MICROWAVE AND FIBER OPTICS COMMUNICATIONS

MICROWAVE AND FIBER OPTICS COMMUNICATIONS

ARTHUR F. WICKERSHAM

PRENTICE HALL
Englewood Cliffs, N.J. 07632

Library of Congress Cataloging-in-Publication Data

Wickersham, Arthur F.
Microwave and fiber optics communications/Arthur F. Wickersham.
p. cm.
Includes index.
ISBN 0-13-582099-5
1. Microwave communication systems. 2. Fiber optics. 3. Optical communications. I. Title.
TK7876.W496 1988
621.38′0413—dc 19 87-29286
CIP

Editorial/production supervision: Editing, Design & Production, Inc.
Cover design: Lundgren Graphics, Ltd.
Manufacturing buyer: Lorraine Fumoso

Smith charts courtesy of Phillip H. Smith. Charts may be obtained from Analog Instrument Co., Box 808, New Providence, NJ 07974.
Fig. 7-12 courtesy of Hewlett Packard, Inc. Used with permission.

10 9 8 7 6 5 4 3 2 1

Printed in the United States of America

ISBN 0-13-582099-5

Prentice-Hall International (UK) Limited, *London*
Prentice-Hall of Australia Pty. Limited, *Sydney*
Prentice-Hall Canada Inc., *Toronto*
Prentice-Hall Hispanoamericana, S.A., *Mexico*
Prentice-Hall of India Private Limited, *New Delhi*
Prentice-Hall of Japan, Inc., *Tokyo*
Simon & Schuster Southeast Asia Pte. Ltd., *Singapore*
Editora Prentice-Hall do Brasil, Ltda., *Rio de Janeiro*

Contents

5 TRANSMISSION-LINE THEORY 123

6 WAVEGUIDE AND THE SPEED OF LIGHT 196

Preface

This combined textbook and laboratory manual is intended for a two-semester, introductory course at the community college or lower-division level; however, it should also be of interest to individual experimenters and radio amateurs. The reader interested in fiber optic and other optic applications may find here a helpful background in wave theory.

Since the student is expected to be familiar with elementary circuit theory, we use circuit concepts to develop topics in wave theory. Rather than attempt a comprehensive overview of current technology, we have developed a select number of topics that are either useful for technicians working in industry or that are fundamental concepts not subject to technological obsolescence.

The text and applications are closely integrated. Occasionally, theory or systems applications are developed in the applications and problems that appear at the end of each chapter. Numerical examples are worked out in both the text and the applications sections. In the first few chapters the numerical examples include instructions for solving the problems with a scientific calculator.

Since small schools often have little microwave or optic equipment, we have tried to present laboratory applications that require a minimum of resources. If a large, well-equipped laboratory is available, the applications can be varied accordingly; for example, if a microwave generator and slotted line are available for each pair of students, there is little incentive for using rope lines to demonstrate standing waves.

Because the mathematical prerequisites are geometry, algebra, and trigonometry, these subjects are reviewed briefly in the text. There is often a gap between mathematics and its application in technology, which can be closed by an instructor familiar with both.

Where some of the material presented is unique or difficult to find in technical literature, we have developed a topic somewhat more extensively than is expected in an introductory course. Such developments can be omitted without loss of continuity. In microwave and optic technology several disciplines overlap, and notation is a problem. We have tried to use symbols that are commonly accepted, redefining them as they take on new meanings in different aspects of the subject.

We thank Tom Hommel and Mike Pieson of Paragon Cable for their enthusiastic consultations.

MICROWAVE AND FIBER OPTICS COMMUNICATIONS

1

Waves

1–1 INTRODUCTION

The higher the frequency of an electromagnetic wave, the shorter its wavelength. At microwave and optic frequencies we shall be working with wavelengths as short as a few centimeters or a few millimeters; thus the length of a connection between components may be a substantial fraction of a wavelength or several wavelengths. This means that phase change and the impedance of the connection must be considered in attempting to transfer energy from one component to another. We cannot simply connect components with a length of wire as we do in dc or low-frequency circuits.

We shall see that when a conductor is about a wavelength in size, it may radiate, that is, act like a broadcasting antenna. This is another reason why we cannot connect components together with wires. Instead of just conducting electromagnetic energy from one component to another, the wires will also radiate energy in the form of electromagnetic waves into surrounding space. We shall have to connect electronic components together with short, carefully designed transmission lines.

Wave effects become so important at high frequencies that we had best start our program of microwave and optical electronics with a study of waves.

1–2 DEFINITION OF A WAVE

A wave is a disturbance moving through a medium; for example, a water wave is a change in height of the water moving across the surface of the water. The surface of the water is the medium. We note that the medium is not transported with the wave: A cork on the surface bobs up and down with the water but does not move across the surface with the wave. Rather than thinking of the disturbance as moving through the medium, it is more accurate to think of the disturbance as reproducing itself at ever-greater distances from its source. For this reason we say that *a wave is a change of state propagating through a medium.*

1–3 WAVELENGTH, FREQUENCY, AND VELOCITY

We speak and hear using sound waves, see with light waves, but can most easily visualize the familiar water wave. The *amplitude* of a water wave is the height of a crest above the mean water surface. *Wavelength* is the horizontal distance between two corresponding points: for example, the distance from crest to crest or trough to trough. These definitions are illustrated in Fig. 1–1.

The *velocity* of a wave is given by the distance s that it travels in an interval of time t,

$$v = \frac{s}{t} \tag{1–1}$$

We can determine velocity by watching a particular wave and measuring the time it takes for it to travel from one point to another; however, an easier way is to count the number of waves that pass a given point in an interval of time t. If n is the number, the total distance traveled by the first wave is $n\lambda$, and its velocity must be $v = n\lambda/t$. Since the number of waves that pass by in an interval of time is defined as the *frequency*, $f = n/t$, we have

$$v = f\lambda \tag{1–2}$$

This derivation is illustrated in Fig. 1–2 for the case in which five waves pass a point P in 1 s.

EXAMPLE 1–1

From Fig. 1–2 and Eq. (1–2) we have

$$\begin{aligned} v &= f\lambda \\ &= 5\ \text{Hz} \times 1\ \text{cm} \\ &= 5\ \text{cm/s} \end{aligned}$$

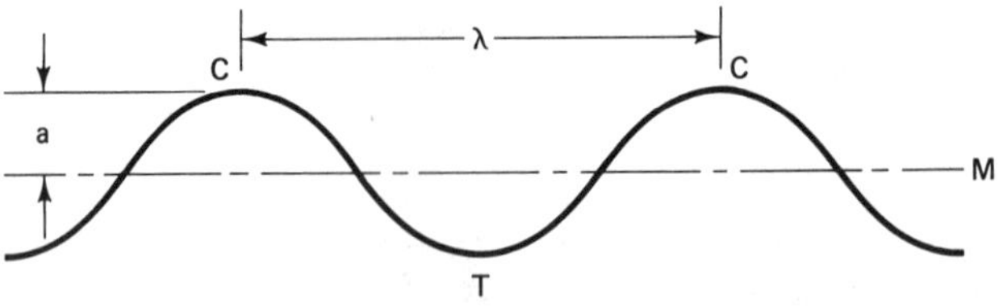

Figure 1–1 Wave parameters: λ, wavelength; a, amplitude; C, crests; T, trough; M, mean water level or reference height.

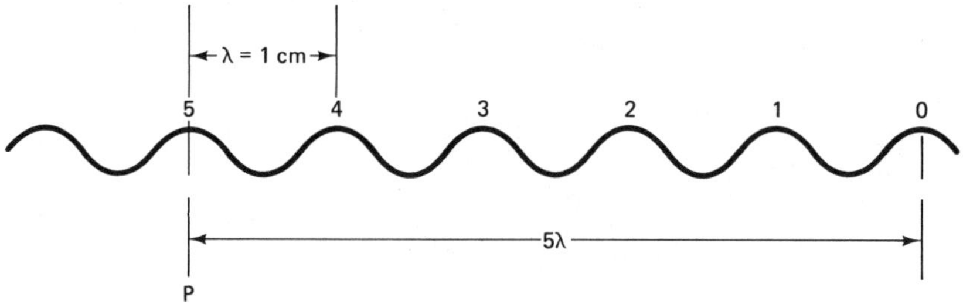

Figure 1–2 Determination of wave velocity from a count of the number of waves that pass a point P in 1 second.

Equation (1–2) is of wide application. It is valid for sound waves, seismic, electromagnetic, and magnetohydromagnetic waves, as well as for water waves. For electromagnetic or light waves in free space it is customary to use the letter c for the speed: $c = \lambda f$. If we know wavelength and frequency, we can immediately determine the wave velocity.

EXAMPLE 1–2

A radio transmitter broadcasts a wavelength of 30 m at a frequency of 10 MHz. What is the wave velocity?

Solution From Eq. (1–2)

$$\begin{aligned} c &= \lambda f \\ &= 30 \times (10 \times 10^6) \\ &= 3 \times 10^8 \text{ m/s} \end{aligned}$$

which is the speed of light in free space.

Another way to derive Eq. (1–2) is to notice that a wavelength passes a fixed point in a time equal to its period. Its speed must be $v = \lambda/T$, where T is the period. Since frequency is the inverse of period, $f = 1/T$, we again obtain Eq. (1–2).

Since the speed of light in free space is a physical constant, we see, from Eq. (1–2), that wavelength must decrease as frequency increases. For each frequency there is only one corresponding wavelength. Electromagnetic waves are given different names depending on their frequency or wavelength bands, as shown in the electromagnetic spectrum in Fig. 1–3. A sound-wave spectrum is shown in Fig. 1–4.

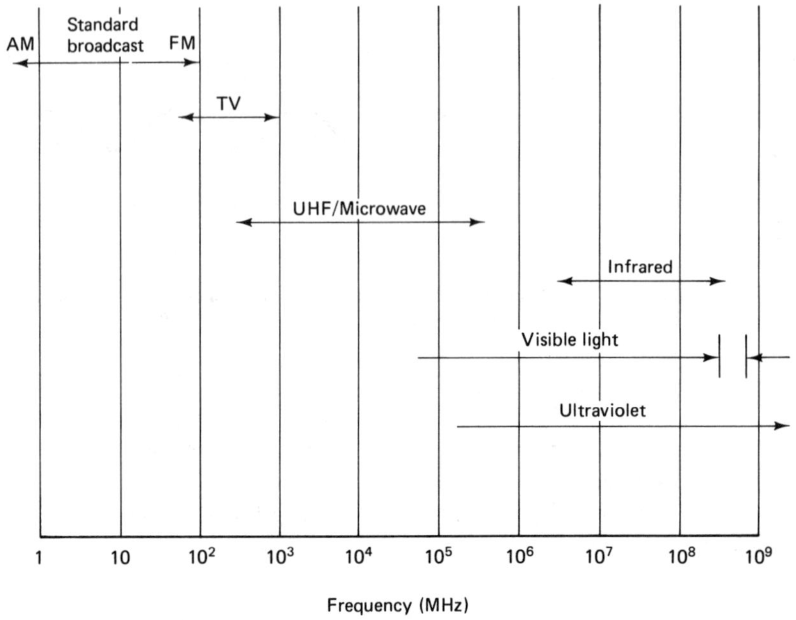

Figure 1–3 Portion of the electromagnetic spectrum.

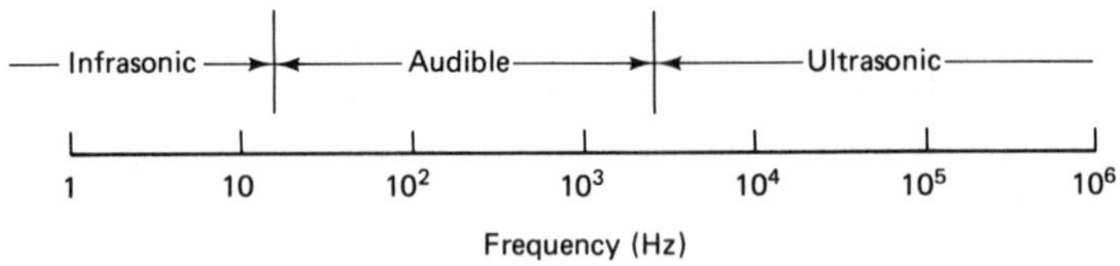

Figure 1–4 Spectrum of sound waves.

1–4 WAVE TYPES

We have been discussing traveling waves; obviously, these are waves that travel or propagate. Another type of wave is the standing wave, a wave that appears to stand still. It is familiar to anyone who has tied a rope to a fixed object and moved the free end of the rope rapidly up and down. The waves reflected from the fixed end combine with the incident waves to form standing waves, as shown in Fig. 1–5. A standing wave actually is formed from two traveling waves moving in opposite directions.

Those points on a standing wave that are not displaced by the motion are called *nodes*, and points at which maximum displacement occurs are *antinodes*. If only one antinode occurs on a line, the line is said to be in its *fundamental mode* of oscillation. If the frequency is doubled, two antinodes appear on the line; if

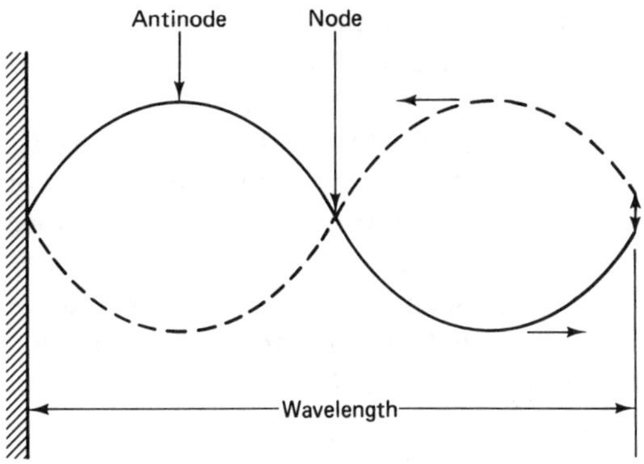

Figure 1–5 Standing wave on a rope line.

tripled, three antinodes; and so on. Frequencies that are integral multiples of the fundamental are called *harmonics*.

If there are two antinodes on a line and the frequency is sufficiently increased, three antinodes will appear, but during the time the frequency is being increased the line appears to be in wild disarray. This is suggestive of the mysterious quantum jump in modern physics. The ultimate nature of matter is dualistic: matter is wavelike as well as corpuscular. If a particle wave is confined, as, for example, an orbital electron is confined in an atom, its wave pattern must contain an integral number of antinodes. The electron can jump to quantum states having fewer or more antinodes, just as can the rope line. Stable or steady states correspond only to integral numbers of antinodes in the standing-wave pattern.

A more complicated wave pattern can be achieved by joining two ropes of different weights, that is, of different linear weight densities. The linear density of a rope is its weight divided by its length. For example, a 100-m line of weight 6 kg has a linear density of 60 g/m. In general, more antinodes will occur on the heavier section of line than on the lighter, as shown in Fig. 1–6.

For the two-rope line there is an increased number of possible modes, each mode defined by the different numbers of antinodes occurring on the heavy and light sections of the line. We are not aware that any system of classification has ever been devised for modes on rope lines, but when we study electromagnetic waves in transmission lines we shall see that an elaborate classification system has

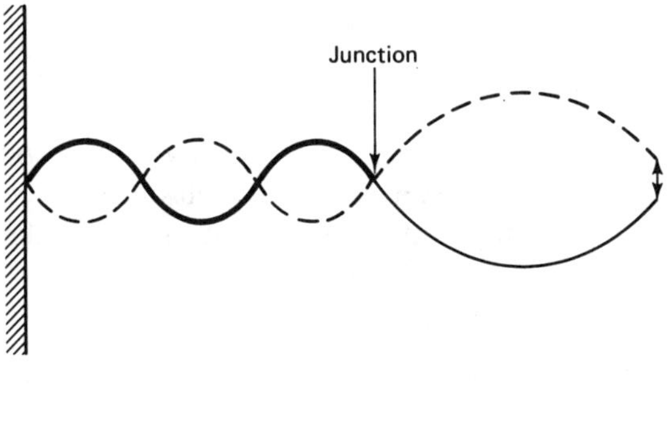

Figure 1–6 Standing waves on rope lines of unequal densities.

been devised in terms of the numbers of antinodes in the electric and magnetic field components of the waves. In summary, a *mode* is a particular configuration of nodes and antinodes of waves in a confined system.

Waves can also be classified in terms of type of motion or displacement. The wave on a rope line is a *transverse* wave, because the displacement of the rope is transverse, or at right angles, to the direction of wave propagation. Electromagnetic waves in free space are transverse; sound waves, however, are *longitudinal.* A longitudinal displacement is a displacement in the direction of propagation. For a sound wave the longitudinal displacements are compressions or dilations of the gas which correspond to variations in pressure and density.

1–5 WAVE MEDIA

Some of the characteristics of a wave, such as frequency and amplitude, are determined by its source, but others, speed for example, are determined by the medium in which the wave propagates. In general, the more rigid the medium, the greater the wave velocity. We see, in Table 1–1, that the speed of sound is 1117 ft/s in air but 16,000 ft/s in steel.

TABLE 1–1 SPEED OF SOUND IN VARIOUS MEDIA

Medium	Speed: English Units	Speed: Metric Units
Air[a]	1,117 ft/s	340.5 m/s
Water	4,800 ft/s	1.463 km/s
Steel	16,000 ft/s	4.88 km/s

[a] For standard atmospheric pressure and a temperature of 59°F.

The formula for the speed of sound in air is, in terms of the physical characteristics of the atmosphere,

$$v = \left(\frac{\gamma kT}{m}\right)^{1/2} \tag{1–3}$$

where, for the earth's atmosphere, the ratio of specific heats is $\gamma = 1.4$; Boltzmann's constant is $k = 1.38 \times 10^{-23}$ joules per kelvin, T is the absolute temperature in kelvin, $T = 273 + {}^\circ\mathrm{C}$, and m is the average molecular mass in a standard atmosphere, 4.78×10^{-26} kg.

EXAMPLE 1–3

Calculate the speed of sound in an atmosphere of normal pressure at a temperature of 20°C.

Solution From Eq. (1–3) we have

$$v = \left(\frac{\gamma kT}{m}\right)^{1/2}$$

$$= \left[\frac{1.4 \times 1.38 \times 10^{-23})(273^\circ + 20^\circ)}{4.78 \times 10^{-26}}\right]^{1/2}$$

$$= 344 \text{ m/s}$$

With a pocket calculator, enter 273 + 20 = 293. Now depress the × key and enter 1.4, again the × key and enter 1.38, depress the EXP key and enter −23, depress the division key and enter 4.78, depress EXP and enter −26, depress the = key and then the key marked $\sqrt{}$ or $x^{1/2}$ to obtain 344 m/s.

For waves on a rope line, the "rigidity" of the line can be increased by increasing the tension. The formula for wave speed on a line is

$$v = \left(\frac{T}{\rho}\right)^{1/2} \tag{1–4}$$

where T is the tension in the line in poundals or newtons and ρ is the linear density, or mass per unit length. In the English system of units, poundals of force are equal to mass in pounds multiplied by the acceleration of gravity, 32.2 ft/s^2; in the MKS or SI system of units, newtons of force are obtained by multiplying the mass in kg by the acceleration of gravity in the metric system, 9.8 m/s^2. These relations follow from Newton's law of motion, which states that force is equal to mass times acceleration:

$$f = ma \tag{1–5}$$

EXAMPLE 1–4

Calculate the speed of a wave on a rope line under 6 lb of tension. The weight of 1000 ft of the rope is 20 lb.

Solution The tension in poundals is

$$T = 6 \text{ lb} \times 32.2 \text{ ft/s}^2 = 193.2 \text{ lb ft/s}^2 = 193.2 \text{ poundals}$$

The linear density is

$$\rho = \frac{20 \text{ lb}}{1000 \text{ ft}} = 0.02 \text{ lb/ft}$$

The wave speed is

$$v = \left(\frac{T}{\rho}\right)^{1/2}$$

$$= \left(\frac{193.2 \text{ lb ft/s}^2}{0.02 \text{ lb/ft}}\right)^{1/2}$$

$$= 98.29 \text{ ft/s}$$

The speed of electromagnetic waves in free space, the speed of light, can be expressed in terms of the "rigidity" of the electromagnetic ether,

$$c = \left(\frac{1}{\epsilon_0 \mu_0}\right)^{1/2} \tag{1–6}$$

where the permittivity is $\epsilon_0 = 8.85 \times 10^{-12}$ F/m and the permeability is $\mu_0 = 4\pi \times 10^{-7}$ H/m. Equation (1–6) is valid in any medium if the permittivity and permeability are appropriate for the medium,

$$v = \left(\frac{1}{\epsilon\mu}\right)^{1/2} \tag{1–7}$$

EXAMPLE 1–5

Calculate the speed of electromagnetic waves in free space and in Teflon. The relative dielectric constant of Teflon is 2.0.

Solution In free space we have

$$c = \left(\frac{1}{\epsilon_0 \mu_0}\right)^{1/2}$$

$$= \left[\frac{1}{(8.85 \times 10^{-12}) \times (4\pi \times 10^{-7})}\right]^{1/2}$$

$$= 3.00 \times 10^8 \text{ m/s}$$

Since relative dielectric constant or relative permittivity is the ratio of the permittivity of the substance to the permittivity of free space, $\epsilon_r = \epsilon/\epsilon_0$, we have

$$\epsilon = \epsilon_r \epsilon_0$$
$$= 2.0 \times (8.85 \times 10^{-12})$$
$$= 1.77 \times 10^{-11} \text{ F/m}$$

For nonmagnetic substances the relative permeability is approximately unity, since $\mu = \mu_0$. The wave speed in Teflon is

$$v = \left(\frac{1}{\epsilon\mu}\right)^{1/2}$$

$$= \left[\frac{1}{(1.77 \times 10^{-11}) \times (4\pi \times 10^{-7})}\right]^{1/2}$$

$$= 2.12 \times 10^8 \text{ m/s}$$

In Example 1–5 you may notice that the wave speed in Teflon is the speed in free space divided by the square root of the relative dielectric constant of Teflon,

$$v = \frac{c}{\epsilon_r^{1/2}}$$

$$= \frac{3.00 \times 10^8}{(2.0)^{1/2}}$$

$$= 2.12 \times 10^8 \text{ m/s}$$

This is a particular case of a general result obtained by dividing Eqs. (1–6) and (1–7),

$$\frac{c}{v} = \frac{(1/\epsilon_0\mu_0)^{1/2}}{(1/\epsilon\mu_0)^{1/2}} = \left(\frac{\epsilon}{\epsilon_0}\right)^{1/2}$$

$$= \epsilon_r^{1/2}$$

The ratio of the speed in free space to the speed in a substance is the square root of the relative dielectric constant.

The result described above was known centuries ago to early workers in physical optics. The ratio of velocity in free space to velocity in a substance was called the *refractive index* η.

$$\eta = \frac{c}{v} = \epsilon_r^{1/2} \qquad (1\text{–}8)$$

The word "refractive" was used because light was observed to be bent, broken, or "refracted" at the interface between two media in which the speed of light is different; thus a stick appears to be bent when it is partly submerged in water because light travels more slowly in water than in air.

1–6 PHASE REPRESENTATION IN SPACE AND TIME

If we take a photograph of a traveling wave, a water wave, for example, we "freeze" the wave in time and display it over a distance, as is indicated in Fig. 1–7. Another way to display a traveling wave is to detect it at a fixed point and display it in

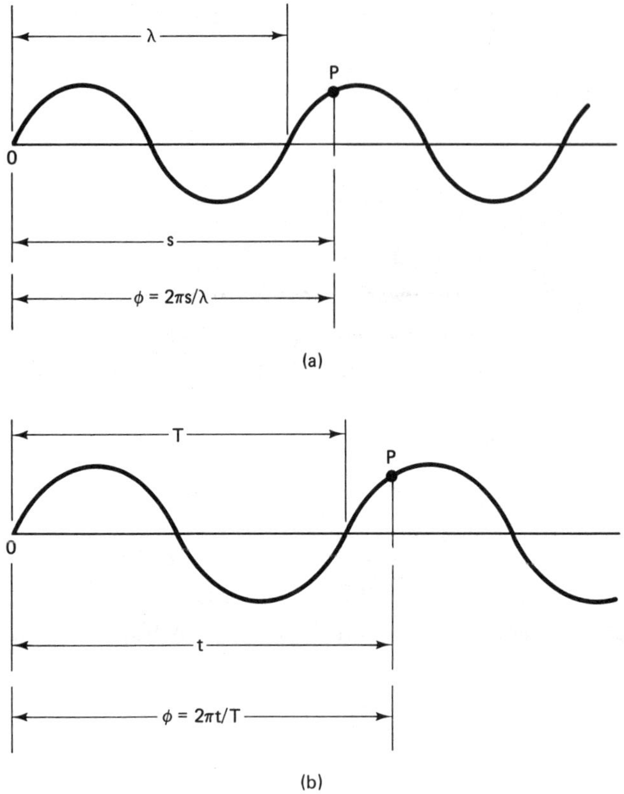

Figure 1–7 (a) A wave frozen in time is displayed over a linear or angular distance. (b) A wave detected at a fixed distance can be displayed over an interval of time.

time. For example, an electromagnetic traveling wave is detected at the input to an oscilloscope and displayed in time according to the horizontal sweep speed of the electron beam.

Space and time representations are shown in Fig. 1–7. In the upper part of the figure, if s is the distance from the origin to a point P, then s/λ is the number of wavelengths from O to P. We shall see later that waves are generated by oscillators. One wavelength is generated by one cycle of an oscillator; thus one wavelength corresponds to one cycle, measured as a *phase angle* of 360° or 2π radians,

$$\phi = \frac{2\pi s}{\lambda} \qquad \text{rad} \tag{1–9}$$

In Section 1–8 we shall review the definition of an angle and show in more detail how the phase at a point on a wave is related to its angular displacement from a reference line.

EXAMPLE 1–6

If the phase of a wave is 0° at the source and its wavelength is 3.0 m, what is the phase at a distance of 5.0 m?

Solution From Eq. (1–9) we have

$$\begin{aligned}\phi &= \frac{2\pi s}{\lambda} \\ &= \frac{2\pi 5}{3} \\ &= 10.47 \text{ rad} \\ &= 600^\circ \\ &= 240^\circ\end{aligned}$$

We may, after reaching 360°, start over at 0°; thus the relative phase can be expressed as 600 − 360 = 240°.

EXAMPLE 1–7

If a voltage wave is generated with a phase of 30° at the source and has a wavelength of 300 km, what is its phase at a distance of 20 km from the source?

Solution In this case we add the phase at the source to the phase computed from Eq. (1–9),

$$\begin{aligned}\phi &= \frac{\phi_s + 360^\circ s}{\lambda} \\ &= \frac{30 + 360 \times 20}{300} \\ &= 54^\circ\end{aligned}$$

To represent a wave in time, for example, a wave detected by an oscilloscope at a fixed distance from the source of the wave, we let t be the total time interval from the instant of detection to that time in the past when the wave was generated. We let T be the period, that is, the time of one cycle. The number of cycles in the total time interval is t/T. Since each cycle corresponds to 360° or 2π rad, the phase interval from the time of detection to the time of generation is

$$\phi = \frac{360°t}{T} \quad \text{deg} \qquad (1\text{–}10)$$

$$= \frac{2\pi t}{T} \quad \text{rad}$$

as shown in the lower part of Fig. 1–7.

EXAMPLE 1–8

If the frequency is 2kHz, what is the phase after 150 μs?

Solution Since period and frequency are reciprocals,

$$T = \frac{1}{f}$$

$$= \frac{1}{2 \times 10^3}$$

$$= 500\ \mu s$$

From Eq. (1–10) we now find

$$\phi = \frac{2\pi t}{T}$$

$$= \frac{2\pi \times (150 \times 10^{-6})}{500 \times 10^{-6}}$$

$$= 1.88 \text{ rad}$$
$$= 108°$$

Equation (1–10) also may be written as

$$\phi = \omega t \qquad (1\text{–}11)$$

where $\omega = 2\pi f$ is the angular frequency, the frequency measured in radians per second rather than in cycles per second or hertz. It is sometimes called the angular velocity since the equation is the angular form of the formula for linear distance traveled at a constant velocity v,

$$s = vt \qquad (1\text{–}12)$$

We shall often use angular variables in place of linear ones. The correspondence between some of the commonly used linear and angular variables is shown in Table 1–2.

If our space and time representations of phase are to be equivalent, we must have $2\pi s/\lambda = 2\pi t/T$ or, after rearranging terms, $s/t = \lambda/T$, which again gives us the velocity formula $v = \lambda f$.

TABLE 1–2 ANGULAR FORM FOR VARIOUS VARIABLES

Variable	Linear Symbol	Angular Form
Frequency	f	$\omega = 2\pi f$
Distance	s	$\phi = \dfrac{2\pi s}{\lambda}$
Wavelength	λ	$\lambdabar = \dfrac{\lambda}{2\pi}$
Wave number	$\dfrac{1}{\lambda}$	$k = \dfrac{2\pi}{\lambda}$

If we drop a rock in a pond, we notice that the waves spread out in circles; that is, if we imagine a curve drawn through the crest, trough, or through any corresponding phase points, the curve is a circle. In general, a *wavefront* is the locus of equiphase points.

1–7 AMPLITUDE, POWER, AND INTENSITY

The amplitude of a wave is its maximum displacement from its rest value or zero point. For a water wave, amplitude is the height of a crest above mean water level. For a voltage wave, amplitude is the peak voltage, V_p, as shown in Fig. 1–8. Peak voltage can be measured directly with an oscilloscope. It is sometimes convenient to measure from maximum to minimum to obtain the peak-to-peak voltage, which is double the peak voltage, $V_{\text{p-p}} = 2V_p$. If a multimeter is used to measure ac voltage, the scale usually is calibrated in root-mean-square average voltage, which is peak voltage divided by the square root of 2, $V_{\text{rms}} = V_p/(2)^{1/2}$. These definitions are illustrated in Fig. 1–8.

EXAMPLE 1–9

The peak-to-peak voltage across a 100-Ω resistor is 100 mV. What are the peak power and average power dissipated in the resistor?

Solution

$$\text{peak power} = \frac{V_p^2}{R}$$

$$= \frac{(50 \times 10^{-3})^2}{100}$$

$$= 25\ \mu\text{W}$$

$$\text{average power} = \frac{V_{\text{rms}}^2}{R}$$
$$= \frac{V_p^2}{2R}$$
$$= \frac{(50 \times 10^{-3})^2}{2 \times 100}$$
$$= 12.5\ \mu\text{W}$$

EXAMPLE 1–10

The rms current through a resistor is 40 mA and the peak voltage across the resistor is 60 mV. What is the peak power dissipated in the resistor?

Solution Since power is the product of voltage and current, we have

$$\begin{aligned} P &= V_p i_p \\ &= (60 \times 10^{-3}) \times (40 \times 10^{-3}) \times (2)^{1/2} \\ &= 3.39\text{mW} \end{aligned}$$

In Examples 1–9 and 1–10 we assume the energy dissipated in a resistor was carried there by electromagnetic waves. To find an expression for the energy carried by the waves we assume that a wave produces a displacement of a parcel of the medium and that the displacement follows Hooke's law; that is, the displacement is proportional to the force that produces it. For the well-known case of a mechanical spring, Hooke's law is a statement that force is proportional to stretch,

$$f = ka \tag{1–13}$$

where a is the stretch or displacement and the constant of proportionality k is called the spring constant. The bigger the k, the "stiffer" the spring.

If f is the maximum force, corresponding to maximum displacement a, and if the displacement starts from a rest position of equilibrium, the average force during the displacement is $f/2$, and the average work done by the wave on the parcel is given by the average force times the distance through which it moves, $W = (f/2)a$. Combining this with Eq. (1–13), we obtain

$$W = \tfrac{1}{2}ka^2 \tag{1–14}$$

and we see that wave energy is proportional to the square of the amplitude.

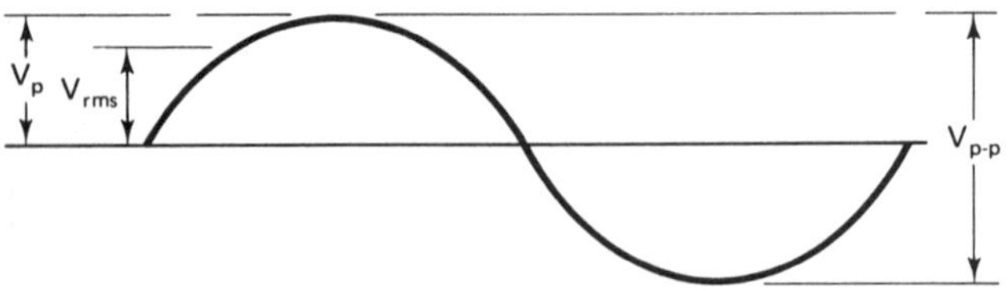

Figure 1–8 Measures of wave displacement: V_p, amplitude or peak voltage; $V_{\text{p-p}}$, peak-to-peak voltage; V_{rms}, root-mean-square average voltage.

For an electric wave we introduce E, the force per unit charge, also called the electric field intensity, measured in newtons per coulomb or volts per meter. The electric displacement D is measured in coulombs per square meter. Hooke's law becomes $E = (1/\epsilon)D$, where the "spring constant" is the reciprocal of the permittivity. From Eq. (1–14) we now obtain $W = \frac{1}{2}(1/\epsilon)D^2$, or

$$W = \tfrac{1}{2}\epsilon E^2 \tag{1–15}$$

Since the units of ϵ and E are farads/m and volts/m, Eq. (1–15) gives the work per unit volume, or energy density, in joules/m^3.

Wave *intensity* is defined as the energy delivered through a unit cross-sectional area in 1 s. If the wave speed is v, the energy in a length $v \times 1$ s will pass through a unit cross section in 1 s, as shown in Fig. 1–9. From Eq. (1–15) we now find the intensity to be

$$I = \tfrac{1}{2}\epsilon E^2 v \tag{1–16}$$

The total power delivered through an aperture of A square meters will be $P = \frac{1}{2}\epsilon E^2 A v$ or

$$P = \tfrac{1}{2}\epsilon V_p^2 v \tag{1–17}$$

If we substitute for the velocity from Eq. (1–7), we obtain $P = \frac{1}{2}V_p^2/(\mu/\epsilon)^{1/2}$, or

$$P = \frac{\tfrac{1}{2}V_p^2}{Z_c} \tag{1–18}$$

where $Z_c = (\mu/\epsilon)^{1/2}$ is the characteristic impedance of the medium. For free space its value is $(\mu_0/\epsilon_0)^{1/2} = 377\ \Omega$. The idea that a medium has an impedance may be surprising, especially if the "medium" is empty space, a vacuum. This will become more plausible when we study transmission lines.

If we compare the result we have obtained for the power carried by a wave, $\frac{1}{2}V_p^2/Z_c$, with the power dissipated in a load, $\frac{1}{2}V_p^2/R$, we see that the load resistance can absorb all of the power in the wave if $R = Z_c$. We shall see later that the power carried by a transmission line will be delivered to a resistive load if the load resistance is equal to the characteristic impedance of the line, that is, if the load is matched to the line. If there is mismatch, not all of the energy will be

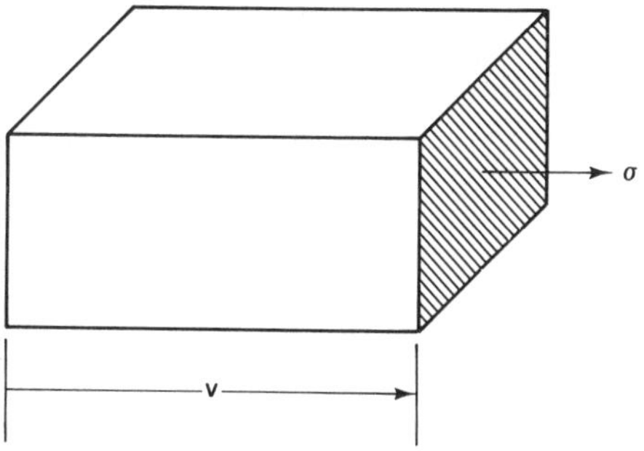

Figure 1–9 In 1 second the energy in a column of length $v \times 1$ is transported through a unit cross section σ.

delivered to the load; some of it will be reflected back to the source where it will be wasted and may cause damage.

1–8 MATHEMATICAL DESCRIPTION OF WAVES

Ordinary household alternating electricity has a sinusoidal waveform. It has a sinusoidal waveform because it is generated by rotating machinery, by the circular motion of a coil of wire in a magnetic field. The mathematical description of a sine wave proceeds in a like manner: a point moving on a circle generates a sinusoidal motion in rectangular coordinates.

On the left side of Fig. 1–10 the points labeled 0, 1, 2, . . . , 11 measure the positions of a point P as it moves around a circle of radius r. We can display the motion of P in rectangular coordinates by measuring its position on a straight line, rather than around a circle, as shown on the right side of Fig. 1–10. At each value of the angle ϕ we plot the height of P above the horizontal axis. This can be done by projecting its height on the circle over to its position in the rectangular system, as shown by dashed lines in the figure. Connecting the projected points with a smooth curve, we obtain a sine wave. To understand why it is called a sine wave, we shall briefly review elementary trigonometry.

An *angle* is defined as the ratio of a circular arc to its radius,

$$\phi = \frac{s}{r} \tag{1–19}$$

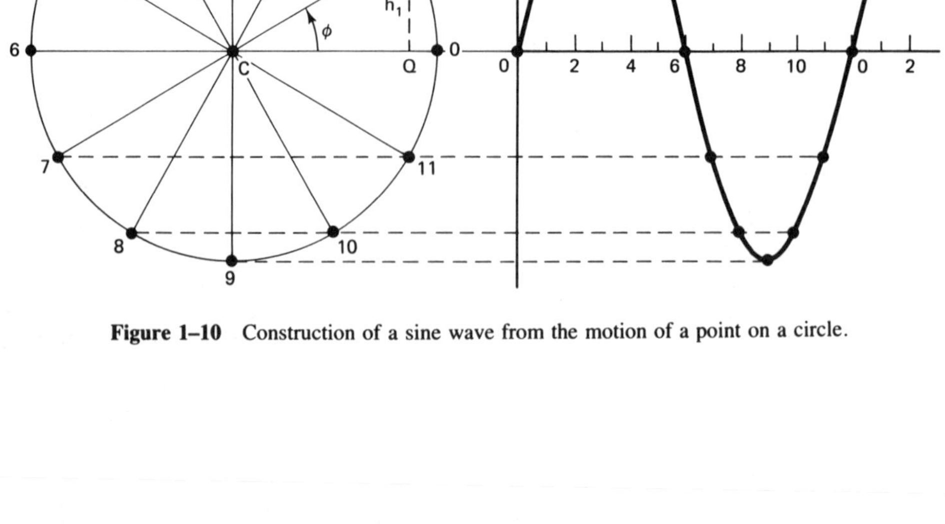

Figure 1–10 Construction of a sine wave from the motion of a point on a circle.

as shown in Fig. 1–11. Since the ratio is dimensionless the units are arbitrary. Three commonly used units are the degree, obtained by dividing the circumference of a circle into 360 equal parts, the mil, produced by dividing a circle into 6400 parts, and the radian, defined as unity when the length of the arc equals the radius. Since the circumference of a circle is $2\pi r$, the angle of a complete circle is $\phi = 2\pi r/r = 2\pi$ radians. Evidently, 2π radians is equivalent to 360° and 1 rad = 57.3°. By custom, angles are measured as positive in the counterclockwise direction.

A triangle is a plane figure formed from three straight lines. A right triangle has one of its three interior angles equal to a right angle, that is, equal to 90°. The two shorter sides of a right triangle are called *sides* or *legs*. The longest side, the side opposite the right angle, is called the *hypotenuse*. The sum of the interior angles of any triangle is 180°.

The sine of an angle α is abbreviated $\sin \alpha$ and is defined in a right triangle as the ratio of the side opposite the angle to the hypotenuse. In Fig. 1–12 we have

$$\sin \alpha = \frac{a}{c} \tag{1–20}$$

The cosine of an angle is the ratio of the adjacent side to the hypotenuse,

$$\cos \alpha = \frac{b}{c} \tag{1–21}$$

and the tangent is the ratio of the sides,

$$\tan \alpha = \frac{a}{b} \tag{1–22}$$

The *rule of Pythagoras* is that the square of the hypotenuse equals the sum of the squares of the sides,

$$c^2 = a^2 + b^2 \tag{1–23}$$

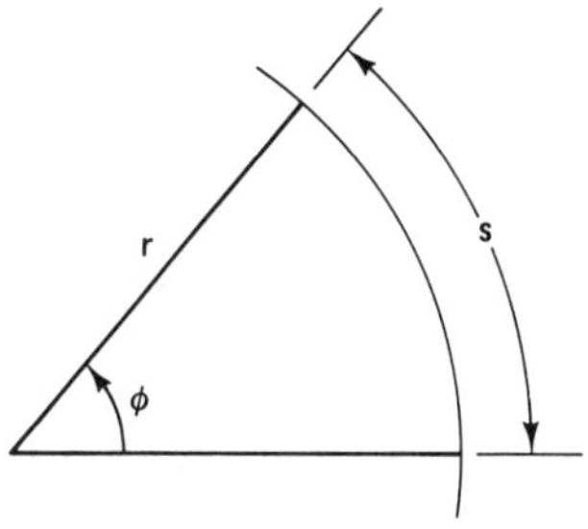

Figure 1–11 Elements of an angle: arc s of radius r.

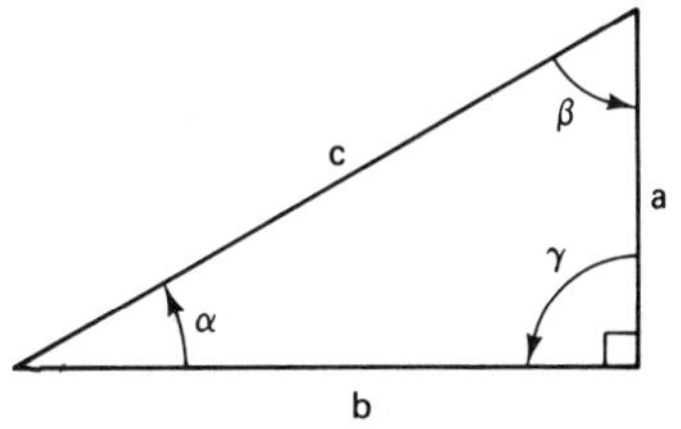

Figure 1–12 Parts of a right triangle: c, hypotenuse; a, b, sides or legs; α, β, and γ, interior angles, with $\gamma = 90°$.

By dividing the last equation by c^2 we obtain $1 = a^2/c^2 + b^2/c^2$, or

$$\sin^2 \alpha + \cos^2 \alpha = 1 \quad (1\text{–}24)$$

a useful trigonometric relation.

EXAMPLE 1–11

What is the sine of 30°?

Solution Turn on a pocket calculator, put it in the degree mode, and enter 30. Depress the key marked SIN and read 0.5. We have $\sin 30° = 0.5$. This tells us that the side opposite a 30° angle in a right triangle is one-half the length of the hypotenuse.

The expression $\sin \alpha$ can be regarded as an operation, sin, performed on an operand, α. An inverse operation undoes the original operation; thus just as dividing by 3 undoes the operation of multiplying by 3, $(3 \times 6)/3 = 6$, so does the operation INV sin or $\sin^{-1}$ undo the operation sin: $\sin^{-1}(\sin \alpha) = \alpha$.

EXAMPLE 1–12

Find the value of the angle α in the right triangle shown in Fig. 1–13.

Solution The side opposite α divided by the hypotenuse gives the sine of α, $\sin \alpha = 1.6\ \text{km}/2.5\ \text{km}$. On a pocket calculator enter 1.6 and divide by 2.5 to obtain 0.64. Now depress the inverse key, usually labeled INV, and then SIN to obtain 39.79°: you have performed the operation $\sin^{-1} 0.64 = 39.79°$. In other words, the angle whose sine is 0.64 is 39.79°.

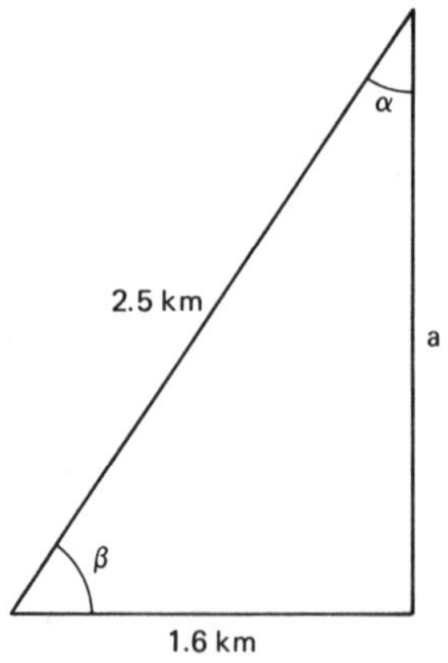

Figure 1–13 Right triangle for Examples 1–12 and 1–13.

EXAMPLE 1–13

Find the length of side a in Fig. 1–13.

First solution By the rule of Pythagoras, $a^2 = c^2 - b^2$ or

$$\begin{aligned} a &= (c^2 - b^2)^{1/2} \\ &= (2.5^2 - 1.6^2)^{1/2} \\ &= 1.92 \text{ km} \end{aligned}$$

Second solution From Example 1–12 we know $\alpha = 39.79°$. Since $\tan \alpha = 1.6/a$, we have

$$\begin{aligned} a &= \frac{1.6}{\tan \alpha} \\ &= \frac{1.6}{\tan 39.79°} \\ &= 1.92 \text{ km} \end{aligned}$$

On a pocket calculator enter 39.79 and depress the key labeled TAN to obtain 0.8329. Now depress the key marked $1/x$ to obtain 1.201 and multiply this result by 1.6 to obtain 1.92 km.

Third solution The sum of the interior angles is 180°; therefore, we have $\beta = 180 - 39.79 - 90 = 50.21°$. Since $\sin \beta = a/2.5$, we now have

$$\begin{aligned} a &= 2.5 \sin 50.21° \\ &= 1.92 \text{ km} \end{aligned}$$

EXAMPLE 1–14

If $\tan \alpha = 0.5$, what is the value of α?

Solution On a pocket calculator enter 0.5. Depress the INV key and then the key marked TAN to obtain

$$\begin{aligned} \alpha &= \tan^{-1}(\tan \alpha) \\ &= \tan^{-1}(0.5) \\ &= 26.57° \end{aligned}$$

After this brief review of trigonometry we can return to the waveform shown in Fig. 1–10 to see why it is called a sine wave. On the left side of the figure the line $\overline{PQ} = h$ is the height, or vertical displacement, of the point P and also is one leg of the right triangle CPQ. From the definition of the sine function we have $\sin \phi = h/r$ or

$$h = r \sin \phi \tag{1–25}$$

We see that for a circle of unit radius the vertical displacement of the wave at any time is the sine of ϕ, and this is one reason for calling it a sine wave. For a voltage wave, the instantaneous voltage is

$$V = V_p \sin \phi \tag{1–26}$$

EXAMPLE 1–15

If the peak voltage is 60 mV, what is the voltage when the phase angle is 240°?

Solution From Eq. (1–26) we have

$$\begin{aligned} V &= V_p \sin \phi \\ &= 60 \times 10^{-3} \sin 240° \\ &= -52.0 \text{ mV} \end{aligned}$$

On a pocket calculator enter 240 and depress the SIN key, depress the multiplication key and enter 60, depress EXP and enter −3, and depress the = key to obtain −0.05196 or −52.0 mV.

If the phase angle is not zero at the source or reference point, we must include the initial value in our calculations,

$$V = V_p \sin (\phi + \phi_0) \tag{1–27}$$

EXAMPLE 1–16

If the angular frequency is 6800 rad/s, the initial phase $\pi/2$ rad, and the peak voltage is 6.0 V, what is the voltage at 1.2 ms?

Solution From Eqs. (1–27) and (1–11) we have

$$\begin{aligned} V &= V_p \sin (\phi + \phi_0) \\ &= 6.0 \sin \left(6800 \times 1.2 \times 10^{-3} + \frac{3.14}{2}\right) \\ &= -1.80V \end{aligned}$$

Put a calculator in radian mode and calculate $6800 \times 1.2 \times 10^{-3} + 3.14/2$, depress the SIN key and multiply the result by 6.0 to obtain −1.80V.

EXAMPLE 1–17

If the phase at the origin is 135°, the wavelength 1.0 cm, and the peak-to-peak voltage 80 mV, what is the instantaneous voltage at a distance of 3.5 cm from the source?

Solution In Eq. (1–27) we must include in the argument of the sine function the phase at the origin; thus

$$V = V_p \sin\left(360\frac{s}{\lambda} + \phi_0\right)$$

$$= (40 \times 10^{-3}) \sin\left(360 \times \frac{3.5}{1.0} + 135\right)$$

$$= -28.28 \text{ mV}$$

We have displayed the sine wave either as frozen in space or fixed in time. If we allow both time and space coordinates to vary simultaneously, we obtain the equation for a traveling wave,

$$V = V_p \sin(ks - \omega t + \phi_0) \tag{1–28}$$

where k is the wave number, $2\pi/\lambda$, and ω is the angular frequency, $2\pi f$. In this case the equation describes a wave traveling from left to right in the positive s-direction.

EXAMPLE 1–18

For a wave with a peak amplitude of 10.0 V, an initial phase of 30°, a frequency of 1.0 MHz, and a wavelength of 0.30 km, sketch the waveform at $t = 0$, 0.05, and 0.10 μs.

Solution We can sketch one wavelength on a total horizontal distance of 300 m, as shown in Fig. 1–14. At $t = 0$, $s = 0$, we find from Eq. (1–28),

$$V = V_p \sin\phi_0$$
$$= 10 \sin 30°$$
$$= 5.0 \text{ V}$$

For $t = 0$ we now find additional points of the waveform by taking $s = 50$ m, 100 m, ... We have, for $s = 50$ m,

$$V = V_p \sin\left(360\frac{s}{\lambda} + \phi_0\right)$$

$$= 10 \sin\left[360\left(\frac{50}{300}\right) + 30°\right]$$

$$= 10.0 \text{ V}$$

Combining the calculations, we obtain enough points to sketch in the solid sine wave shown in Fig. 1–14.

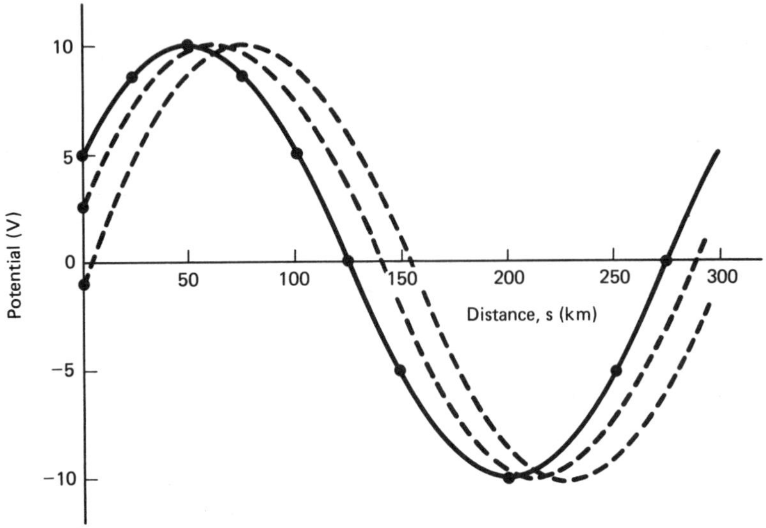

Figure 1–14 Sine wave showing progression from left to right with increasing time, from Example 1–18.

We now repeat the calculations for $t = 0.05\ \mu s$. For $s = 0$, we find

$$V = V_p(-360ft + \phi_0)$$
$$= 10.0 \sin [(-360 \times 10^6) \times (0.05 \times 10^{-6}) + 30°]$$
$$= 2.08\ \text{V}$$

For $s = 50$ m we have

$$V = V_p \sin\left(360\frac{s}{\lambda} - 360ft + \phi_0\right)$$

$$= 10.0 \sin\left[360 \times \frac{50}{300} - (360 \times 10^6) \times (0.05 \times 10^{-6}) + 30°\right]$$

$$= 9.51\ \text{V}$$

Repeating the calculations, we obtain sufficient points to sketch the dashed sine wave shown in Fig. 1–14. Repeating the entire process for $t = 0.1\ \mu s$, we obtain the second dashed wave shown in the figure. We see that as time increases the wave propagates from left to right.

1–9 DIRECTION FINDING

The measurement of phase intervals between a wave and a reference or baseline can be used to find the direction of the source of the waves, a measurement that has many applications. Acoustic arrays of microphones, for example, can be used

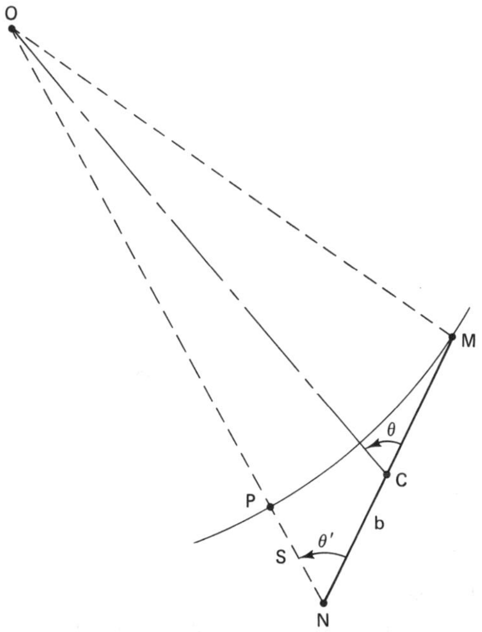

Figure 1–15 Acoustic detection array. The baseline is $\overline{MN} = b$, angle θ is the direction of the source, arc MP is the wavefront of a sound wave from the source, and $\overline{PN} = S$ is the extra path length the wave will travel to reach the microphone at N.

to determine the direction of sources of sound, such as artillery fire or missile launchings.

In Fig. 1–15 let O be the location of a missile launching site and let M and N be the locations of two remote detectors, in this case microphones. A line through the detectors is called the baseline of the array of detectors, and our problem is to find the direction of the source, measured from the center C of the baseline.

Let us assume that the received signals are displayed on a dual-channel oscilloscope and that the display is as shown in Fig. 1–16. In Fig. 1–15 we see that the sound waves must travel a greater distance to reach N than to reach M, the difference being S; thus the waves arrive later at N, the time difference being t as shown on the oscilloscope. The extra distance S can be calculated from a measurement of t on the oscilloscope display, $S = vt$, where v, the local sound speed, is known. We next assume that the baseline b has been measured and that the source is so far away that the angle it makes to the center of the baseline is approximately equal to the angle it makes at N, in Fig. 1–15, $\theta \approx \theta'$. If this condition is met, then MNP is approximately a right triangle and we have $\cos\theta \approx \cos\theta' = S/b = vt/b$, or

$$\theta = \cos^{-1}\frac{vt}{b} \tag{1–29}$$

EXAMPLE 1–19

The baseline of an acoustic array is 600 m long in a compass direction 12° east of north. Artillery sounds are detected with a time difference of 1.2 s between channels. The local speed of sound is 342 m/s. Find the compass bearing of the source.

Solution From Eq. (1–29) we have

$$\theta = \cos^{-1}\frac{vt}{b}$$

$$= \cos^{-1}\frac{342 \times 1.2}{600}$$

$$= 46.84^\circ$$

With a calculator enter 342 × 1.2/600 to obtain 0.684. Now depress the INV and COS keys to find 46.84°.

The compass bearing of the artillery is 46.84° − 12° = 34.84° west of north.

In Example 1–19 there is an ambiguity in that the same result would be obtained if the source were on the other side of the array at −46.84°. Enter either +46.84° or −46.84° on a calculator and depress the COS key to obtain 0.684. The cosine

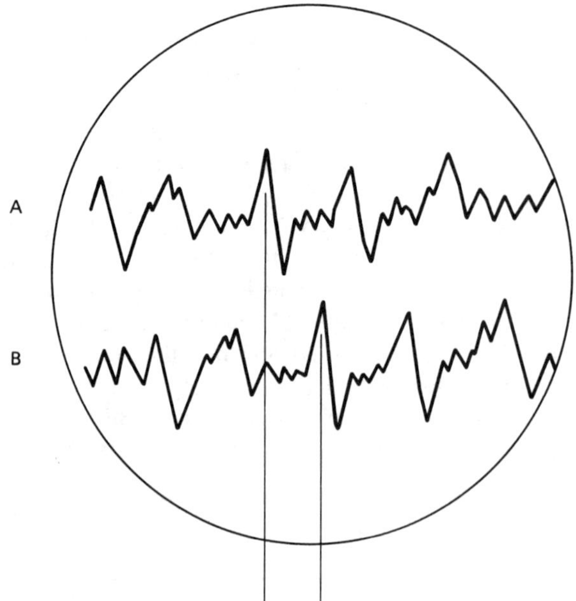

Figure 1–16 Oscilloscope display of infrasonic signals received by acoustic detectors: channel A, signal received from microphone M; channel B, signal received from N; t, time difference between signals.

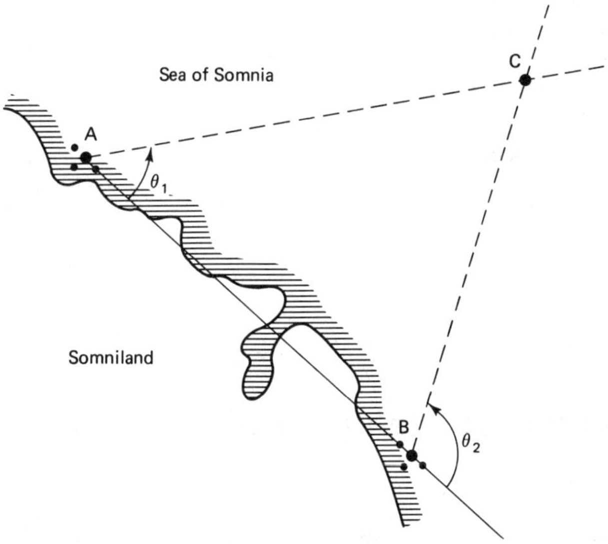

Figure 1–17 Map of coastal installations of underwater acoustic detectors.

function is called an *even function* because it yields the same result for positive or negative arguments. In practice this ambiguity is removed by adding a third microphone to the array. The third microphone cannot be on the original baseline but must be displaced so that the array is triangular.

If two widely separated directional arrays are used to find the directions of a source, the location of the source can be determined by triangulation. In the map shown in Fig. 1–17, *A* and *B* show the locations of coastal installations of underwater acoustic arrays. If sound detections indicate directions θ_1 and θ_2 at the two installations, the lines of direction can be extended on the map and a fix made at *C*. If the sound source is a hostile submarine, the coordinates of *C* can be relayed to a defensive unit for appropriate action.

Directional arrays have many applications besides military: for example, depth sounders for locating underwater objects; ionospheric sounders or "ionosondes" for tracking ionospheric waves; arrays for detecting seismic waves in oil exploration. We later shall study the directional capabilities of electromagnetic arrays or "antenna arrays," but first we must learn more about electromagnetic waves.

1–10 UNITS

The metric MKS system of units, also called the SI system, is used in electronics; however, in practice and in the assignments and laboratory work at the end of this chapter it may be necessary occasionally to use English units. In Table 1–3 we list some of the conversions that may be required.

TABLE 1–3 METRIC AND EQUIVALENT ENGLISH UNITS FOR VARIOUS CONCEPTS

Concept	MKS Units	English Units
Distance	meter	39.37 inches, in.
Distance	kilometer	0.621 mile, mi
Mass	kilogram	2.203 pounds, lb
Force	newton	7.23 poundals, pdl
Time	second	second, s

APPLICATION 1–1: Acoustic Delay Line and Directional Array

In this application an acoustic delay line is used to measure the local speed of sound, and an array of microphones is used to determine the direction of an acoustic source. The results are to be reported as briefly as possible in a laboratory notebook. To achieve brevity and readability, tables and graphs should be used to record data and report results. Every table and graph is to be numbered, titled, and captioned.

Remarks. Tables and graphs are useless if units are omitted. Never connect measured points on a graph with segments of straight lines. Although brevity is extremely important, nothing significant should be omitted. When repetitive calculations are required, one sample calculation is sufficient for the report.

The report should tell what you did, how you did it, and what the results were. For examples of good and bad reports, see the Appendix of *Experiments for Electrical Circuit Analysis with BASIC Programming*, by T. F. Bogart (Chicago: Science Research Associates, 1982).

Equipment and Materials

1 dual-channel oscilloscope
2 microphones
1 loudspeaker
1 variable-frequency signal generator, 4 Hz to 2 MHz
1 audio output transformer, if required
2 rulers
1 protractor
10-ft string

PART 1: SPEED OF SOUND

Assemble the equipment as shown in Fig. 1–18. Observe the signal on channel B while varying the generator frequency and select a suitable range of operating frequencies. For the inexpensive microphones we have used, a recognizable sine wave could be detected only between 2.5 and 4.0 kHz. Outside this range the signal was too badly distorted to permit phase measurements. The transformer can be omitted if the generator-to-speaker mismatch is not too great, or if the signal generator output is sufficiently large, about 15 V.

Procedure

The signal on channel B will show a phase delay because of the extra distance S that it travels as a sound wave. A combination of transducers that causes a signal to travel a distance as a sound wave is called an *acoustic delay line*.

Adjust the distance S until the signals on channels A and B appear to be in phase; that is, until the extra distance S corresponds to exactly an integral number of wavelengths. Call this distance S_1. We now have

$$S_1 = n\lambda \tag{1–30}$$

where n is an integer and λ is the wavelength of the sound wave.

Increase S until the signals again appear to be in phase. Call this distance S_2. We have

$$S_2 = (n + 1)\lambda \tag{1–31}$$

Combining Eqs. (1–30) and (1–31), we obtain

$$\lambda = S_2 - S_1 \tag{1–32}$$

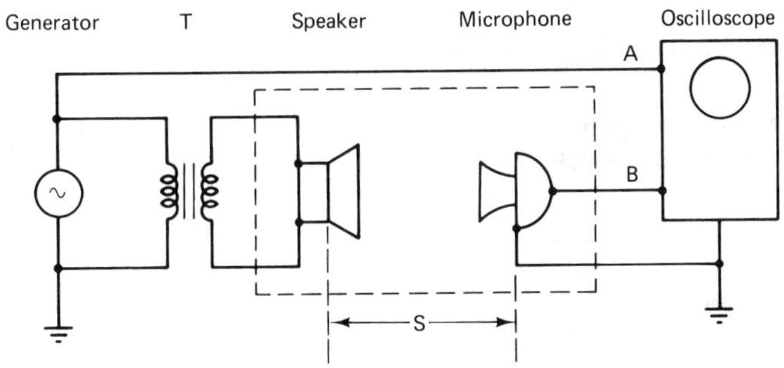

Figure 1–18 Speed of sound apparatus. The components within the dashed lines comprise an acoustic delay line.

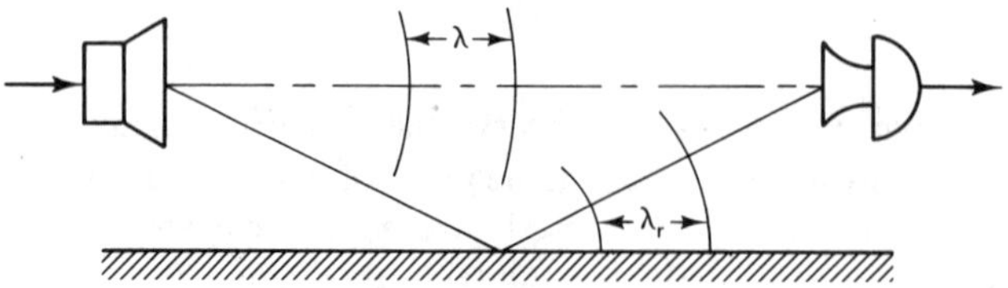

Figure 1–19 Error resulting from the presence of reflected waves. If the reflected wavelength is measured in the horizontal direction, the result will be greater than the correct value, $\lambda_r > \lambda$.

With the determination of the sound wavelength from Eq. (1–32) and the known frequency f of the signal generator, we can calculate the speed of sound from Eq. (1–2).

Repeat the measurements for several operating frequencies and for distance intervals that are one or more multiples of a wavelength. Record the room temperature and calculate the speed of sound from Eq. (1–3). Tabulate data and results and compare the average of your experimental determinations with the speed of sound calculated from Eq. (1–3). If the distances are measured carefully, the error should be 0.1% or less.

Sources of Error and Difficulties

Care should be taken to eliminate reflections. If, for example, speaker and microphone are close to a tabletop, obliquely reflected waves also will be received by the microphone, as shown in Fig. 1–19. Since the reflected wavelength is measured in the horizontal direction rather than in a direction normal to the wavefronts, its length will be greater than the correct value, $\lambda_r > \lambda$, as shown in the figure. If a too-large value of wavelength is used in the calculations, the resulting speed of sound will be too large. Most reflections can be minimized by positioning the microphone vertically above the loudspeaker.

Weak signal strength is another difficulty. If there is considerable mismatch between signal generator and loudspeaker, use an audio transformer to improve the match. Some inexpensive "button" microphones are too insensitive to use in this experiment. Maximize signal strength by tuning the generator to a frequency that gives maximum response.

PART 2: DIRECTION FINDING

Assemble the equipment as shown in Fig. 1–20. The range $\overline{SC}$ should be at least five times the baseline $\overline{MN}$. Microphones can be held in place at M and N by taping them to a ruler. For the first measurements make the baseline small, say 3 or 4 in. Tape a protractor at S or C to measure the angle θ.

Procedure

After selecting the lowest usable operating frequency, measure the time difference between the signals on channels A and B and measure the baseline distance $\overline{MN}$. Use these measurements and the measured speed of sound from Part 1 to determine

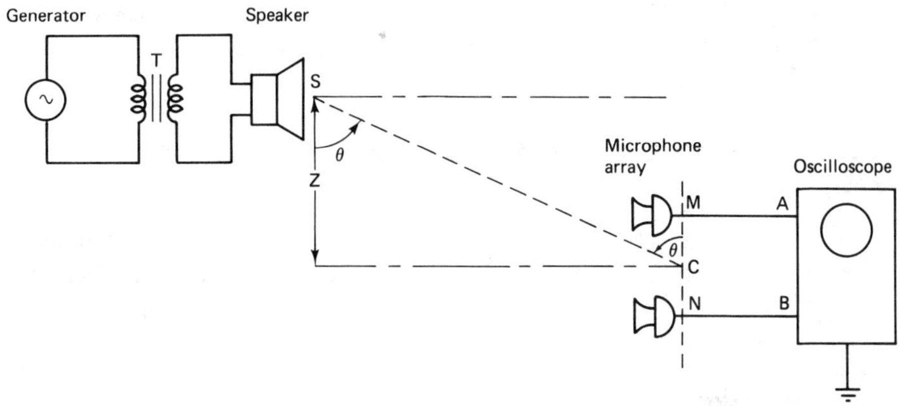

Figure 1–20 Acoustic directional array: $\overline{SC}$, range; $\overline{MN}$, baseline; θ, direction angle; z, offset.

the direction angle θ in Eq. (1–29). Start a series of four measurements with the equipment initially in direct line, θ = 90°, and decrease to about θ = 70°. At each position, also determine the angle θ with the protractor by stretching a string from S to C. When you write your report, compare the electrical determinations of θ with those made with the protractor.

Increase the offset of the array from the centerline until the angle θ becomes so small that the signals on channels A and B again appear to be in phase. Measure the angle electrically and with a protractor; also, measure the range $\overline{SC}$. The electrical and mechanical measurements should now be in disagreement.

In your report use a scale drawing of the experiment to show the positions of the source, microphones, and acoustic wavefronts. Explain why the disagreement occurred, and show how to modify Eq. (1–29) so that it can be used correctly when the path difference S, in Fig. 1–15, is greater than one wavelength.

Sources of Error and Difficulties

Inexpensive microphones and loudspeakers vary in sensitivity by as much as a factor of 10. Imbalance can be compensated by adjusting the amplifier gains in a dual-channel oscilloscope. If an assortment is available, test several microphones and speakers and select those with the greatest sensitivity.

Some oscilloscopes are not provided with a trigger switch to select one channel or the other for triggering. If both channels trigger on a predetermined signal level, it may not be possible to measure phase difference; however, if an external trigger input is provided, it can be used to effect measurement of phase difference.

If an external trigger is not provided, or if dual-channel oscilloscopes are not available, or no oscilloscopes at all are available, do one of the modified experiments shown in Part 3.

At higher frequencies the loudspeakers may be so directive that little power is available at large displacements from the centerline. In other words, the acoustic beamwidths are too narrow. In this case the measurements will have to be restricted to smaller offsets than suggested.

PART 3: DIRECTION FINDING BY SIGNAL STRENGTH MEASUREMENT

If a dual-channel oscilloscope is not available, it is possible to determine the direction of the source by measuring signal level on a single channel. Connect the outputs of the microphones together and to a single-channel oscilloscope or to an ac voltmeter, as shown in Fig. 1–21.

Procedure

Place the microphone array on the centerline, $\theta = 90°$, and record the received signal strength, V_{max}. If the array is moved to one side or the other of the centerline, the received signal should decrease, indicating a maximum signal strength on the centerline. If the maximum occurs to one side or the other, either the microphones are of unequal sensitivity or there are reflections from nearby objects. Obtain matched microphones and remove possible sources of reflections.

Displace the microphone array from the centerline, as shown in Fig. 1–21, until the received signal voltage decreases to $0.7V_{max}$. Measure the angle θ with a string and protractor. Repeat these measurements on the other side of the centerline and average the results.

The direction of the source can be calculated from

$$\theta = \cos^{-1} \frac{\lambda}{4b} \tag{1–33}$$

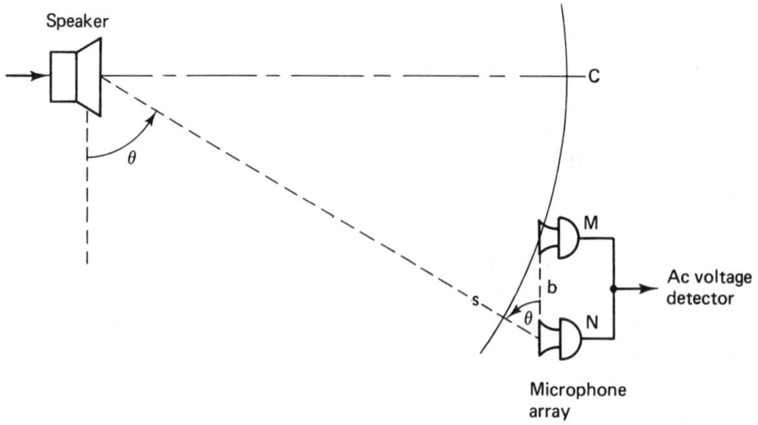

Figure 1–21 Determination of signal source direction by single-channel measurements.

where b is measured directly and λ is determined from the ratio of the speed of sound and the operating frequency.

In your report compare the angle θ calculated from Eq. (1–33) with the result of measurement with a protractor. If the beamwidth of the horn is not too narrow, the error should be less than 5°.

Analysis

To derive Eq. (1–33), we let V_0 be the voltage detected by either microphone; then the total voltage at the detector will be

$$\overline{V}_r = V_0 + V_0 \angle \phi \tag{1–34}$$

where ϕ is the phase angle corresponding to the extra distance the signal must travel to reach the microphone at N,

$$\phi = 2\pi \frac{s}{\lambda} \tag{1–9}$$

If we expresss the voltage $V_0 \angle \phi$ in rectangular form, we obtain

$$\overline{V}_r = V_0 + V_0 \cos \phi + jV_0 \sin \phi,$$

or

$$\begin{aligned} V_r &= V_0[(1 + \cos \phi)^2 + \sin^2 \phi]^{1/2} \\ &= V_0(1 + 2 \cos \phi + \cos^2 \phi + \sin^2 \phi)^{1/2} \\ &= V_0[2(1 + \cos \phi)]^{1/2} \end{aligned} \tag{1–35}$$

If the procedure above is unfamiliar, refer to the complex number summary given in Section 2–3.

When the array is on the centerline, we have $\phi = 0^0$ and maximum signal strength is received,

$$V_{\max} = 2V_0 \tag{1–36}$$

By dividing Eqs. (1–35) and (1–36), we obtain

$$\frac{V_r}{V_{\max}} = (1 + \cos \phi)^{1/2}(2)^{-1/2} \tag{1–37}$$

In the experimental procedure we determined $V_r/V_{\max}$ to be 0.7 or $(2)^{-1/2}$; thus Eq. (1–37) becomes

$$(2)^{-1/2} = (1 + \cos \phi)^{1/2}(2)^{-1/2}$$

and

$$\cos \phi = 0$$

$$\phi = \frac{\pi}{2}$$

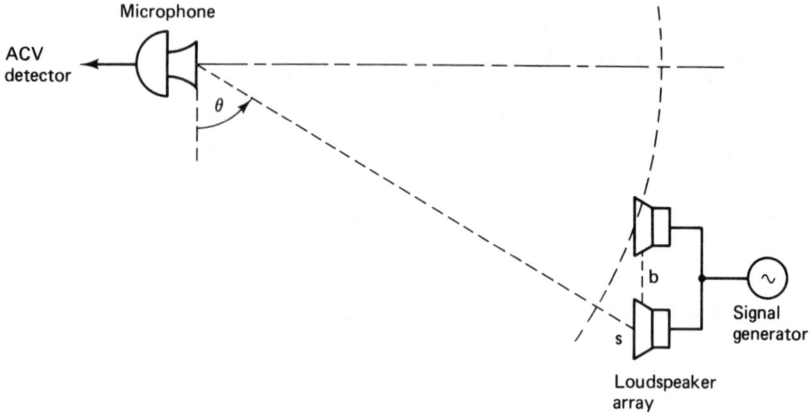

Figure 1–22 Alternative form of the direction-finding experiment, using only one microphone and one channel of detection.

The geometry shown in Fig. 1–21 is the same as that shown in Fig. 1–15, so we again have

$$\theta = \cos^{-1}\frac{s}{b} \tag{1–38}$$

From $\phi = 2\pi s/\lambda = \pi/2$ we have $s = \lambda/4$ and Eq. (1–38) becomes

$$\theta = \cos^{-1}\frac{\lambda}{4b} \tag{1–33}$$

which was to be shown.

Alternative Experiment

If only one microphone is available, modify the experiment to that shown in Fig. 1–22. The geometry, analysis, and procedure are the same as before.

SUMMARY

A wave is defined as a change of state propagating through a medium. It is characterized by its length, a wavelength being the distance covered on one period, its frequency, which is the reciprocal of the period, $f = 1/T$, and its speed, which is a wavelength per period, $v = \lambda/T = \lambda f$.

Waves are classified by the change they cause in a medium: A sound wave is a change in pressure and density in a gas; a seismic wave is a twisting, compression, or bending of the earth's crust; a light wave or electromagnetic wave is a change in electric and magnetic field strengths. Waves also are classified as traveling or standing, and a standing or confined wave is further classified by the number of its antinodes, its "mode." A wave may be *longitudinal* if the change of state is in the direction of propagation, as in a sound wave, or *transverse*, if the change is across the direction of propagation, as on a rope line.

The speed of a wave can be related to the physical properties of the medium in which it propagates:

Sound wave $$v = \left(\frac{\gamma kT}{m}\right)^{1/2}$$

Wave on a rope line: $$v = \left(\frac{T}{\rho}\right)^{1/2}$$

EM wave: $$v = \left(\frac{1}{\mu\epsilon}\right)^{1/2}$$

The ratio of the speed of an EM wave in free space to its speed in a dielectric medium is the refractive index of the medium, $\eta = c/v$, and the square of the refractive index is the relative permittivity or "dielectric constant" of the medium, $\epsilon_r = \eta^2$.

The amplitude of a wave is a measure of the magnitude of the change of state it produces in a medium; for example, the amplitude of an EM wave can be measured as a change in electric field strength E, in volts per meter. The *energy density* transported by a wave is proportional to the square of its amplitude,

$$W = \tfrac{1}{2}\epsilon E^2$$

The *power* of a wave is the energy density delivered per unit time,

$$P = \frac{\tfrac{1}{2}V_p^2}{Z_c}$$

where Z_c is the characteristic impedance of the medium,

$$Z_c = \left(\frac{\mu}{\epsilon}\right)^{1/2}$$

$$Z_0 = \left(\frac{\mu_0}{\epsilon_0}\right)^{1/2} = 377\ \Omega$$

and the *wave intensity* is

$$I = \tfrac{1}{2}\epsilon E^2 v$$

The phase of a wave is the portion of its cycle that has been completed, measured as an angle. At any instant of time the angle is measured spatially,

$$\phi = \frac{2\pi s}{\lambda} = ks$$

At a fixed distance, phase can be measured temporally,

$$\phi = \frac{2\pi t}{T} = \omega t$$

If both distance and time are varied, the phase is

$$\phi = ks - \omega t + \phi_0$$

where ϕ_0 is the phase at the origin at $t = 0$. The amplitude of a wave is related to its phase by the sine function,

$$V = V_p \sin (ks - \omega t + \phi_0)$$

By measuring the phase of a wave at three different locations, we can determine the direction and range of its source.

A review of trigonometric definitions: An angle is the ratio of a circular arc to its radius, $\phi = s/r$; the circular functions can be defined by the sides of a triangle; the rule of Pythagoras relates the hypotenuse to the sides of a right triangle, from which we have $\sin^2 \phi + \cos^2 \phi = 1$.

PROBLEMS

1–1. Look up the permittivity and permeability of free space. Try to obtain as many significant figures as possible. Calculate the speed of light to as many significant figures as is justifiable. Cite references used.

1–2. The atomic mass of helium is 4, and one atomic mass unit (amu) is 1.66×10^{-27} kg. The atomic mass density of helium is accordingly, $m = 4 \times (1.66 \times 10^{-27}) = 6.64 \times 10^{-27}$ kg. Calculate the speed of sound in an atmosphere of He at 20°C. The ratio of specific heats for a monatomic gas is $\gamma = 1.66$.

1–3. Calculate the wave speed on a rope line under a tension of 6 lb force. A 40-ft length of the line weighs 3 lb. Express the answer in ft/s and m/s.

1–4. A rope line 30 ft long and fixed at both ends has a standing-wave pattern with two antinodes when the frequency is 20 Hz. Use Eq. (1–2) to calculate the wave speed. At what frequency will it have three antinodes?

1–5. A radio station broadcasts at two frequencies, 100.00 and 110.00 kHz. If the two signals leave the broadcasting antenna in phase, at approximately what distance from the station will they be 180° out of phase?

1–6. The voltage across a 2.2-kΩ resistor is $V = 5 \sin (377t + 0.7)$ volts, where the argument of the sine function is in radians. What is the average power dissipated in the resistor? What is the peak current in the resistor? What is the voltage across the resistor at $t = 0.1$ s?

1–7. A model airplane is tethered to a 60-ft rope and is traveling at a speed of 35 ft/s. What is its angular velocity in rad/s?

1–8. For the acoustic array shown in Fig. 1–15, we have $b = 1.61$ km, a sound frequency of 4 Hz in a standard atmosphere at 58°F, and a time difference of 0.9 s between signals received at M and N. Assuming no ambiguity, find the angle θ.

1–9. For the right triangle shown in Fig. 1–23, find the distances a and c.

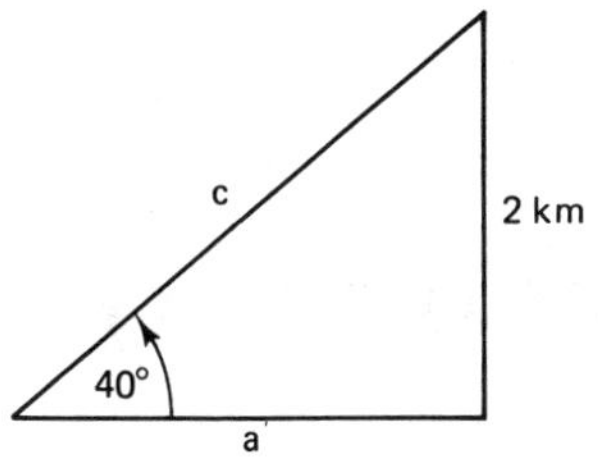

Figure 1–23 Right triangle for Problem 1–9.

1–10. In the analysis given in Part 3 of Application 1–1 it was shown that the sum of two sine waves fell to 0.7 of the maximum in-phase sum when the phase difference was increased to $\pi/2$ radians or 90°. Verify this result by plotting

$$a_1 = \sin \frac{360°s}{\lambda}$$

$$a_2 = \sin \left(\frac{360°s}{\lambda} + 90°\right)$$

for $\lambda = 10$ cm and $s = 0$ to 20 cm. Find $a_1 + a_2$ graphically and show that it is equal to $0.7(a_1 + a_2)_{max}$, where $(a_1 + a_2)_{max}$ is the sum when there is no phase difference.

2

The Reflection of Waves

2–1 INTRODUCTION

We may be aware that waves are reflected whenever we look into a mirror or hear an echo. It appears as if a wave is reflected from a surface much as a billiard ball bounces from a rail; however, if an eavesdropper places an ear to a wall, he can hear speech in another room. Evidently, not all of the sound is reflected by the wall. Unlike the billiard ball, some of the sound wave penetrates into and through the material of the wall. The portions of a wave that are transmitted or reflected depend on the difference in physical characteristics of the media on either side of the surface or *interface*, in this case the difference in elasticity and density of air on one side and plasterboard on the other.

In Chapter 1 we found that the speed of a wave is determined by the physical properties of the medium in which it propagates. If the properties change abruptly, the propagation speed changes abruptly and part of the wave is reflected. The greater the change, the greater the portion reflected. By "abrupt" we mean that the change occurs in a distance small compared to a wavelength. We shall see that we can express quantitatively the portion of a wave reflected in terms of the magnitude of an abrupt change in speed.

A transmission line can be regarded as a one-dimensional medium. A load is said to be "matched" to a transmission line if there is no reflection, if all of the incident wave energy is transmitted into the load. An understanding of wave reflections will enable us to design efficient circuits; also, we shall find that by measuring the phase and amplitude of a reflected wave we can determine the impedance of a load.

The ratio of the reflected to incident amplitudes of a wave is called a *reflection coefficient*. If the reflection occurs from a device connected to transmission lines, the reflection and transmission coefficients are called S-parameters. Since manufacturers use S-parameters to describe the performance of high-frequency devices, a knowledge of reflection theory will help us to understand device specifications in manufacturers' catalogs.

2–2 TYPES OF REFLECTIONS

When a wave propagates from one medium into another in which the propagation speed is different, part of the wave is reflected at the interface and only the remaining part is transmitted. For example, the velocity of light in glass is less than in air; accordingly, there is a *partial reflection* at the air–glass interface. If the change in velocity is extreme, all of the wave energy may be reflected, and this is called *total reflection*. A standing wave on a rope line fastened to a rigid wall is an example of total reflection. The wave cannot propagate into the wall, all of the wave energy is reflected.

In general, both the amplitude and phase of the reflected and transmitted waves differ from that of the incident wave. For the special case of total reflection the phase change is either 0° or 180°. Returning once again to our example of a rope line tied to a wall, we see that at the point of reflection the reflected wave is out of phase with the incident wave, producing a node at the wall. If the end of the rope were free, the reflected wave would be in phase with the incident, producing an antinode at the end of the line. This can be demonstrated by hanging a rope from a balcony, generating standing waves, and observing the antinode at the free end of the line.

Electromagnetic waves can be reflected obliquely, as shown in Fig. 2–1. In this figure the direction of propagation of a wave is represented by a straight line or arrow. This is called a *ray*, and in isotropic media the rays are perpendicular to the wavefronts. In anisotropic media, such as certain crystals, the rays may not be perpendicular to the wavefronts.

The vertical dashed line in Fig. 2–1 is called the *normal* to the surface. The *angle of incidence* θ_1 lies between the normal and the incident ray; similarly, the *angle of reflection* θ_3 lies between the normal and the reflected ray. The *law of reflection* is that the angle of incidence is equal to the angle of reflection,

$$\theta_1 = \theta_3 \tag{2–1}$$

The *angle of refraction* θ_2 is measured from the transmitted ray to the normal extended and is related to the angle of incidence by Snell's law,

$$\sin \theta_2 = \frac{\eta_1}{\eta_2} \sin \theta_1 \tag{2–2}$$

where η_1 and η_2 are the refractive indices of the two media shown in Fig. 2–1.

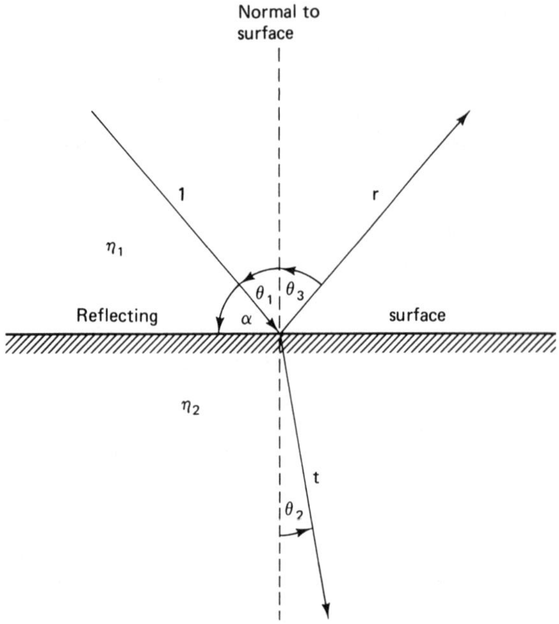

Figure 2–1 Reflection at oblique incidence: 1, incident ray; *r*, reflected ray; *t*, transmitted or refracted ray; θ_1, angle of incidence; θ_2, angle of refraction; θ_3, angle of reflection; α, grazing angle.

EXAMPLE 2–1

An airborne search radar transmits its beam at an angle of 45° to the surface of the sea. The transmission frequency is 0.5 GHz, and the index of refraction of water is 8.88 at this frequency. What is the direction of propagation beneath the surface of the sea?

Solution From Snell's law we have

$$\theta_2 = \sin^{-1}\left(\frac{\eta_1}{\eta_2}\sin\theta_1\right) = \sin^{-1}\left(\frac{1}{8.88}\sin 45^\circ\right)$$
$$= 4.57^\circ$$

The radar waves propagate almost directly downward in the water.

On a calculator enter 45 and depress the SIN key, divide the result by 8.88 to obtain 0.0796, and press the SIN^{-1} key to obtain 4.567°.

EXAMPLE 2–2

A beam of light is incident on Crown glass at an angle of 25° and is refracted at an angle of 10.3°. What is the relative dielectric constant of the glass?

Solution We solve Snell's equation for η_2 and obtain

$$\eta_2 = \eta_1 \frac{\sin \theta_1}{\sin \theta_2} = 1 \times \frac{\sin 25°}{\sin 10.3°}$$
$$= 2.36$$
$$\epsilon_2 = \eta_2^2 = (2.36)^2$$
$$= 5.59$$

The angles in Fig. 2–1 are geometric angles that show the directions in which the waves propagate; they should not be confused with phase angles. Phase angles relate to the cyclic motion that produces a wave, not to the direction of propagation. Geometric angles will be of interest when we study radio-wave propagation. For the present we shall mostly consider only reflection at *normal incidence*, in which $\theta_1 = \theta_3 = 0$ and the wave is reflected directly back on itself. This is the case of interest in transmission-line applications.

2–3 REFLECTION COEFFICIENTS AND IMPEDANCE

If the amplitude of an incident wave is a_0, and if a_r, a_t are the amplitudes of the reflected and transmitted waves, the *reflection* and *transmission coefficients* are

$$r = \frac{a_r}{a_0} \qquad (2\text{–}3)$$

$$t = \frac{a_t}{a_0} \qquad (2\text{–}4)$$

These might just as well have been called the relative or normalized amplitudes, but microwave electronics is an outgrowth of physics, optics, and electrical engineering, and each discipline has its own custom. Since the normalized incident wave is $a_0/a_0 = 1$, it is customary to indicate the incident amplitude as unity, as we did in Fig. 2–1.

When there is a partial reflection at an interface, the reflected wave combines with the incident to form a standing wave (Fig. 2–2). Since the reflected amplitude is not as large as the incident, the subtraction $1 - r$ at the out-of-phase points produces minima, not zeros or complete nodes as in the case of total reflection. At those points where the reflected and incident wave are in phase we obtain maxima of amplitude $1 + r$.

The *standing-wave ratio* is the ratio of the maximum to minimum amplitude,

$$\text{SWR} = \frac{1 + r}{1 - r} \qquad (2\text{–}5)$$

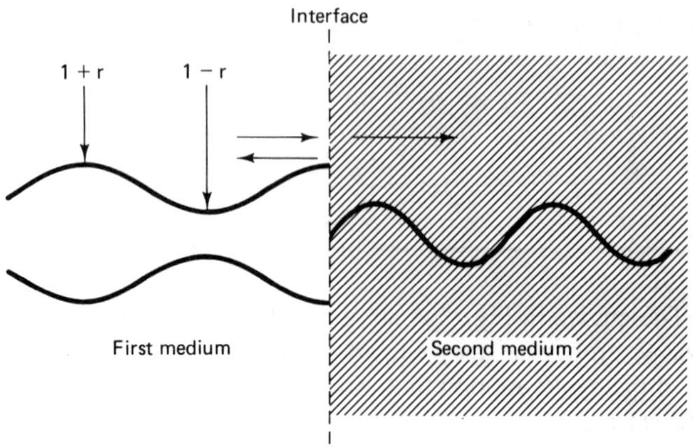

Figure 2–2 Standing-wave pattern produced by the partial reflection of a wave at an interface between two media.

The results we are about to obtain are applicable to many types of waves, but we are interested primarily in electromagnetic waves detected by a voltage probe; accordingly, we shall speak of the *voltage standing-wave ratio* or VSWR. In this case the reflection coefficient is the ratio of the reflected wave voltage to the incident voltage, V_r/V_0. The VSWR is equal to the ratio of the propagation speeds in the two media,

$$\text{VSWR} = \frac{1 + r}{1 - r} = \frac{v_1}{v_2} \tag{2–6}$$

If we divide the numerator and denominator of the fraction v_1/v_2 by c, we have $(v_1/c)/(v_2/c) = \eta_2/\eta_1$; thus we can write

$$\frac{1 + r}{1 - r} = \frac{\eta_2}{\eta_1} \tag{2–7}$$

If we solve this equation for r, we have

$$r = \frac{\eta_2 - \eta_1}{\eta_2 + \eta_1} \tag{2–8}$$

or

$$r = \frac{v_1 - v_2}{v_1 + v_2} \tag{2–9}$$

We see that the reflection coefficient increases with the velocity difference in the two media; also, we can calculate the reflection coefficient if we know the refractive indices of the media.

EXAMPLE 2–3

The index of refraction of water is 8.1 at 3.70 GHz. If a radar signal of intensity 6 μW/m^2 is at normal incidence on a calm lake, what is the amplitude of the reflected wave?

Solution Solving Eq. (1–16) for E_0, we find an incident amplitude of

$$E_0 = \left(\frac{2I}{\epsilon_0 c}\right)^{1/2}$$

$$= \left[\frac{2 \times (6 \times 10^{-6})}{(8.85 \times 10^{-12}) \times (3 \times 10^8)}\right]^{1/2}$$

$$= 67.2 \text{ mV/m}$$

The reflection coefficient is

$$r = \frac{\eta_2 - \eta_1}{\eta_2 + \eta_1}$$

$$= \frac{8.1 - 1}{8.1 + 1}$$

$$= 0.780$$

The reflected amplitude is

$$E_r = E_0 r$$

$$= (67.2 \times 10^{-3}) \times 0.780$$

$$= 52.4 \text{ mV/m}$$

EXAMPLE 2–4

Two long ropes of different linear densities are tied together and a short-wave disturbance or *pulse* is sent down the line (Fig. 2–3). The wave speed on the lighter line is 26 ft/s and on the heavier line the speed is 12 ft/s.

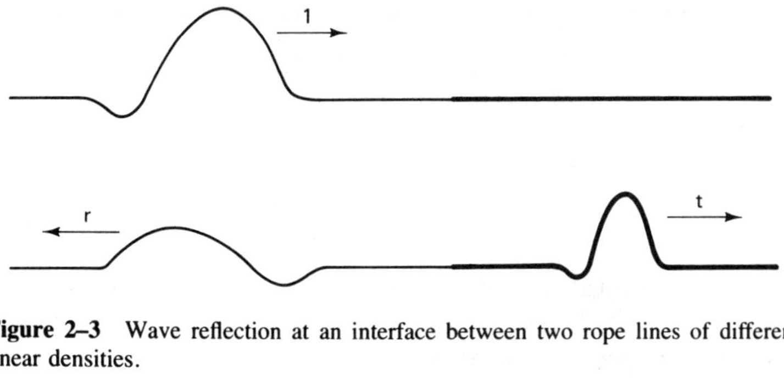

Figure 2–3 Wave reflection at an interface between two rope lines of different linear densities.

What is the reflection coefficient at the knot? If the incident amplitude is 10 in., what is the reflected amplitude?

Solution From Eq. (2–9) we have

$$r = \frac{v_1 - v_2}{v_1 + v_2}$$

$$= \frac{26 - 12}{26 + 12}$$

$$= 0.368$$

From the definition of reflection coefficient we find the reflected amplitude to be

$$a_r = ra_0$$
$$= 0.368 \times 10$$
$$= 3.68 \text{ in.}$$

For electromagnetic waves we can substitute Eq. (1–7) for the velocities on the right side of Eq. (2–6) and so obtain an expression for VSWR in terms of the permittivities and permeabilities of the media: $(1 + r)/(1 - r) = (\mu_2\epsilon_2/\mu_1\epsilon_1)^{1/2}$. For nonmagnetic materials we have $\mu_1 = \mu_2 = \mu_0$; therefore, we can interchange μ_1 and μ_2 to obtain $(1 + r)/(1 - r) = (\mu_1/\epsilon_1)^{1/2}/(\mu_2/\epsilon_2)^{1/2}$, or

$$\text{VSWR} = \frac{1 + r}{1 - r} = \frac{Z_1}{Z_2} = z \tag{2–10}$$

where z is the *normalized impedance* and is the ratio of the impedances of the media. Solving this equation for the reflection coefficient, we have

$$r = \frac{z - 1}{z + 1} \tag{2–11}$$

EXAMPLE 2–5

In Example 2–3, what is the characteristic impedance of the water?

Solution Since we know from Example 2–3 that the reflection coefficient is 0.780, we can use Eq. (2–10) to find the normalized impedance,

$$z = \frac{1 + r}{1 - r}$$

$$= \frac{1 + 0.780}{1 - 0.780}$$

$$= 8.09$$

Since the impedance of air is approximately the same as the impedance of free space (377 Ω), we can reverse the normalization process to find the absolute impedance of the water:

$$Z_2 = \frac{Z_1}{z}$$

$$= \frac{377}{8.09}$$

$$= 46.6\ \Omega$$

Check: The index of refraction of water is 8.10 and the relative permittivity is the square of the index; therefore, from the definition of relative permittivity we have

$$\begin{aligned} \epsilon_2 &= \epsilon_0 \eta_2^2 \\ &= 8.85 \times 10^{-12}(8.10)^2 \\ &= 5.806 \times 10^{-10} \end{aligned}$$

We now can find the answer by using the expression for the characteristic impedance of a medium,

$$Z_2 = \left(\frac{\mu_2}{\epsilon_2}\right)^{1/2} = \left(\frac{\mu_0}{\epsilon_2}\right)^{1/2}$$

$$= \left(\frac{4\pi \times 10^{-7}}{5.806 \times 10^{-10}}\right)^{1/2}$$

$$= 46.5\ \Omega$$

EXAMPLE 2–6

The characteristic impedance of a coaxial line is 50 Ω and it is terminated in an 80-Ω load. What percentage of the incident power will be reflected from the load?

Solution The normalized impedance is

$$z = \frac{Z_1}{Z_2}$$

$$= \frac{80}{50}$$

$$= 1.6$$

and the reflection coefficient is

$$r = \frac{z - 1}{z + 1}$$

$$= \frac{1.6 - 1}{1.6 + 1}$$

$$= 0.231$$

The *power reflection coefficient* is defined as the ratio of the reflected to incident power or intensity,

$$\Gamma = \frac{P_r}{P_0} = \frac{I_r}{I_0} \tag{2–12}$$

From Eq. (1–16) or (1–17) we know that power and intensity are proportional to the square of the amplitude reflection coefficient:

$$\Gamma = \frac{P_r}{P_0} = r^2$$
$$= 0.231^2$$
$$= 0.0533$$

5.33% of the incident power is reflected.

EXAMPLE 2–7

The VSWR on a 50-Ω transmission-line terminated in a resistive load is 2.2. What is the impedance of the load?

Solution From Eq. (2–10) we have

$$z = \text{VSWR}$$
$$= 2.2$$

We also have $z = Z_1/Z_2$ by definition; therefore,

$$Z_2 = \frac{Z_1}{z}$$
$$= \frac{50}{2.2}$$
$$= 22.73\ \Omega$$

EXAMPLE 2–8

A 50-Ω line is terminated in a 110-Ω load. What is the VSWR?

Solution From Eq. (2–10) we have

$$\text{VSWR} = \frac{Z_2}{Z_1}$$
$$= \frac{110}{50}$$
$$= 2.2$$

In this case we have reversed the procedure in that we have normalized the load to the line rather than the line to the load. This amounts to turning Z_1/Z_2 upside down and is done because the VSWR is customarily taken to be greater than unity.

In comparing Examples 2–7 and 2–8, we see that either load, a load of 22.73 or 110 Ω, would produce a VSWR of 2.2; thus the answer to Example 2–7 is ambiguous. This ambiguity is removed by observing that the smaller answer, 22.73 Ω, corresponds to a change in phase of 180° on reflection, the same as the change in phase at a short or fixed end. The larger answer corresponds to the case of 0° phase change on reflection, the same as for an open or free end. We need to pay attention to the phase change on reflection, even for the simple case of reflection from a resistive load. To understand this more fully, we must consider the general case in which the load can be inductive or capacitive, as well as resistive.

The preceding discussion and examples have been limited to real quantities and resistive loads; however, Eq. (2–11) also is valid for complex reflection coefficients resulting from inductive or capacitive loads. Before applying this equation to complex coefficients and impedances, we shall briefly review elementary complex number theory. Section 2–4 may be skipped if the reader is proficient in this subject.

2–4 COMPLEX NUMBER SUMMARY

We owe much of our number system and geometry to the ancient Greek philosophers of 300 to 400 B.C. One of their accomplishments was the establishing of a correspondence between algebra and geometry. This correspondence is often used in elementary grades to teach arithmetic by associating numbers with points on a number line.

About five centuries ago mathematicians started using the number $(-1)^{1/2}$. It was discovered that theorems could be proven and results calculated that were correct, even though $(-1)^{1/2}$ had been used in the process. Since this number did not "exist" in the sense that there is no real number which when multiplied by itself yields -1, the quantity $(-1)^{1/2}$ was called "imaginary." This was unfortunate because it is not imaginary, but the name persists to this day, to the discomfort of many students. As mathematicians became more confident in the use of $(-1)^{1/2}$, they abbreviated it with the letter i. In electronics this would be confused with ac current, so we use the letter j.

A geometric representation of imaginary and complex numbers was given at the end of the eighteenth century by Carl F. Gauss. In this representation imaginary numbers $j, j2, j3, \ldots$ are associated with points on a number line drawn perpendicular to the number line for "real" numbers. The concept of "number" now is given a greater abstraction: instead of associating real numbers with points on a line, we can associate complex numbers with points on a plane. A complex number plane is shown in Fig. 2–4. Here the x-axis is the real number line, the y-axis is the imaginary number line, and the point P represents a complex number. In the case shown the number is $4 + j2$.

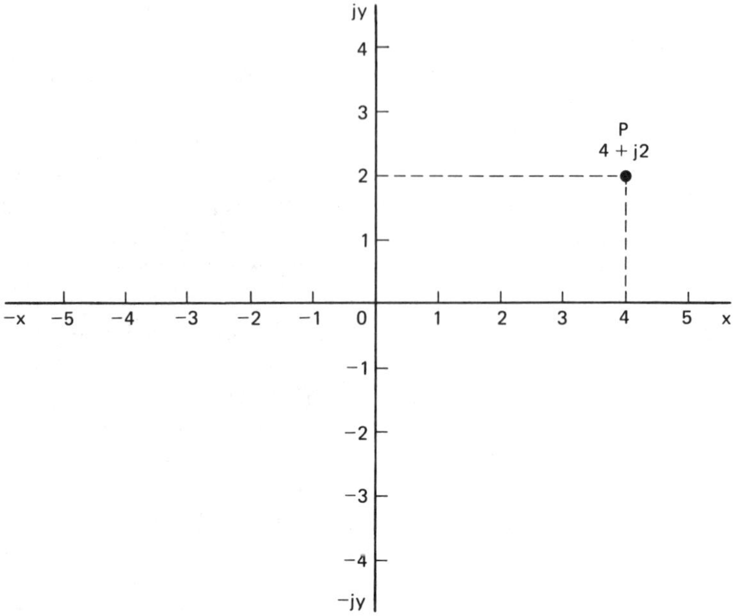

Figure 2–4 Complex number plane: *jy*, imaginary axis; *x*, real axis; point *P*, complex number $4 + j2$.

While we have provided a geometrical representation of imaginary and complex numbers, the student may feel uneasy and suspect that we have not really defined an imaginary number. This difficulty perhaps can be traced back to the original number-line concept. The elementary student may readily grasp the association of positive numbers with points on a line, but may have more difficulty with associating negative numbers with points to the left of zero. Fortunately for our teachers, young pupils have much faith and usually can be talked into accepting the idea of a negative number. The positive number 3, for example, can be associated with 3 dollars, which the pupil can hold in his hands. He cannot hold -3 dollars in his hands. Negative numbers are *operations*; in this example, one that may have resulted from the act of writing an IOU for 3 dollars.

To associate $(-1)^{1/2}$ with an operation, we consider the following series of multiplications of the number j:

$$\begin{aligned} & 1 \\ j \times 1 &= j \\ j \times j &= -1 \\ j \times (-1) &= -j \\ j \times (-j) &= 1 \end{aligned}$$

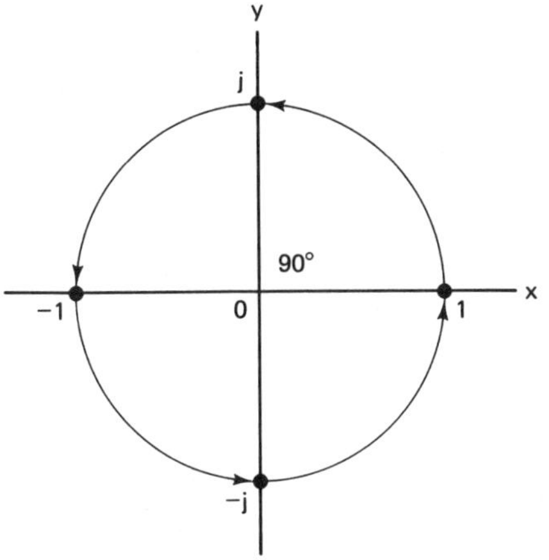

Figure 2–5 Representation of the numbers $1, j, -1$, and $-j$ on the complex plane.

We see that after four multiplications we have returned to our starting point, unity. The results of these multiplications are shown graphically in Fig. 2–5. It is evident that each multiplication by j is equivalent to a rotation of 90°.

Evidently, a mathematical tool such as j is useful for describing current and voltage relations in an impedance. We know, for example, that for a resistor and capacitor in series, the voltage across the resistor will be in phase with the current through the resistor, but the voltage across the capacitor will lag the current by 90°. We have, for a resistor and capacitor in series,

$$Z = R - jX_C$$

where $X_C = 1/\omega C$ is the capacitive reactance. From Ohm's law we have

$$\begin{aligned} V &= iZ \\ &= iR - jiX_C \end{aligned}$$

The first term on the right side of the equation is a voltage drop which is in phase with the current, but the second term contains a factor $-j$ which indicates the current leads the voltage drop across the capacitor by 90°, as shown in Fig. 2–6.

The addition and subtraction of complex numbers is accomplished through the separate addition and subtraction of the real and imaginary components:

$$\begin{array}{c} a + jb \\ \underline{c + jd} \\ a + c + j(b + d) \end{array} \qquad \begin{array}{c} a + jb \\ \underline{-(c + jd)} \\ a - c + j(b - d) \end{array}$$

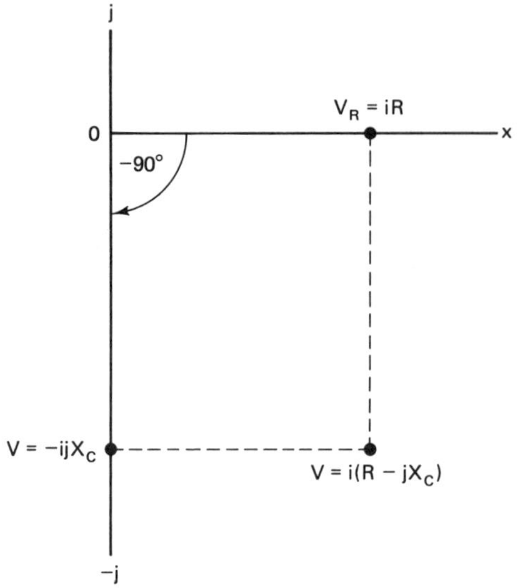

Figure 2–6 Voltage diagram on the complex plane showing the use of $-j$ to represent a phase advance of 90°. The voltage drop across the resistor is $V_R = iR$; the drop across the capacitor is $V_C = -jiX_C$, where $-j$ indicates the current is leading the voltage V_C by 90°; the total voltage across the circuit is represented by the complex number $V = iR - jiX_C$.

EXAMPLE 2–9

Add the complex numbers $6 + j2$ and $3 + j3.6$.

Solution

$$\begin{array}{r} 6 + j2 \\ \underline{3 + j3.6} \\ 9 + j5.6 \end{array}$$

EXAMPLE 2–10

Add $36.3 - j27.2$ and $20.1 + j13.0$.

Solution

$$\begin{array}{r} 36.3 - j27.2 \\ \underline{20.1 + j13.0} \\ 56.4 - j14.2 \end{array}$$

EXAMPLE 2–11

Subtract $10 - j6$ from $6 + j12$.

Solution

$$\begin{array}{r} 6 + j12 \\ \underline{-(10 - j6)} \\ -4 + j18 \end{array}$$

The multiplication and division of complex numbers is accomplished by first transforming the complex number from rectangular to polar form. In Fig. 2–7 we

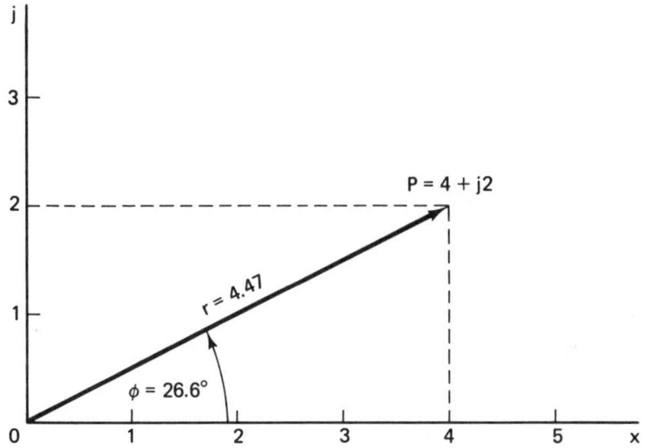

Figure 2–7 Rectangular (x, jy) and polar (r, ϕ) coordinates of a point P on the complex plane.

show the complex number $4 + j2$ represented by the point P in *rectangular coordinates*; that is, by its components on the (x, jy)-plane. Evidently, the point P also can be represented by its distance r from the origin and the angle ϕ at which such distance is measured. This alternative description of P is $r\angle\phi$, where r and ϕ are called the *polar coordinates* of P.

EXAMPLE 2–12

Find the polar coordinates of the point P at $4 + j2$, shown in Fig. 2–7.

Solution Enter 4 on a calculator and depress the rectangular to polar conversion key usually marked R → P. On some calculators this may require an initial depressing of the INV key. Various manufacturers have used different designations, such as REC − POL, X − ρ, and so on. Consult the instruction manual for your calculator. Now enter the number 2 and depress the = key to obtain 4.472. Depress the X − Y key to obtain 26.57. The answer is $4.472\angle 26.57°$, as illustrated in Fig. 2–7.

If a calculator is not available, the transformation from rectangular to polar coordinates can be accomplished through the following trigonometric formulas:

$$\phi = \tan^{-1}\frac{y}{x} \tag{2–13}$$

$$r = (x^2 + y^2)^{1/2} \tag{2–14}$$

EXAMPLE 2–13

Transform $4 + j2$ into polar form using Eqs. (2–13) and (2–14).

Solution The calculation of the angle is

$$\phi = \tan^{-1}\frac{y}{x}$$

$$= \tan^{-1}\frac{2}{4}$$

$$= \tan^{-1}(0.5)$$
$$= 26.57°$$

If a calculator is available, enter 0.5 and depress keys INV and TAN to obtain 26.57°. If a calculator is not available, look through a table of tangents to find 0.5 and read the corresponding angle, 26.57°.

From Eq. (2–14), which is the Pythagorean rule, we have

$$r = (x^2 + y^2)^{1/2}$$
$$= (4^2 + 2^2)^{1/2}$$
$$= 4.472$$

The answer is $4.472\underline{/26.57°}$.

EXAMPLE 2–14

Transform the complex number $60\underline{/35°}$ into rectangular form.

Solution Enter 60 on the calculator and depress the P → R key, now enter 35 (after checking to see if the calculator is in the DEGREE mode) and depress the = key to obtain 49.15. Depress the X – Y key to obtain 34.41. The answer is $49.15 + j34.41$.

The calculator uses a memory in making these transformations. Always clear the memory before starting, or the answers will be incorrect.

One of the reasons for transforming into polar coordinates is to take advantage of the simple rules for multiplication and division of complex numbers in polar form. To multiply two complex numbers, multiply the amplitudes and add the angles:

$$(a_1\underline{/\phi_1})(a_2\underline{/\phi_2}) = a_1a_2\underline{/\phi_1 + \phi_2} \tag{2–15}$$

EXAMPLE 2–15

Multiply $4 + j2$ and $3 - j3$.

Solution In polar form we have

$$4 + j2 = 4.472\underline{/26.57°}$$

and

$$3 - j3 = 4.243\angle{-45^\circ}$$

The resultant amplitude is 4.472 × 4.243 = 18.97, and the resultant phase is 26.57 − 45 = −18.43°. The answer is $18.97\angle{-18.43^\circ}$. In rectangular form the answer is 18.00 − j6.00.

Check:

$$\begin{aligned}(4 + j2)(3 - j3) &= 12 + 6 + j6 - j12\\ &= 18 - j6\end{aligned}$$

To divide complex numbers, divide the amplitudes and subtract the angle of the divisor from the angle of the dividend:

$$\frac{a_1\angle{\phi_1}}{a_2\angle{\phi_2}} = \frac{a_1}{a_2}\angle{\phi_1 - \phi_2} \tag{2–16}$$

EXAMPLE 2–16

Find the quotient (4 + j2)/(3 − j3).

Solution We know the polar forms from Example 2–15; thus

$$\begin{aligned}\frac{4 + j2}{3 - j3} &= \frac{4.472\angle{26.57^\circ}}{4.243\angle{-45^\circ}}\\ &= 1.054\angle{26.57^\circ + 45^\circ}\\ &= 1.054\angle{71.57^\circ}\end{aligned}$$

While multiplication and division are easier in polar form, addition and subtraction are easier in rectangular form.

EXAMPLE 2–17

Subtract $3\angle{-20^\circ}$ from $6\angle{30^\circ}$.

Solution We first convert to rectangular form:

$$\begin{aligned}6\angle{30^\circ} &= \quad 5.196 + j3.000\\ -(3\angle{-20^\circ}) &= \underline{-(2.819 - j1.026)}\\ &\quad\ \ 2.377 + j4.026\\ &= 4.675\angle{59.44^\circ}\end{aligned}$$

EXAMPLE 2–18

A 6-kHz signal generator is connected to a 15-Ω resistor in series with a 2-μF capacitor. What is the phase angle between the current and the total voltage across the circuit?

Solution The reactance is

$$X_C = \frac{-j}{2\pi f C}$$

$$= \frac{-j}{2\pi \times (6 \times 10^3) \times (2 \times 10^{-6})}$$

$$= -j13.26\ \Omega$$

The total impedance is

$$Z = 15 - j13.26$$
$$= 20.02\underline{/-41.48^\circ}$$

The current leads the voltage by 41.48°.

If a calculator is not available, the transformation from polar to rectangular form can be accomplished through the equations

$$x = r \cos \phi \qquad (2\text{–}17)$$
$$y = r \sin \phi \qquad (2\text{–}18)$$

EXAMPLE 2–19

Express $27\underline{/-30^\circ}$ in rectangular form.

Solution From Eqs. (2–17) and (2–18) we have

$$x = 27 \cos -30^\circ$$
$$= 23.38$$
$$y = 27 \sin -30^\circ$$
$$= -13.50$$

The answer is $23.38 - j13.50$. If a calculator is not available, the sine and cosine of −30° will have to be found in a handbook of trigonometric functions.

The *modulus* or *amplitude* of a complex number is its magnitude r, when it is expressed in polar form as $r\underline{/\phi}$. If the number is in rectangular form $x + jy$, its modulus can be found by taking the square root of the product of the number and its complex conjugate:

$$\begin{aligned} r &= [(x + jy)(x - jy)]^{1/2} \\ &= (x^2 + jxy - jxy + y^2)^{1/2} \\ &= (x^2 + y^2)^{1/2} \end{aligned} \tag{2–19}$$

The *complex conjugate* of a number is obtained by changing the sign of its imaginary terms; for example, the complex conjugate of $6 - j14 + 3 + j2$ is $6 + j14 + 3 - j2$. The complex conjugate is usually indicated by an asterisk: if the number is z, its conjugate is z^*. Equation (2–19) is the rule of Pythagoras [Eq. (1–23)].

In circuit theory, as well as microwave electronics, it is necessary to be able to manipulate imaginary numbers, and the following should be understood.

$$\begin{aligned} j^2 &= -1 \\ j(-j) &= +1 \\ \frac{1}{j} &= \frac{1}{j}\left(\frac{j}{j}\right) = \frac{j}{-1} = -j \\ jb &= b\angle 90^\circ \end{aligned} \tag{2–20}$$

2–5 COMPLEX REFLECTION COEFFICIENTS

If normalized impedance has reactive as well as resistive components, $z = r \pm jX$, the reflection coefficient given by Eq. (2–11) will be a complex number. As we noted earlier, the reflection will produce a change in phase as well as a change in amplitude.

EXAMPLE 2–20

A 50-Ω transmission line is connected to an inductive load of $50 + j50$ Ω. Find the reflection coefficient.

Solution We start by normalizing the load impedance,

$$z = \frac{50 + j50}{50} = 1 + j$$

From Eq. (2–11) we now have

$$r = \frac{z - 1}{z + 1}$$

$$= \frac{1 + j - 1}{1 + j + 1}$$

$$= \frac{j}{2 + j}$$

$$= \frac{j}{2 + j}\left(\frac{2 - j}{2 - j}\right)$$

$$= \frac{1 + j2}{5} = 0.2 + j0.4$$

$$= 0.4472\angle 63.43°$$

Alternative procedure:

$$r = \frac{j}{2 + j}$$

$$= \frac{\angle 90°}{2.236\angle 26.57°}$$

$$= 0.4472\angle 63.43°$$

On a calculator enter 2, depress the R → P key and enter 1, depress "=" to obtain 2.236, key 1/X to obtain 0.4472; now depress key X − Y to obtain 26.57°, key +/− to obtain −26.57°, the addition key +, enter 90, key = to obtain 63.43°.

2–6 THE SMITH CHART

If two-figure accuracy is acceptable, the preceding calculations for finding complex reflection coefficients can be avoided by using a type of nomograph called a *Smith chart* (Fig. 2–8). The chart is formed of three families of circles. The first family of circles has centers on the upper right side of the figure, and the intersections of these circles with the upper edge of the chart provide a scale for measuring inductive reactances. This scale is labeled "inductive reactance component ($+jX/Z_0$)" on the upper left edge of the chart. A second family of circles, with centers somewhere to the right and below the chart, provides a capacitive reactance scale, labeled on the lower left edge of the chart. For ease in reading, the reactance scale is duplicated on a circle through the center of the chart, the center being the point 1.0 on the horizontal diameter. A third family of circles, with centers on the horizontal diameter, intersects the horizontal diameter, and these intersections provide a scale of normalized resistances. This scale is labeled "resistance component (R/Z_0)" at the left center of the chart. Thus we have three families of circles providing for the representation of inductive reactance, capacitive reactance, and resistance. The "radially scaled parameters" at the bottom of the chart will be described as they are used.

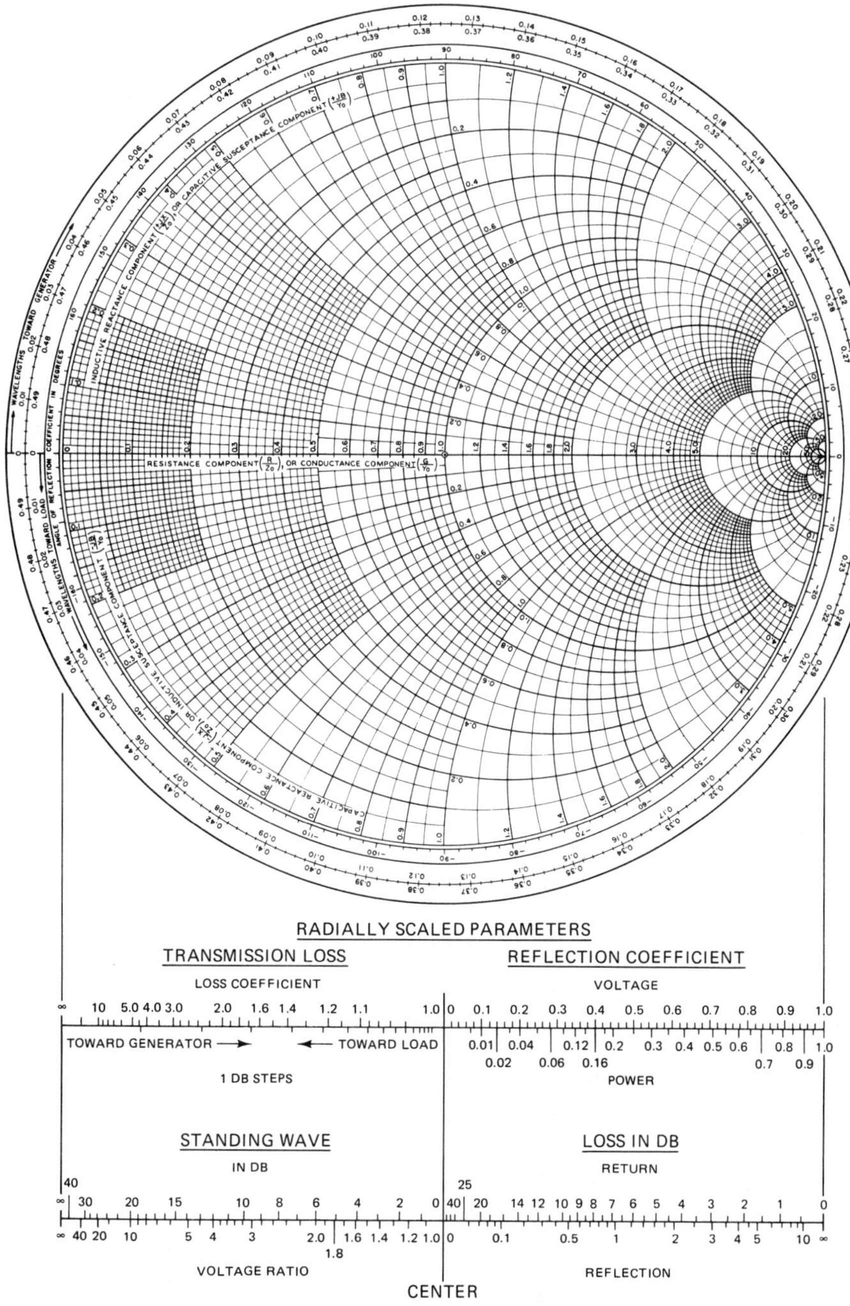

Figure 2–8 Smith chart, impedance normalized to unity.

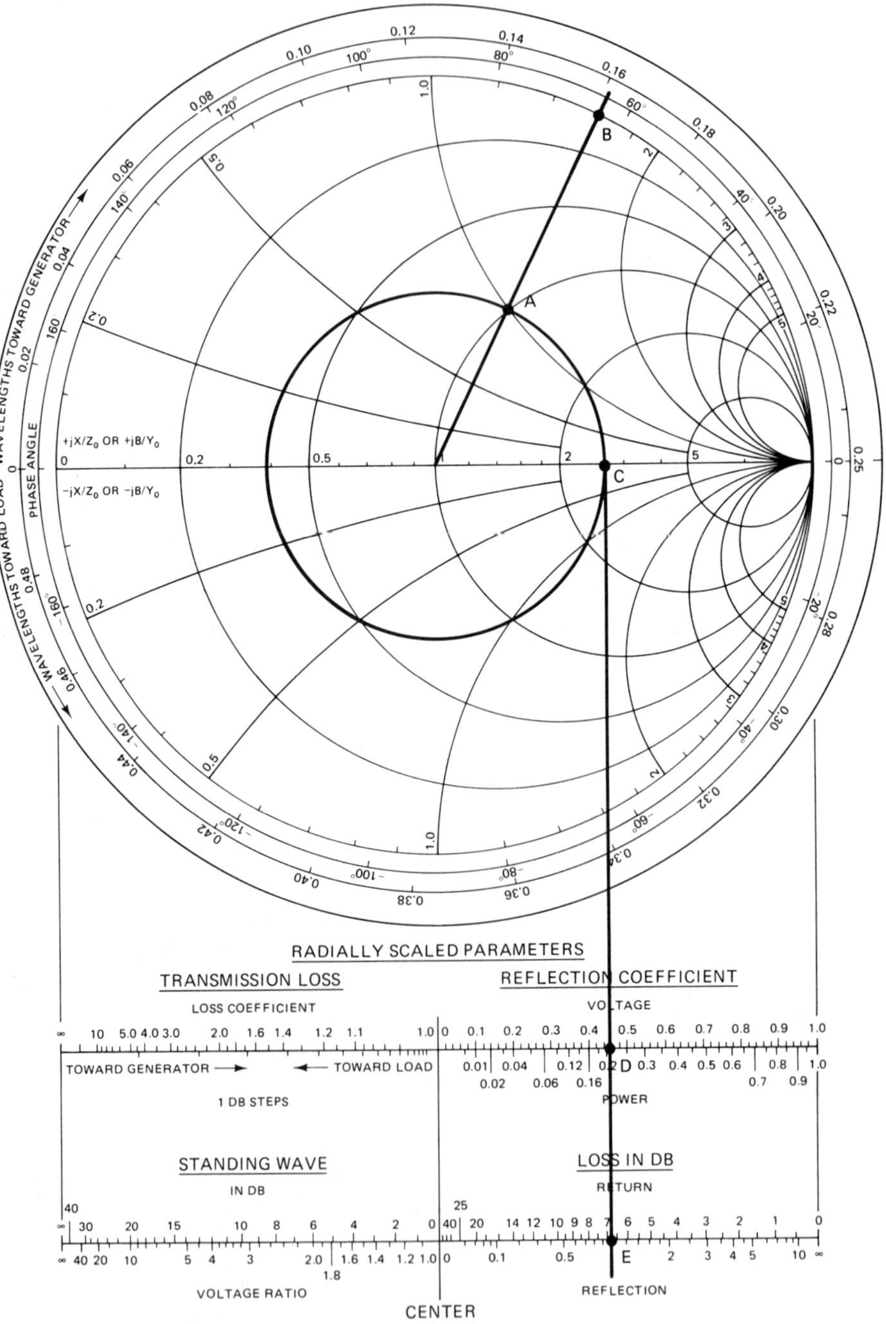

Figure 2–9 Smith chart solution to the problem of finding the complex reflection coefficient for a line terminated in an inductive load.

EXAMPLE 2–21

Repeat Example 2–20 using a Smith chart.

Solution Plot the normalized impedance $1 + j$ at the point A (1, 1), as shown in Fig. 2–9. Draw a radial line from the center of the chart through point A and extend it until it intersects the second circular scale at the edge of the chart at point B. At the left edge of the chart this scale is labeled "angle of reflection coefficient in degrees." Reading the scale at point B we find $\phi_r = 64°$.

To find the amplitude of the reflection coefficient we use a compass at the center of the chart and draw a circle through point A. At the intersection of this circle and the horizontal diameter, at the point C, draw a line, parallel to the edge of the paper, down to the scales at the bottom of the page. The first of these scales, labeled "ref'l coeff vol"—which stands for the voltage reflection coefficient—gives the amplitude of the reflection coefficient. Reading the intersection at point D we find 0.45.

The answer is $r = 0.45\angle 64°$, in agreement with our result for Example 2–20.

Additional information is available from the Smith chart. For Example 2–21 we can obtain the VSWR by reading the scale at point C: VSWR = 2.6.

Check: From Eq. (2–6) and Example 2–20 we have

$$\begin{aligned} \text{VSWR} &= \frac{1 + r}{1 - r} \\ &= \frac{1 + 0.4472}{1 - 0.4472} \\ &= 2.618 \end{aligned}$$

Because VSWR is always a real number, we used only the amplitude of the reflection coefficient in the right side of Eq. (2–6).

We can find the power reflection coefficient on the second scale at the bottom of the Smith chart. It is labeled "ref'l coeff pwr," and at point D we find its value to be 0.20.

Check: From Eq. (2–12) and Example 2–20 we have

$$\begin{aligned} \Gamma &= r^2 \\ &= (0.4472)^2 \\ &= 0.2000 \end{aligned}$$

The lowest of the scales on the right side of the Smith chart, labeled "loss in dB retn" and "loss in dB refl," give the return loss and reflection loss owing to a reflection. These losses are always expressed in decibels.

Return loss is defined as

$$10 \log \frac{1}{r^2} \tag{2–21}$$

If there is total reflection, $r = 1$, all the energy is returned, there is no loss in the return, $10 \log (1) = 0$. If there is no reflection, $r = 0$, nothing is returned, the return loss is infinite, $\log (0) = \text{infinity}$.

EXAMPLE 2–22

Find the return loss in the example shown in Fig. 2–9.

Solution We extend the vertical line until it intersects the lowest scale at point E, and at the intersection we find 7.0 dB.
Check: In expression (2–21), enter the reflection amplitude found in Example 2–20

$$\text{return loss} = 10 \log \frac{1}{r^2}$$

$$= 10 \log \frac{1}{0.4472^2}$$

$$= 6.99 \text{ dB}$$

On a calculator enter 0.4472 and depress the key X^2, depress 1/X to obtain 5.000, depress the key LOG to obtain 0.698996 bel, multiply the answer by 10 to obtain the answer in decibels, 6.990 dB.

Reflection loss also is measured in decibels and is defined as

$$-10 \log (1 - r^2) \tag{2–22}$$

If no power is reflected, $r^2 = 0$, there is no loss in delivering power to the load, $-10 \log (1) = 0$. If all the power is reflected, $r^2 = 1$, there is infinite loss in attempting to deliver power to the load, $-10 \log (0) = \text{infinity}$.

EXAMPLE 2–23

For the same problem (Fig. 2–9), find the reflection loss.

Solution At point E in Fig. 2–9 we read the reflection loss scale and find the value 1.0 dB.

Check: From expression (2–22) we have

$$\begin{aligned}\text{reflection loss} &= -10 \log (1 - r^2)\\ &= -10 \log (1 - 0.4472^2)\\ &= 0.09690 \text{ bel}\\ &= 0.9690 \text{ dB}\end{aligned}$$

2–7 MEASUREMENT OF IMPEDANCE

We shall conclude with the problem mentioned at the beginning of this chapter, how to determine, on a transmission line, a load impedance by measuring the wave reflected from the load. We shall assume that a way has been found to measure the phase and amplitude of the reflected wave and that the characteristic impedance of the transmission line is known. Let the measurement result be $r = r_0\underline{/\phi_r}$. We can find the VSWR from Eq. (2–6), VSWR $= (1 + r_0)/(1 - r_0)$. We can plot this value of the VSWR on the horizontal diameter of the Smith chart. We can now draw a circle through the VSWR point, draw a radius through the phase of the reflection coefficient, and at the intersection of the radius and the circle we find the normalized impedance, z.

EXAMPLE 2–24

Find the impedance of a load terminating a 100-Ω transmission line if the reflection coefficient is $0.4\underline{/-120^\circ}$.

Solution We start by marking 0.4 on the reflection amplitude scale as shown at point A in Fig. 2–10. Extending a line, parallel to the edge of the paper, to its intersection with the resistive axis at point B we read a VSWR of 2.5. Alternatively, we could have calculated VSWR from Eq. (2–6), as explained above, and plotted it directly on the resistive axis. We now draw a circle through point B, as shown. This circle is called the *VSWR circle*. We next locate the phase angle -120° at point C on the chart and draw a radial line from the center of the chart to point C. The radial intersects the VSWR circle at D, and the coordinates of D are the components of the normalized impedance.

In reading the coordinates of D we follow the resistance circle through D to its intersection with the resistive scale at 0.55. Similarly, we follow the other circle through D to its intersection with the reactance scale at -0.45. The normalized impedance is $z = 0.55 - j0.45$.

We now find the load impedance by reversing the normalization procedure,

$$\begin{aligned}Z_L &= Z_0 z\\ &= (100)(0.55 - j0.45)\\ &= 55 - j45\ \Omega\end{aligned}$$

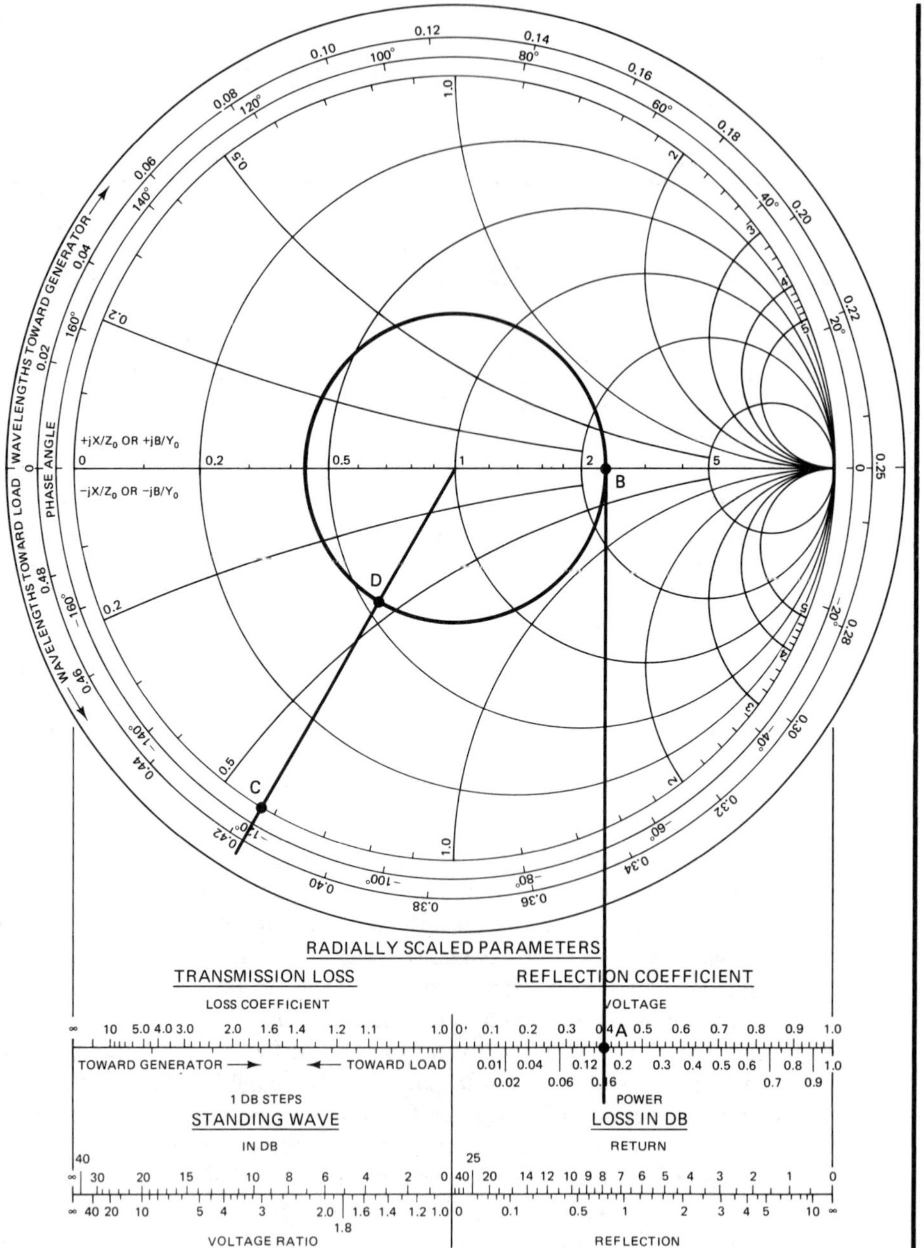

Figure 2–10 Smith chart used to find the impedance from a complex reflection coefficient: *A*, amplitude of the reflection coefficient; B, VSWR from which the VSWR circle is drawn; *C*, phase of the reflection coefficient; *D*, normalized impedance.

The technique of measuring the VSWR and phase of a reflected wave is routine in industry and is an important part of the training of an electronics technician. It will be considered in detail in Application 2–1.

DEMONSTRATION 2–1

Application 2–1 requires either the measurement of standing waves in waveguide or the measurement of standing waves on a rope line. It is helpful if an instructor demonstrates the production and detection of electromagnetic standing waves in the laboratory so that the students can observe the following:

- How to turn on a microwave generator and adjust the modulation frequency to match the operating frequency of a tuned amplifier or VSWR meter
- How to identify a pad, slotted line, waveguide short, and a VSWR meter
- How to adjust the depth of a slotted-line probe
- How to replace the diode in the probe detector
- How to read the vernier scale on the slotted-line carriage (this may require a short lecture by the instructor)
- How to use a VSWR meter as a null detector

A schematic diagram of the equipment required is shown in Fig. 2–11. Turn on the generator and after it has warmed up adjust the modulation frequency to approximately 1 kHz, or whatever frequency is appropriate for the amplifier being used. Adjust the probe depth, slotted-line carriage position, and the carrier frequency until a signal is detected on the VSWR meter. The VSWR meter should initially be on a sensitive scale range. After a signal is detected, maximize the signal by readjusting the carrier frequency, modulation frequency, carriage position, and the barrel of the probe; adjust the VSWR meter to less sensitive scale ranges as the signal is increased. When the equipment is adjusted,

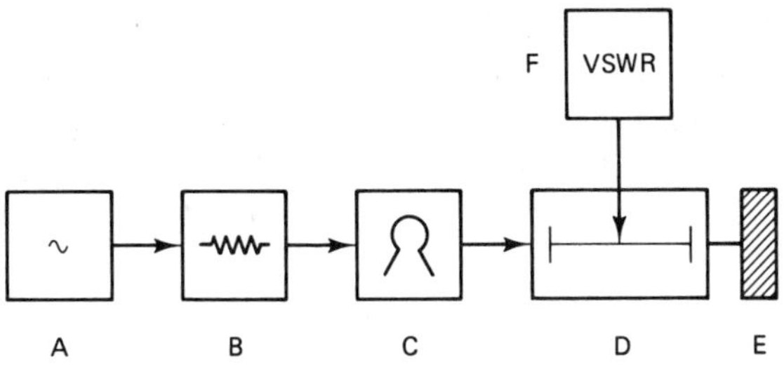

Figure 2–11 Apparatus for the demonstration of electromagnetic standing waves: *A*, signal generator; *B*, pad to provide 5 to 10 dB of attenuation for the suppression of multiple reflections; *C*, optional, a resonant cavity frequency meter; *D*, a slotted line with carriage and probe; *E*, a short or reflecting plate; *F*, a VSWR meter, including amplifier and detector. An oscilloscope can be used in place of the VSWR meter.

the signal-to-noise ratio should be 20 to 40 dB in order to make useful measurements.

Demonstrate the appearance of standing-wave maxima and minima as the carriage is moved. Measure the waveguide wavelength by measuring the distance between standing-wave nulls. Explain, qualitatively, why the guide wavelength is greater than the free-space wavelength (Fig. 1–19). Demonstrate the use of the cavity frequency meter.

APPLICATION 2–1: Measurement of Standing Waves

PART 1: STANDING WAVES IN WAVEGUIDE

Our objective is to learn how to measure the amplitude of a reflected wave and to determine SWR. If microwave equipment is not available, the principles can be learned by making measurements on standing waves on a rope line, as described in Part 2.

Equipment and Materials

In addition to the equipment described in Demonstration 2–1, a waveguide variable attenuator and a waveguide short are required.

Procedure

Set up the equipment shown in Fig. 2–12. Choose an operating frequency suitable to the size of the waveguide (Table 6–1). Terminate the waveguide in a load consisting of a short-circuited variable waveguide attenuator. The variable attenuator will be used to control the amplitude of the reflected wave, the greater the attenuation the smaller the amplitude.

VSWR can be measured by moving the slotted-line probe from a maximum

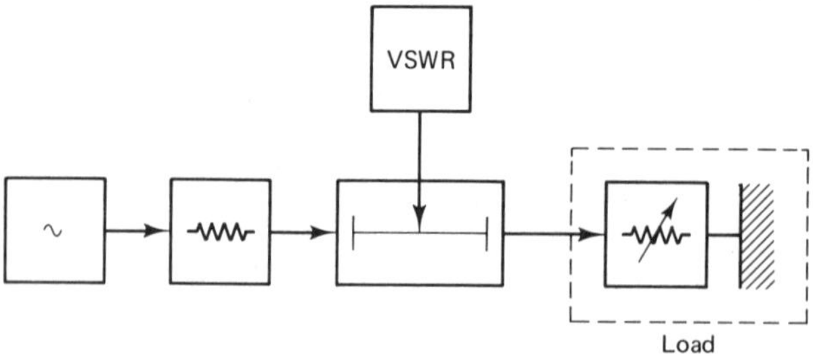

Figure 2–12 Short-circuited variable attenuator used as a resistive load for producing reflected waves of various amplitudes.

to a minimum. At the maximum, adjust the gain of the VSWR meter so that the scale pointer is at unity. The scale on the meter is calibrated to read VSWR directly when the probe is moved to a minimum. Additional scales usually are provided for reading the logarithm of the VSWR in decibels.

Notice that if the load were a perfect short, the reflected amplitude would equal the incident, the denominator of Eq. (2–10) would vanish, and the VSWR would be infinite. In practice, the difference, $1 - r$, is limited to the noise level, and the VSWR for a shorted line becomes a measure of the *dynamic range* of the microwave system. A dynamic range of at least 20 dB is needed for accurate VSWR measurements.

Before taking a set of measurements, slide the probe along the slotted line to see if there is more than one set of standing waves. A series of nonuniformly spaced maxima and minima at different amplitudes is an indication of two or more standing-wave patterns. Check the generator to be sure that power is not being delivered at two or more frequencies; change the operating frequency if necessary. Always use a pad in the generator output to attenuate possible multiple reflections.

Make a series of 10 VSWR measurements, for various settings of the variable attenuator, at an operating frequency near the center of the guide bandwidth. Repeat the measurements at a frequency near the guide cutoff (see Table 6–1). Use Eqs. (2–10) and (2–11) to determine the corresponding reflection coefficients.

$$r = \frac{\text{VSWR} - 1}{\text{VSWR} + 1}$$

You can check your measurements by also determining the reflection coefficients from the dial settings of the attenuator, which of course, you recorded in your notebook. A dial setting, N, is an expression for the reduction of the incident power in decibels. Since, in this case, the energy incident on the short circuit passes through the attenuator twice before emerging as a reflected wave, we have to multiply the dial setting by 2 to find the total power attenuation. We divide $2N$ by 10 to change decibels to bels and take the inverse logarithm to find the ratio of the incident to reflected power: $\log^{-1}(2N/10) = P_0/P_r$. The power or energy carried by a wave is proportional to the square of the amplitude, $P_0/P_r = a_0^2/a_r^2 = 1/r^2$, and we have

$$r = \left(\log^{-1}\frac{N}{5}\right)^{-1/2} \tag{2–23}$$

EXAMPLE 2–25

Find the reflection coefficient when the attenuator is set to 3 dB.

Solution From Eq. (2–23) we have

$$r = \left(\log^{-1}\frac{3}{5}\right)^{-1/2} = 0.5$$

On a calculator enter 3, divide by 5 to obtain 0.6. Press INV and LOG keys to obtain 3.98, then press square-root key to obtain 1.995 and $1/x$ key to obtain 0.5.

Check: The power reflection coefficient is $\Gamma = r^2 = (0.5)^2 = 0.25$, or -6 dB. The attenuator setting is half of this, or 3 dB.

Present your results in tables, compare amplitude measurements made by using the slotted line to those determined from the attenuator settings, and compute differences, averages, and percentage errors. If the difference between average values, determined from the two methods of measurement, appreciably exceeds 5%, there may be systematic error in one or both of your measurement techniques. This is a scientific way of saying you are making a mistake.

PART 2: STANDING WAVES ON ROPE LINES

The observation of standing waves on rope lines is quite instructive; however, it requires some manual dexterity as well as ample room. It may be advisable, especially if the class is large, for the instructor to demonstrate the experiments, making measurements with the help of two or three students. The remainder of the class can record the data and use the results to write their own reports. The equations for wave velocity and reflection coefficient will be verified by measuring the frequency, wavelength, and amplitude of standing waves on rope lines of different linear densities.

Remark. We specify definite values for the hardware to be used; however, these values are for guidance only. Within reason, any convenient or available ropes may be used.

Equipment and Materials

30-ft rope, 1.188 oz/ft
30-ft rope, 0.500 oz/ft
stopwatch
12-ft tape measure
beam balance, or any small weight scale

Procedure

Weigh the lines, measure their lengths, and determine their linear densities. Join the light and heavy lines and fasten the other end of the heavy line to a fixed object.

While moving the free end of the light line rapidly up and down, adjust the length until the mode shown in Fig. 2–13 is generated. A slight circular motion

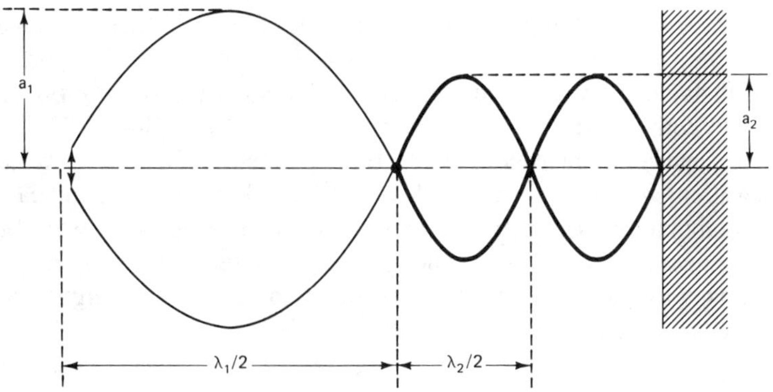

Figure 2–13 Standing-wave mode on rope lines of different linear densities.

may be helpful. The length can be adjusted by starting with a fairly short length and allowing the line to slip through the fingers while backing away from the fixed end. If the mode shown is difficult to obtain, any mode will do, provided that there is a distinct null at the junction of the two lines.

The measurements are best made by two or three students working together. The frequency can be determined by count by the person generating the waves, while the count is clocked by a second person. Amplitudes and wavelengths can be measured by holding a tape, vertically or horizontally, behind the line while a third person sights the amplitude or wavelength.

Measure frequency, standing wavelengths, and amplitudes on the light and heavy lines for three trials. If possible, repeat the measurements for a different mode.

Analysis

From Eq. (1–4) we know that the wave speeds on the light and heavy lines are $v_1 = (T/\rho_1)^{1/2}$ and $v_2 = (T/\rho_2)^{1/2}$. Since the tension must be the same in both sections of line, we can divide the equations to obtain

$$\frac{v_1}{v_2} = \left(\frac{\rho_2}{\rho_1}\right)^{1/2} \tag{2–24}$$

From Eq. (1–2) we have $v_1 = \lambda_1 f$ and $v_2 = \lambda_2 f$; again dividing equations, we obtain

$$\frac{v_1}{v_2} = \frac{\lambda_1}{\lambda_2} \tag{2–25}$$

We find, by combining Eqs. (2–25) and (2–24),

$$\frac{\lambda_1}{\lambda_2} = \left(\frac{\rho_2}{\rho_1}\right)^{1/2} \tag{2–26}$$

and we see that the ratio of the standing wavelength measurements should agree with the square root of the density measurements. In your report tabulate the wavelength and density data and compare the left and right sides of Eq. (2–26). Agreement can be considered to be a partial verification of Eq. (1–4).

Determine the velocities on the two sections of line from Eq. (1–2) and your measurements of frequency and wavelength. Tabulate data and calculated velocities, then tabulate values of the reflection coefficient calculated from Eq. (2–9).

Include in your report calculations of the tension, using Eq. (1–4). Most people perform this experiment using a tension of about 2 to 4 lb weight (64 to 128 poundals). Are your results reasonable?

SUMMARY

A knowledge of the theory of wave reflections helps us to understand the formation of standing waves, measure impedance, and match loads to transmission lines.

A reflection coefficient is defined as the ratio of reflected and incident amplitudes of a wave and can be related to the change in propagation speed on either side of an interface between two media,

$$r = (v_1 - v_2)(v_1 + v_2)$$

or the change in refractive index,

$$r = (\eta_2 - \eta_1)(\eta_2 + \eta_1)$$

or the change in impedance,

$$r = \frac{z - 1}{z + 1}$$

The last two expressions are also valid if η or z are complex, in which case the reflection coefficient is complex.

That portion of a wave that is not reflected is transmitted into the second medium, and the transmitted portion is expressed as a transmission coefficient, the ratio of the transmitted amplitude to the incident amplitude,

$$t = \frac{a_t}{a_0}$$

The voltage standing-wave ratio is the ratio of maximum and minimum values in the voltage standing-wave pattern, that is, the ratio of the sum of the incident and reflected amplitudes and their difference,

$$\text{VSWR} = \frac{1 + r}{1 - r}$$

$$= |z|$$

Reflection coefficients, real or complex, can be calculated with the aid of a Smith chart. The accuracy of such determinations is sufficient for most practical applications. The ability to read data on a Smith chart is also useful in reading manufacturers' catalogs, since device performance often is presented on Smith charts.

Because reflection and transmission coefficients may be complex we need to be able to make calculations with complex numbers. A scientific calculator is very useful for converting complex numbers to and from rectangular and polar representations. Such conversions simplify the operations of adding, subtracting, multiplying, and dividing complex numbers.

The relative power reflected from a mismatched load is proportional to the square of the amplitude reflection coefficient, and the power reflection coefficient can be expressed as

$$\Gamma = \frac{P_r}{P_0}$$

$$= \frac{I_r}{I_0}$$

$$= r^2$$

If all of the incident power is reflected, there is no loss in the "return"; the return loss,

$$10 \log \frac{1}{r^2} \qquad \text{dB}$$

is zero. If no power is reflected, all of the power is delivered to the load and the reflection loss,

$$-10 \log (1 - r^2) \qquad \text{dB}$$

is zero.

In this chapter we learned how to use a Smith chart to find the impedance of a load from measurements of the amplitude and phase of the wave reflected from the load, assuming that the measurements had already been made. Later we learn how to make such measurements.

PROBLEMS

2–1. A laser beam is incident on a slab of dielectric at an angle of 30°. A receptor on the lower surface of the slab detects the beam at an angle of refraction of 19.47°. What are the refractive index and dielectric constant of the dielectric? Assume that the slab is in air and the refractive index of air is unity.

2–2. An electromagnetic wave in air enters at normal incidence into a slab of glass of dielectric constant 5.5. If the amplitude in air is 150 *v*/m, what will it be in the glass?

2–3. In Problem 2–2, what is the reflected amplitude? What VSWR would be detected in front of the glass?

2–4. Teflon has a relative dielectric constant of 2.0 and is, of course, nonmagnetic. What is the speed of light in Teflon? What is the characteristic impedance of Teflon?

2–5. What is the power reflection coefficient at an air–Teflon interface?

2–6. The VSWR is 3.6 on an 80-Ω transmission line terminated in a resistive load. What is the impedance of the load if the phase change on reflection is 180°?

2–7. Perform the following additions and subtractions.

(a) $6 + j18 + 3 + j6$
(b) $25 - j20 + 18 + j10$
(c) $3.5\ 10^4 - j6\ 10^3 + 2700 + j4200$
(d) $-126.3 - j75.0 + (-150.0 + j4200)$
(e) $12.6 + j15.0 - (25 + j12.5)$
(f) $-182 + j53 - (-80 - j26)$

2–8. Transform the following into polar form.

(a) $6 + j18$
(b) $25 - j20$
(c) $(3.5 \times 10^4) - (j6 \times 10^3)$

2–9. Transform into rectangular form.

(a) $26\angle -30°$
(b) $126\angle 270°$
(c) $8\angle 145°$
(d) $10\angle 60°$

2–10. Perform the following operations and express the answers in polar form:

(a) $\dfrac{6 + j18}{3 + j6}$

(b) $\dfrac{25 - j20}{18 + j10}$

(c) $(25 + j20) \times (18 + j10)$
(d) $(6 + j18) \times (3 + j6)$

(e) $\dfrac{4\angle 35°}{1\angle 270°}$

2–11. Find the voltage across the capacitors in Fig. 2–14.

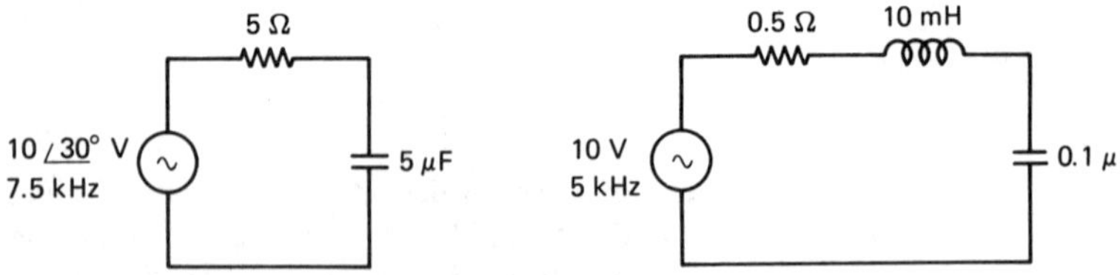

Figure 2–14 Circuit diagrams for Problem 2–11.

2–12. A 50-Ω line is terminated in a load of $100 - j150\ \Omega$. Calculate the amplitude and power reflection coefficients. Repeat the problem, using a Smith chart, and turn in the chart, clearly marked, with your assignment.

2–13. The normalized impedance on a transmission line is $0.5 + j0.35$. Use a Smith chart to find the complex reflection coefficient and the VSWR.

2–14. For Problem 2–13, calculate the return loss and reflection loss. Use a Smith chart to check your answers.

2–15. Find the impedance of a load terminating a 50-Ω line if the reflection coefficient is $0.4\angle 35^\circ$.

2–16. Find the impedance of a load at the end of an 80-Ω transmission line if the reflection coefficient is $0.6\angle 0^\circ$.

2–17. Repeat Problem 2–16 for a reflection coefficient of $0.6\angle 180^\circ$.

3

Impedance-Matching Techniques

3–1 INTRODUCTION

Impedance mismatch causes reflections and inefficiency. Not only is the reflected power not delivered to the load, it may produce standing waves with dangerously high voltages and unwanted radiation. Reflection loss or mismatch loss is a measure of power loss due to reflections and is defined by Eq. (2–22). Another measure of mismatch is return loss, defined by Eq. (2–21).

There are several ways to reduce reflection loss, and one of the most frequently used is the quarter-wave matching section. An optical lens, for example, is made more efficient by coating it with a thin, quarter-wavelength layer of dielectric. Because of the quarter-wave thickness, light reflected from the dielectric–glass interface must travel a half-wavelength farther than light reflected from the air–dielectric interface, and the two reflected waves emerge 180° out of phase. The reflected waves thus cancel. If the reflected amplitudes are equal, the cancellation is complete, and there is no net reflection. All of the energy is transmitted into the lens.

Quarter-wave layers are also used at microwave frequencies. A radar target, for example, if covered with an absorber and a quarter-wave matching layer, would be "invisible" to a search radar. Of course, the matching layer would be effective only at that frequency for which the layer is a quarter-wavelength in thickness, but it is not too difficult to calculate the manner in which the refractivity of the dielectric must vary with frequency to provide a broadband matching layer.

A matching layer can be a section of transmission line between two lines of different characteristic impedance, or a quarter-wavelength section between a transmission line and its terminating load, or a quarter-wavelength section between a source and a load. In any case the section is called a quarter-wave section or a quarter-wave transformer. At lower frequencies the same matching technique can be implemented with lumped-parameter components, and here the name quarter-wave transformer seems more appropriate.

3–2 WAVELENGTH IN A DIELECTRIC

A quarter-wave matching section must have a thickness equal to a quarter of a wavelength *in the matching layer*. A change in velocity, as a wave crosses an interface between two media, not only causes a reflection but also causes a change in wavelength. This may have been noticed in the rope-line experiment, where the standing wavelength in the heavier line is shorter than in the lighter line. Another familiar example is the shoaling of water waves at shallow depths. The speed of a water wave is proportional to the depth of the water. At a shoal the waves move more slowly, the wavelengths are shorter, and the reflected waves form standing waves of amplitude greater than that of the incident waves. The greater amplitude and shorter wavelengths can be hazardous for small boats.

The shortening of the wavelengths follows from the speed equation, $\lambda = v/f$. Since the frequency is the same in both media, when the velocity is smaller, the wavelength must be smaller. If we let $\lambda_0 = c/f$ be the wavelength in free space and divide this equation by the previous equation, we obtain $\lambda_0/\lambda = c/v = \eta$, and we see that refractivity not only is a measure of the speed ratio but also of the wavelength ratio. The wavelength in the medium is

$$\lambda = \frac{\lambda_0}{\eta} \tag{3–1}$$

The thickness of a quarter-wave section is a quarter of the wavelength in the medium, as given by Eq. (3–1).

3–3 QUARTER-WAVE SECTION DESIGN

There are two conditions to be satisfied in quarter-wave section design: a phase difference of 180° between the waves reflected from the first and second faces and, for complete cancellation, waves of equal amplitude. The first condition has already been discussed. To satisfy the second condition we have, from Eq. (2–8), $r_1 = (\eta_1 - \eta_2)/(\eta_1 + \eta_2)$ and $r_2 = (\eta_2 - \eta_3)/(\eta_2 + \eta_3)$, where r_1 and r_2 are the reflection coefficients from the first and second interfaces and η_1, η_2, and η_3 are the refractive indexes of the three media, as shown in Fig. 3–1. The amplitudes of the two reflected waves will be equal if the reflection coefficients are equal, or

$$\frac{\eta_1 - \eta_2}{\eta_1 + \eta_2} = \frac{\eta_2 - \eta_3}{\eta_2 + \eta_3} \tag{3–2}$$

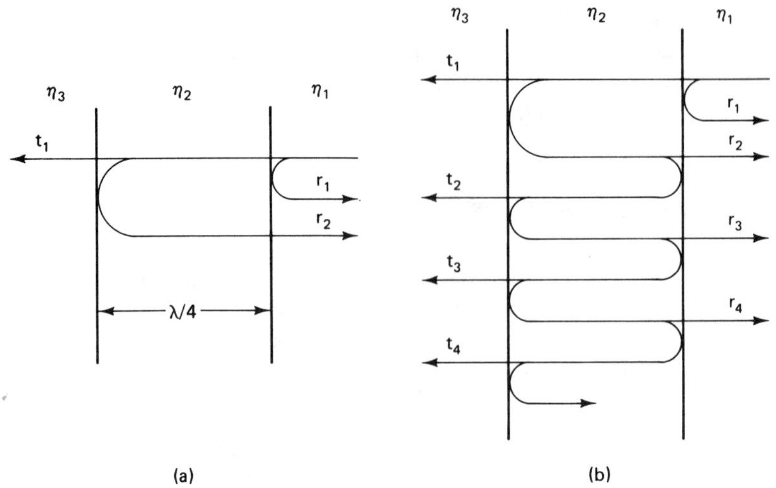

Figure 3–1 (a) Transmitted ray and first and second reflected rays in a quarter-wave matching layer; (b) multiple reflections in a matching layer.

We solve Eq. (3–2) for η_2 and find

$$\eta_2 = \sqrt{\eta_1 \eta_3} \tag{3–3}$$

and we see that the refractivity of the matching layer must be equal to the geometric mean of the refractivities of the media being matched.

In deriving Eq. (3–3) we ignored the multiple reflections that are shown in the right side of Fig. 3–1. If the first and second reflections cancel, the third and fourth, the fifth and sixth, . . . , will cancel pair by pair, and Eq. (3–3) remains valid. The problem of multiple reflections is considered in more detail in Chapter 7.

EXAMPLE 3–1

Use a quarter-wave section to match air to Teflon, $\eta_1 = 1$, $\eta_3 = 1.5$, at an operating frequency of 10 GHz.

Solution We need a matching layer with a refractivity of $(1 \times 1.5)^{1/2} = 1.225$. At 10 GHz the wavelength in free space is $\lambda_0 = (3 \times 10^{10})/(10 \times 10^9) = 3$ cm. The wavelength in the matching layer is $\lambda_0/\eta_2 = 3/1.225 = 2.45$ cm; therefore, the thickness of the matching layer should be $2.45/4 = 0.61$ cm.

We can design a transmission-line transformer by recalling that Eqs. (2–8) and (2–11) relate refractive index to normalized impedance. Our design equation becomes $z_2 = \sqrt{z_1 z_3}$, where the impedances have been normalized to the impedance of free space. We now unnormalize the equation by multiplying it by Z_0, and we obtain

$$Z_2 = \sqrt{Z_1 Z_3} \tag{3–4}$$

EXAMPLE 3–2

Design a quarter-wave transformer to match a 600-Ω generator to a 37.5-Ω load. Assume an operating frequency of 500 MHz.

Solution The required characteristic impedance of a transmission-line matching section is $Z_c = \sqrt{600 \times 37.5} = 150\ \Omega$. If we use twin-conductor line, the wavelength on the line will be the same as the wavelength in free space, $\lambda_0 = c/f = (3 \times 10^{10}\ \text{cm/s})/(500 \times 10^6\ \text{s}^{-1}) = 60$ cm. We need $60/4 = 15$ cm of 150-Ω twin line to form the transformer. We discuss the problem of constructing transmission lines to a specified impedance in Chapter 5. For now, we can do a makeshift job by connecting two pieces of standard 300-Ω twin-conductor line in parallel. Standard 300-Ω line is used for television receivers and is readily available.

EXAMPLE 3–3

Design a quarter-wave section to match 100-Ω coaxial line to a 50-Ω load at 500 MHz.

Solution The characteristic impedance of the matching line is $Z_c = \sqrt{100 \times 50} = 70.7\ \Omega$. The operating wavelength is $\lambda_0 = c/f = (3 \times 10^{10})/(500 \times 10^6) = 60$ cm. If 70-Ω coax is available with an insulation of dielectric constant 2.25, the required length is $\lambda/4 = \lambda_0/4\sqrt{\epsilon_r} = 60/4\sqrt{2.25} = 10$ cm.

So far we have considered only resistive loads. If we want to use a quarter-wave section to match a line to a complex impedance, we must find a point on the line where the impedance is a pure resistance and insert the quarter-wave section at that point. First, we need to know how impedance varies on a transmission line.

3–4 THE IMPEDANCE CONCEPT

Impedance is defined as the ratio of a force to a velocity, $Z = F/v$. In mechanics it is also called drag or the drag coefficient. Anyone who has ridden a bicycle at speed knows that most of the force is exerted to overcome wind drag. When Georg Ohm formulated his law, he presented it as

$$\rho = \frac{E}{i} \tag{3–5}$$

where ρ is the resistivity, or resistance per unit volume, E is the electric force per unit charge, and i is the current density—the amount of current flowing through a cross section of unit area. If we think of current flow as a velocity, we see that Ohm's law conforms to the mechanical definition of impedance.

If we multiply each side of Eq. (3–5) by volume, L^3, where L is length, we have $L^3\rho = EL/(i/L^2)$, or

$$R = \frac{V}{I} \tag{3–6a}$$

which is the form of Ohm's law we use in circuit theory. Here R and I are the total resistance and current, and V is the *work* or energy per unit charge. Work, remember, is force times distance, and work per unit charge is potential or "voltage," $V = EL$. We see from this last equation that force per unit charge can be measured as volts per meter, $E = V/L$; thus electric field strength or *electric field intensity* can be measured as newtons per coulomb or volts per meter.

We can extend the impedance concept to the electromagnetic field in space by once again multiplying Eq. (3–5) by volume, but this time we will arrange the terms differently, $L^3\rho = E/(i/L^3)$, or

$$Z = \frac{E}{I/L} = \frac{E}{H} \tag{3–6b}$$

The last step follows from the definition of the magnetic field strength as ampere turns per meter. Equation (3–6b) defines impedance at any point in space as the ratio of the electric and magnetic field strengths at that point. If the last step seems vague—the assumption of a current in empty space—remember that an alternating current can flow through the empty space between the plates of a capacitor. In the nineteenth century James Maxwell formulated his celebrated theory of electromagnetism on the assumption that such a *displacement current* can exist in space.

If a wave propagates through a medium or a transmission line, the average value of the ratio of the electric and magnetic fields determines the *characteristic impedance* of the medium or the line. Of course, everything changes if there are reflections, for then there are two or more waves present and forming standing waves. The characteristic impedance is appropriate only when there are no reflections, when the line is matched to a load. There also are no reflections when power is first connected to a long line, whether the line is matched or not. For this reason the power engineer calls the characteristic impedance the *surge impedance*.

When there are reflections and standing waves, the impedance varies from point to point along the line. The Smith chart is a convenient way to visualize such variation. In the chart shown in Fig. 3–2, a load impedance of $50 + j50$ ohms has been normalized to a line impedance of $Z_c = 50$ ohms, plotted at point $1 + j$ (point A), and a VSWR circle drawn through A. Let us imagine that we are measuring impedance along the line as we journey away from the load toward the generator. At first the reactance increases to a maximum of about $j1.6 \times 50 = j80$

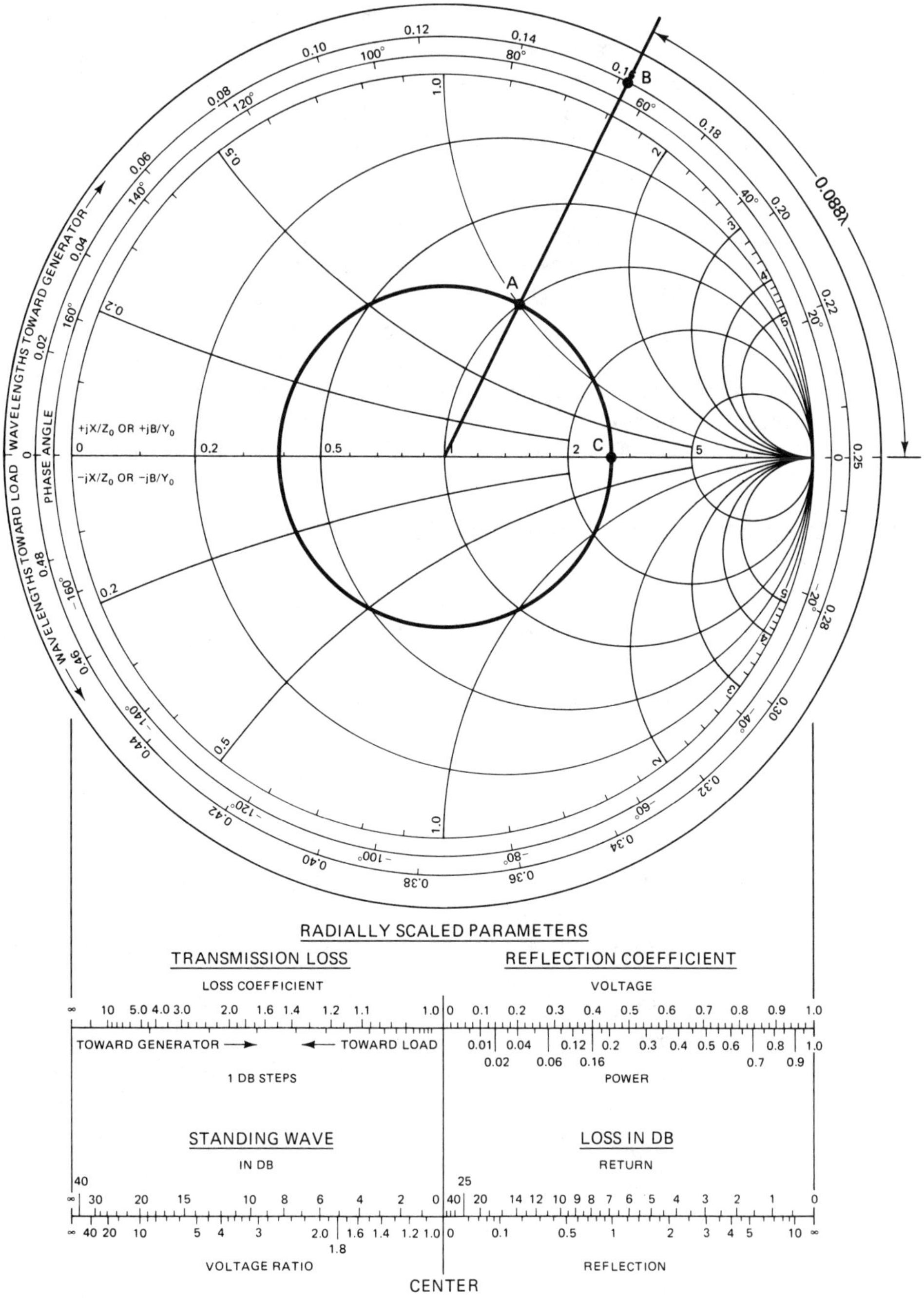

Figure 3–2 Smith chart procedure for designing a quarter-wave transformer to match a transmission line to a complex load impedance.

Ω, then decreases abruptly until it vanishes at point *C*, where the impedance is a pure resistance, 2.6 × 50 = 130 Ω. As we continue our journey the reactance becomes negative (capacitive) and reaches a maximum of $-j80$ Ω. We pass through the point $1 - j$ and again cross the real axis, this time at 0.38 × 50 = 19 Ω. After we cross the real axis we find the reactance is again positive (inductive) and increases as we return to point *A*. By referring to the wavelength scale at the edge of the chart we see that the complete trip corresponds to a journey of half a wavelength down the line. For every point on a transmission line there is another point a half-wavelength away that has the same impedance. We have to be careful to start our journeys in the right direction. If we start at the load we have to travel toward the generator; otherwise, we would run into the load.

When we use a Smith chart we usually normalize the impedance by dividing by the characteristic impedance of the transmission line. The characteristic impedances of most transmission lines can be calculated from formulas based on the "geometry" or configuration of the line, and this will be discussed in Chapter 4.

3–5 MATCHING TO COMPLEX IMPEDANCES

Now that we know how impedance varies along the length of a mismatched line, we can insert a quarter-wave transformer at a resistive point and match the line to the load.

EXAMPLE 3–4

Design a quarter-wave transformer to match a 50-Ω coaxial line to a load impedance of $50 + j50$ Ω. The operating frequency is 500 MHz and the dielectric constant of the insulation in the coax is 2.25.

Solution We start by plotting on a Smith chart the normalized impedance $(50 + j50)/50 = 1 + j$. This is shown as point *A* in Fig. 3–2. We draw a VSWR circle through *A* and extend a radial line from the center of the chart through *A* to point *B* on the circumference.

The VSWR circle intersects the real axis at point *C*. Here the impedance has only a resistive component, and its value is 2.6 × 50 = 130 Ω. We can put a quarter-wave transformer at *C* and match 130 Ω to 50 Ω. The characteristic impedance of the transformer should be $\sqrt{Z_1 Z_2} = \sqrt{50 \times 130} = 80.6$ Ω.

To determine the length of the transformer we have to find the wavelength in the insulation in the coax. The wavelength in free space is, from Eq. (1–2), $(3 \times 10^{10})/(500 \times 10^6) = 60$ cm, and the wavelength in the insulation is, from Eq. (3–1), $\lambda = \lambda_0/\sqrt{\epsilon_r} = 60/2.25 = 40$ cm. The length of the transformer is one-fourth of this, or 10 cm.

We determine the location of the transformer by subtracting the difference between points *A* and *C* on the wavelength scale. As shown in the figure, the difference is 0.088 λ, which is equal to 0.088 × 40 = 3.52 cm.

Our design is complete: The transformer has a characteristic impedance of 80.6 Ω, a length of 10 cm, and is located 3.52 cm from the load, as shown in Fig. 3–3. This design is not the most elegant. We discuss in Chapter 7 a method that eliminates the extra piece of transmission line between the load and the transformer.

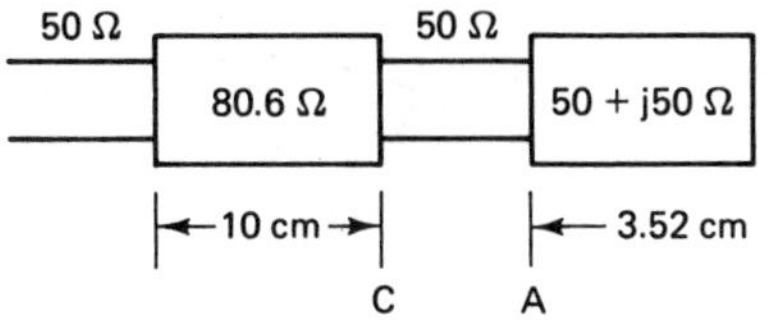

Figure 3–3 Quarter-wave transformer for matching a complex impedance to a transmission line.

There is another way to use a quarter-wave section to match to a complex impedance. Instead of looking for a resistance point on the transmission line, we use a stub to cancel the reactance of the load and match the remaining resistance with a quarter wave transformer.

3–6 MATCHING WITH STUBS

A *stub* is a short length of transmission line connected in parallel with a main line, as indicated in Fig. 3–4. If one end of the stub is short-circuited, its impedance is

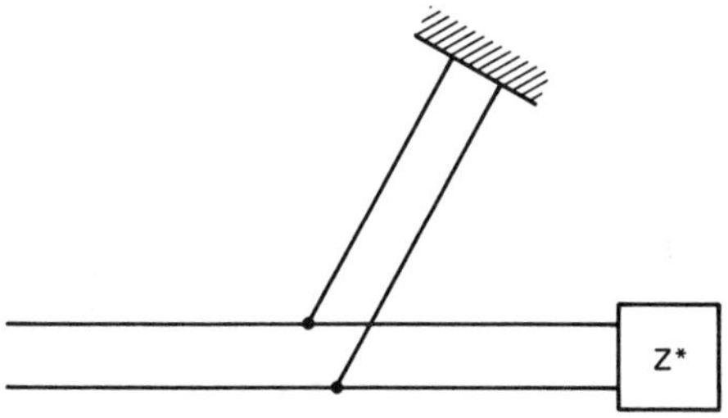

Figure 3–4 Short-circuited stub for matching a load to a transmission line.

zero and is represented by a point at the extreme left end of the real axis of a Smith chart, or at the extreme right end if we plot admittance. As distance from the short is increased, the point on the chart moves around the circumference, through various values of normalized inductive and capacitive reactance. We see that a stub can be regarded as a device for obtaining reactance, the value of the reactance being determined by the length of the stub and its characteristic impedance.

EXAMPLE 3–5

Find the admittance of a short-circuited stub 0.15 wavelength long. Its characteristic impedance is 75 Ω.

Solution We consider the short circuit to be the load and plot its infinite admittance at point *A* in Fig. 3–5. We now move away from the load a

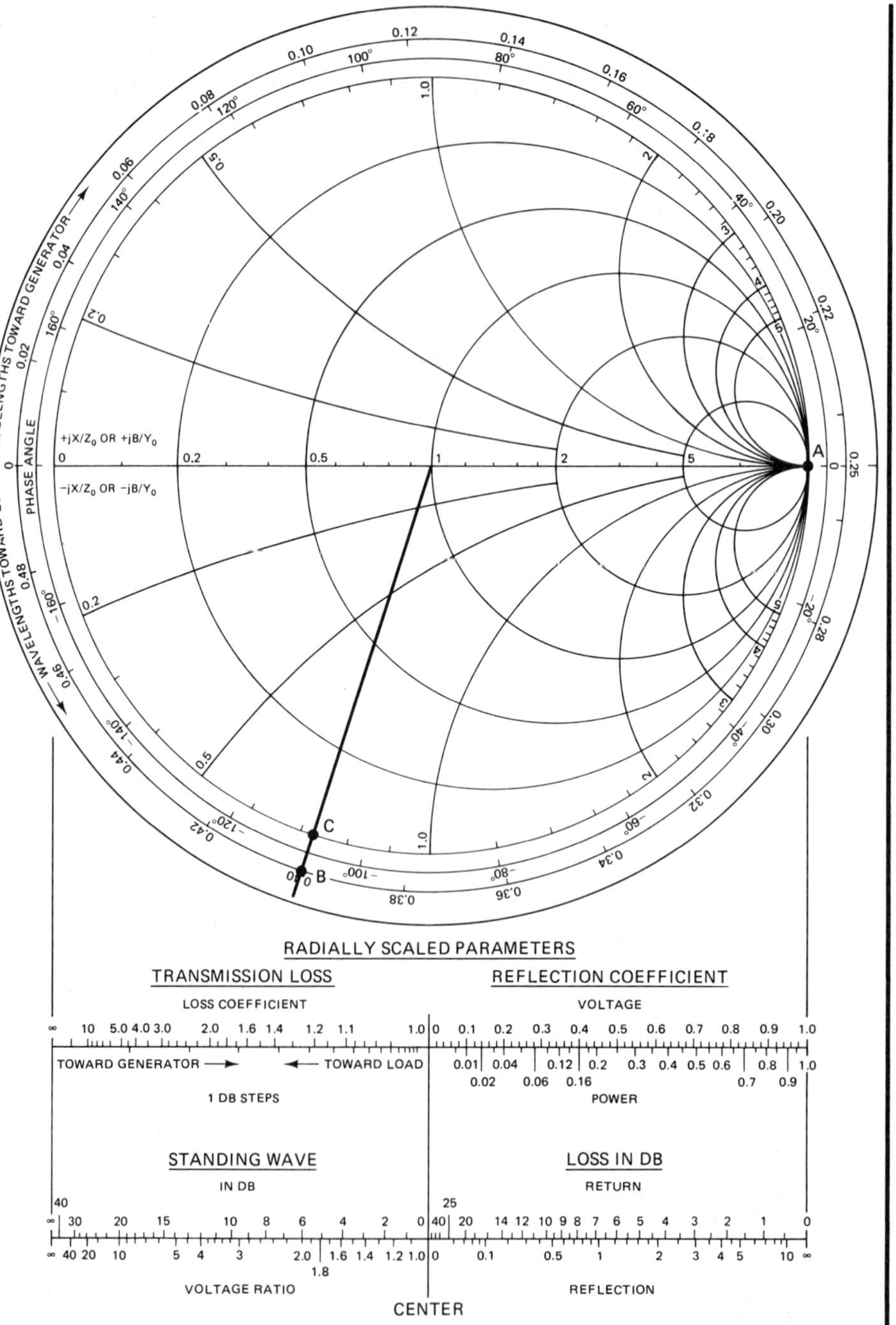

Figure 3–5 Smith chart procedure for finding the admittance of a short-circuited stub.

distance of 0.15 wavelength to point B and draw a radial line from the center and through B. The radial line intersects the VSWR circle at point C, where the normalized susceptance is $-j0.73$. The unnormalized admittance is $-j0.73/75 = -j9.73$ mS.

Short lengths of transmission line can be used as inductors or capacitors. In Example 3–5 we have a short-circuited stub that is equivalent to $1/(-j9.73) = j102\ \Omega$ of shunt inductance.

The Smith chart can be used to convert impedance to admittance. On a VSWR circle, points that are separated by a quarter-wavelength are reciprocals; that is, they are related as impedances and admittances. We shall find this a convenience in the next example, where we shall use both a stub and a quarter-wave section to match to a complex impedance.

EXAMPLE 3–6

The output impedance of a transistor is $61 - j83\ \Omega$ at 1 GHz. Use a stub and quarter-wave transformer to match it to a 50-Ω load.

Solution We plot the normalized impedance, $(61 - j83)/50 = 1.22 - j1.66$, at point A in Fig. 3–6. After drawing a VSWR circle, we draw a diameter through A and find the normalized admittance, $0.29 + j0.39$, at point B.

Check: Normalized admittance is the reciprocal of normalized impedance: $y = 1/z = 1/(1.22 - j1.66) = 1/2.06\angle{-53.7^\circ} = 0.485\angle{53.7^\circ} = 0.287 + j0.391$.

We can reduce the admittance to a conductance if we add, in parallel, a susceptance of $-j0.391$. We mark $-j0.391$ at point C and draw a radial line through it. The length of the stub is the distance from the short ($y = \infty$) to point C: $0.250\ \lambda - 0.059\lambda = 0.191\lambda$.

The last step is the matching of the remaining resistance, $r = (1/0.287) \times 50 = 174\ \Omega$, to the 50-Ω load. For this we shall need a quarter-wave section with a characteristic impedance of $\sqrt{174 \times 50} = 93.3\ \Omega$. A sketch of the design is shown in Fig. 3–7.

An alternative solution is to use a one-eighth wavelength stub with a characteristic impedance of $(1/0.391) \times 50 = 128\ \Omega$. The susceptance at a distance of $\lambda/8$ from the short is $-j1.0$, point D on the chart. Since now we have normalized to 128 Ω, point D corresponds to a susceptance of $-j1.0/128 = -j7.82 \times 10^{-3}$ S, which will cancel the transistor susceptance of $j0.392/50 = +j7.82 \times 10^{-3}$ S. This solution provides a shorter stub but requires the manufacture of a transmission line with a characteristic impedance different than that of the main 50-Ω line.

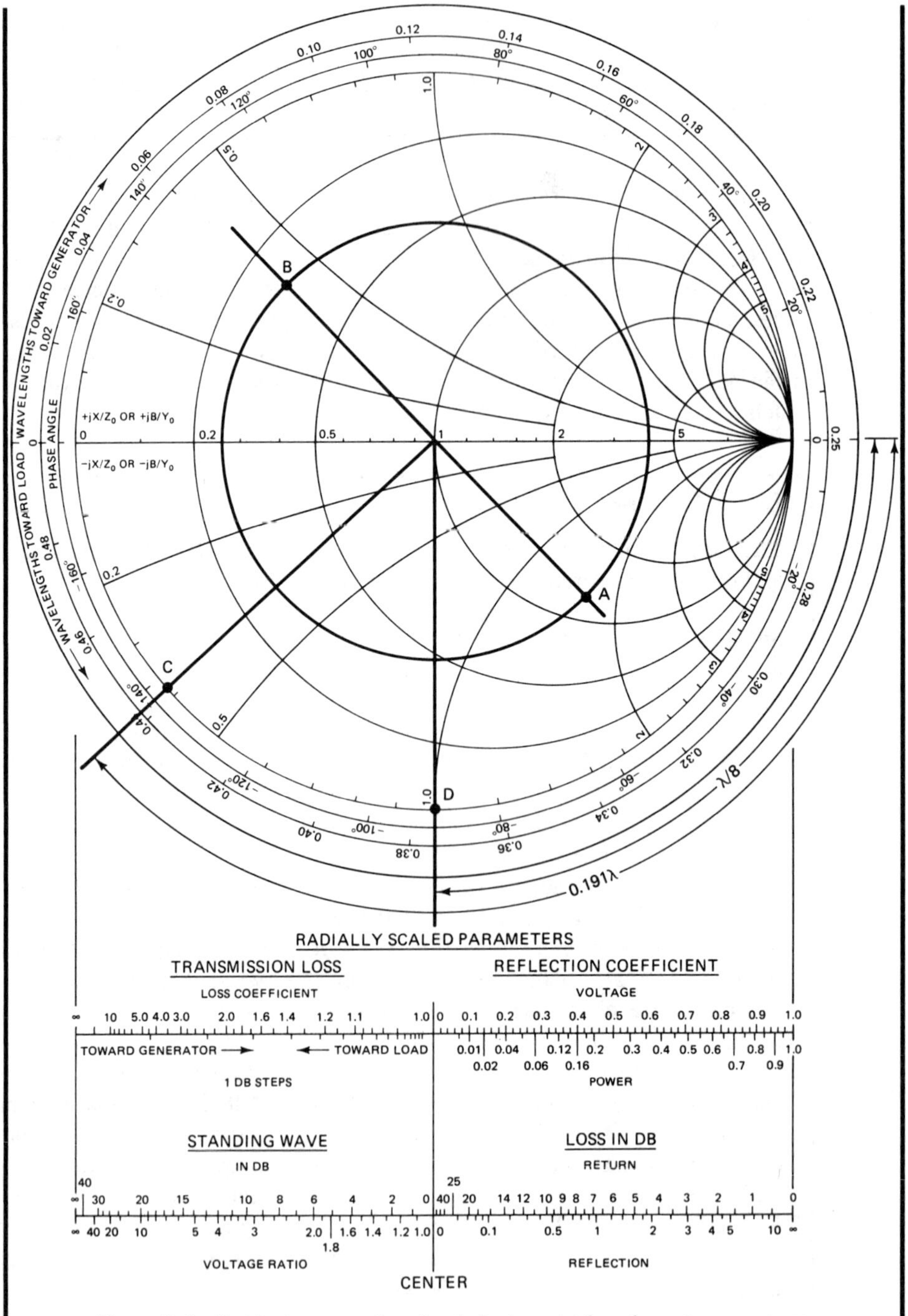

Figure 3–6 Smith chart procedure for designing a stub and quarter-wave transformer, Example 3–6.

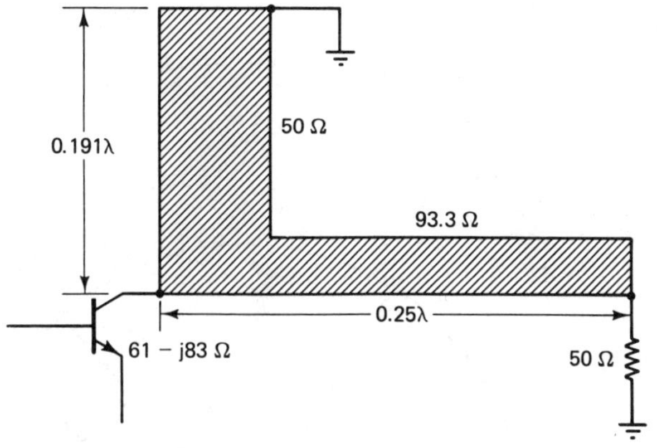

Figure 3–7 Stub and quarter-wave transformer design, Example 3–6, executed in microstrip line (Chapter 4).

Before continuing, we review, in terms of circuit elements, what we have done on the Smith chart. The output impedance of the transistor, $r - jX_c$, is shown in Fig. 3–8. As shown in the figure, this impedance is transformed into an equivalent parallel circuit $g + jB$, a stub is used to provide $-jB$, and the result, g, is then transformed to 50 Ω.

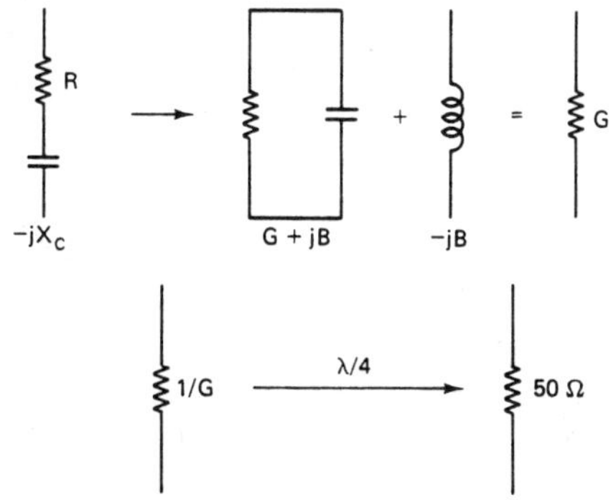

Figure 3–8 Schematic representation of the stub and quarter-wave transformer design procedure, Example 3–6.

A stub by itself can be used to match a transmission line to a load; the technique is more complicated but the hardware easier to build. The idea is to find a point on a transmission line where the real part of the impedance is equal to the load resistance, then use a stub to cancel the reactance.

EXAMPLE 3–7

Design a stub to match a 200-Ω transmission line to a load impedance of 50 + j100 Ω.

Solution We plot the normalized impedance of the load, 0.25 + j0.5, on a Smith chart at point A (Fig. 3–9). We draw a VSWR circle and a diameter

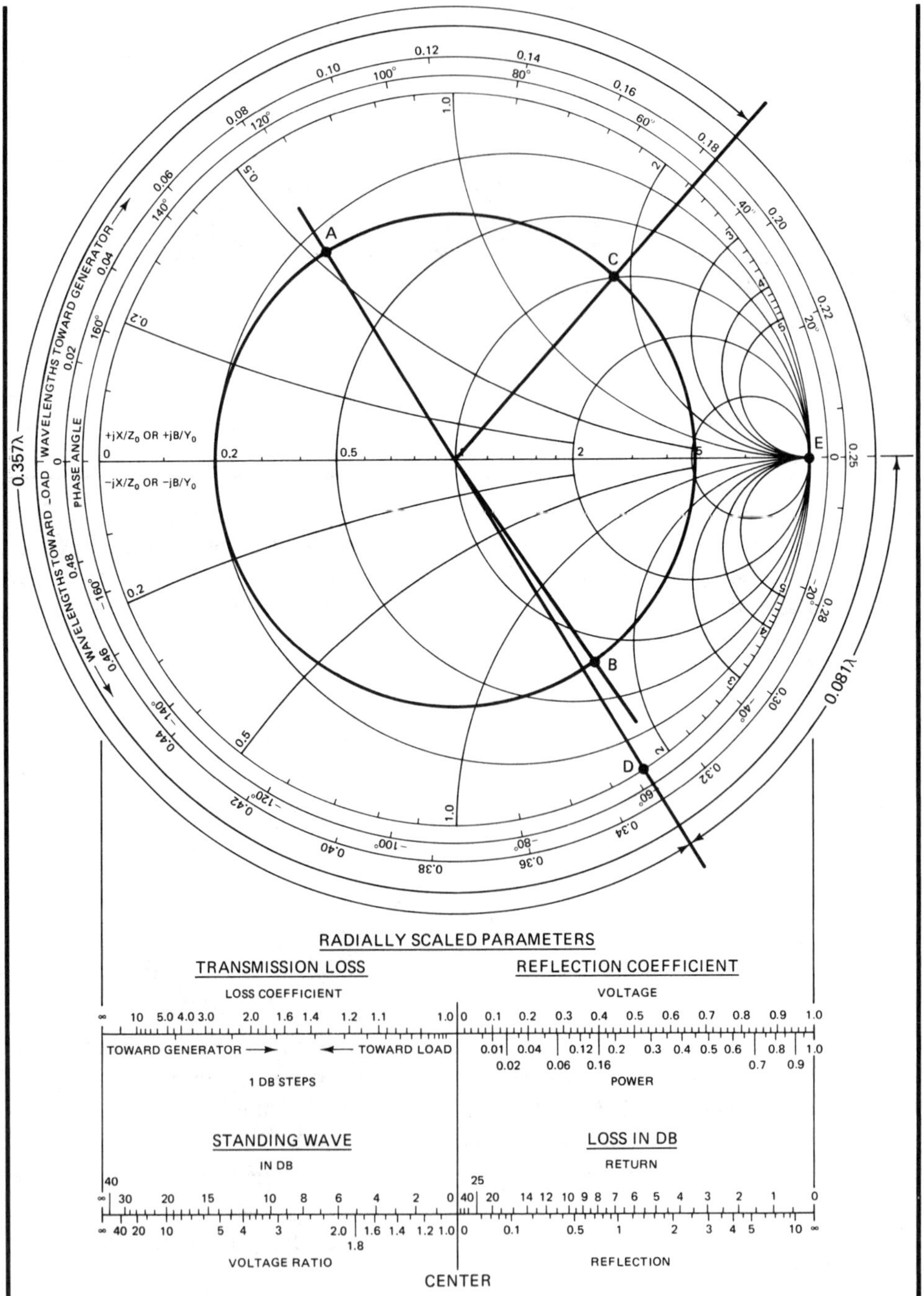

Figure 3–9 Design procedure for using a stub to match a transmission line to a complex load impedance, Example 3–7.

through point A to determine the normalized admittance of $0.8 - j1.6$ at point B. It is convenient to work with admittances rather than impedances because the stub is parallel to the main line, and admittances of parallel elements can be added or subtracted directly.

The crux of the technique is to rotate from the load point B to the point C, where the VSWR circle intersects the $g = r = 1$ circle. At point C the normalized admittance is $y = 1 + j1.8$, and the real part of the admittance is equal to the characteristic conductance of the transmission line. At this point we need only to connect a stub to cancel the susceptance. The susceptance to be canceled is $+j1.8$, so we locate $-j1.8$ at point D on the chart. The distance from point D to the short circuit at E is the required stub length, 0.081λ, and the distance from the load to the stub is the distance, toward the generator, from B to C, 0.357λ. A sketch of the solution is shown in Fig. 3–10.

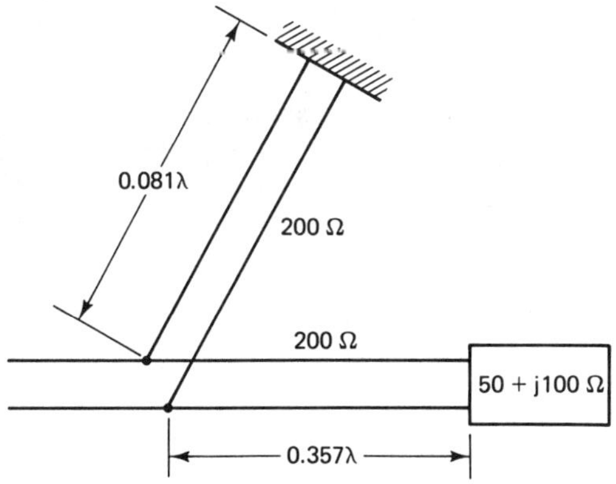

Figure 3–10 Stub matching design, Example 3–7.

3–7 STUBS AND BALUNS

Stubs are sometimes used in balanced-to-unbalanced transformers, called *baluns*. Because of its symmetry, twin-conductor line is an example of a *balanced* transmission line, whereas coax is an example of *unbalanced* line. Owing to the different field configurations at a balanced–unbalanced interface, reflections and radiation may occur even if the impedances are matched. The electric field distributions in coax and twin line are shown in Fig. 3–11. If we were to connect coax directly to twin line, as shown at the bottom of the figure, the contortion required of the field in propagating from one line to the other would cause reflections. At microwave frequencies currents would be induced on the outer shield of the coax and radiation would occur.

In Fig. 3–12 we show a balun sometimes used to connect a balanced load to an unbalanced line. Here the balanced load is a dipole antenna which is to be connected to a coaxial feed line. We assume the impedance of the dipole is 70 Ω

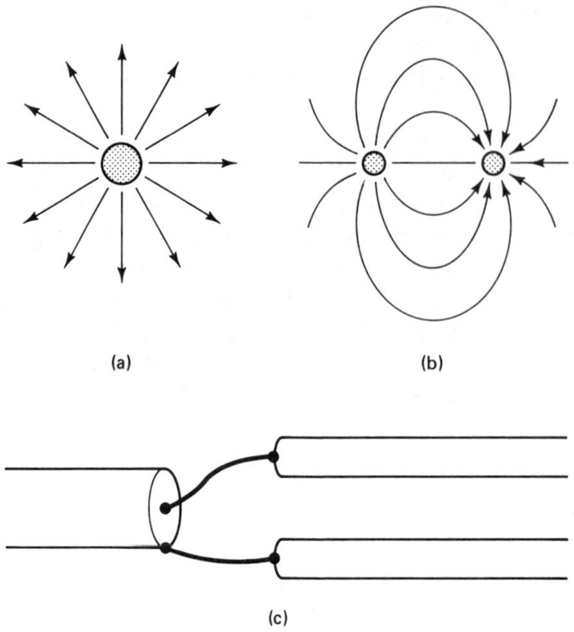

Figure 3–11 Electric field distributions in (a) coaxial and (b) twin-conductor transmission lines; (c) improper unbalanced-to-balanced transition.

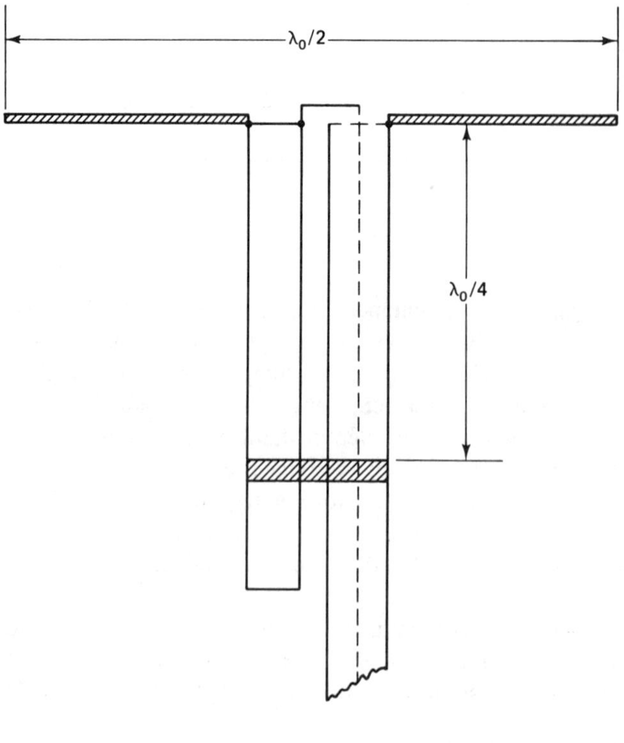

Figure 3–12 Balun for connecting an unbalanced transmission line to a balanced load.

of radiation resistance and the characteristic impedance of the coax also is 70 Ω, so we have no impedance matching problem. The transformation is achieved through the use of a short-circuited quarter-wavelength stub. The stub is in twin line formed by the outer conductors of two coaxial cables, one of the cables being the feed line.

We see, by referring to a Smith chart, that the impedance of a short-circuited, quarter-wavelength stub is infinite; thus the stub has no effect on the impedance at the point of excitation of the antenna. The characteristic impedance of the twin line is immaterial: the dipole appears to be driven by the twin line. We have a balanced line to a balanced load.

3–8 LUMPED-PARAMETER TRANSFORMERS

We noted, in deriving Eq. (1–18), that the characteristic impedance of a medium is $\sqrt{\mu/\epsilon}$, where μ is the permeability in henries per meter and ϵ is the permittivity in farads per meter. This can be demonstrated by assuming the electric and magnetic energies in a wave are equal. From Hooke's law we showed that the electric field energy is $\frac{1}{2}\epsilon E^2$ [Eq. (1–15)]. By the same argument the magnetic field energy is $\frac{1}{2}\mu H^2$. Equating the two, we have $\epsilon E^2 = \mu H^2$; further, since $Z_c = E/H$, we find $Z_c = \sqrt{\mu/\epsilon}$.

A similar result attains for transmission lines,

$$Z_c = \sqrt{\frac{L}{C}} \tag{3–7}$$

where L and C are the inductance and capacitance in henries and farads per meter. We obtain the above by equating the electric and magnetic energies of a wave on a transmission line, $\frac{1}{2}Li^2 = \frac{1}{2}CV^2$, and substituting Z_c for V/i.

We expect, from Eq. (3–7), that we can substitute a series inductance and a shunt capacitance for a length of transmission line and so obtain a lumped-parameter equivalent of a quarter-wave section. We show a circuit for matching a resistive source, R_1, to a resistive load, R_2, in Fig. 3–13. From Eqs. (3–4) and (3–7) we

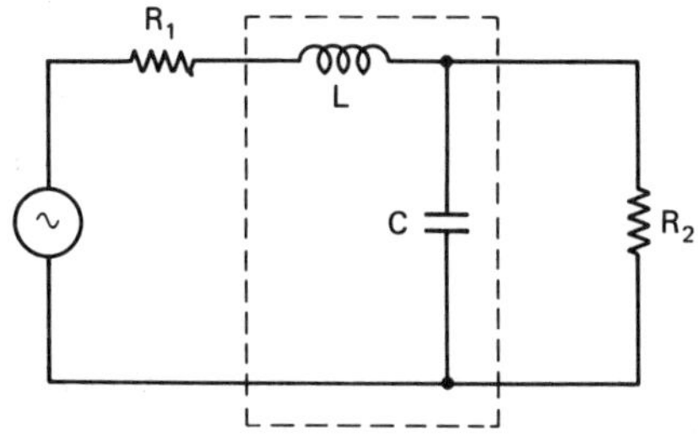

Figure 3–13 Lumped-parameter, step-up, quarter-wave transformer.

find the quarter-wave amplitude condition to be $Z_c = \sqrt{L/C} = \sqrt{\omega L/\omega C} = \sqrt{X_L X_C} = \sqrt{R_1 R_2}$, or

$$X_L = \frac{R_1 R_2}{X_C} \tag{3–8}$$

If the quarter-wave transformer is to match the load to R_1, we must have $R_1 = jX_L + 1/(G_2 + jB_C)$ or, after clearing the equation of fractions,

$$R_1 G_2 + jR_1 B_C = jX_L G_2 - X_L B_C + 1 \tag{3–9}$$

The second quarter-wave condition is that the reflected waves be out of phase and cancel completely. In circuit theory this is expressed as the vanishing of the reactance; only the real terms in Eq. (3–9) remain, and we have $R_1 G_2 = X_L B_C + 1$, or

$$\frac{R_1}{R_2} = \frac{X_L}{X_C} + 1 \tag{3–10}$$

We now substitute for X_L from Eq. (3–8) into Eq. (3–10) and rearrange terms to obtain

$$X_C = R_2 \sqrt{\frac{R_1}{R_2 - R_1}} \tag{3–11}$$

We see that we have a solution if $R_2 > R_1$.

EXAMPLE 3–8

Design a lumped-parameter transformer to match a 50-Ω source to a 200-Ω load at 500 MHz.

Solution From Eq. (3–11) we find

$$X_C = 200 \sqrt{\frac{50}{200 - 50}} = 115.5\ \Omega$$

and from Eq. (3–8),

$$X_L = 50 \times \frac{200}{115.5} = 86.6\ \Omega$$

At 500 MHz the capacitance is

$$C = \frac{1}{2\pi f X_C} = \frac{1}{2\pi(500 \times 10^6) \times 115.5} = 2.75\ \text{pF}$$

and the inductance,

$$L = \frac{X_L}{2\pi f} = \frac{86.6}{2\pi(500 \times 10^6)} = 0.0276\ \mu\text{H}$$

Check:

$$Z_c = \sqrt{R_1R_2} \quad = \sqrt{X_LX_C} \quad = \sqrt{\frac{L}{C}}$$

$$= \sqrt{50 \times 200} = \sqrt{86.6 \times 115.5} = \sqrt{\frac{0.0276 \times 10^{-6}}{2.75 \times 10^{-12}}}$$

$$= 100 \qquad = 100 \qquad = 100$$

If we are faced with a "step-down" problem, $R_1 > R_2$, the obvious thing is to try the inductor and capacitor in reverse order (Fig. 3–14). The amplitude condition [Eq. (3–8)] is again applicable, but the equation for the input admittance now is $G_1 = jB_C + 1/(R_2 + jX_L)$. Clearing the fraction, we have $R_2G_1 + jG_1X_L = jR_2B_C - B_CX_L + 1$. We again require that the reactive terms cancel, which leaves

$$\frac{R_2}{R_1} = -\frac{X_L}{X_C} + 1 \tag{3–12}$$

We combine this result with Eq. (3–8) to obtain

$$X_C = R_1\sqrt{\frac{R_2}{R_1 - R_2}} \tag{3–13}$$

and we see what we wanted, a solution for $R_1 > R_2$.

EXAMPLE 3–9

Design a transformer to match a 200-Ω source to a 50-Ω load.

Solution From Eq. (3–13) we have

$$X_C = 200\sqrt{\frac{50}{200 - 50}} = 115.5\ \Omega$$

and from Eq. (3–8),

$$X_L = 200 \times \frac{50}{115.5} = 86.6\ \Omega$$

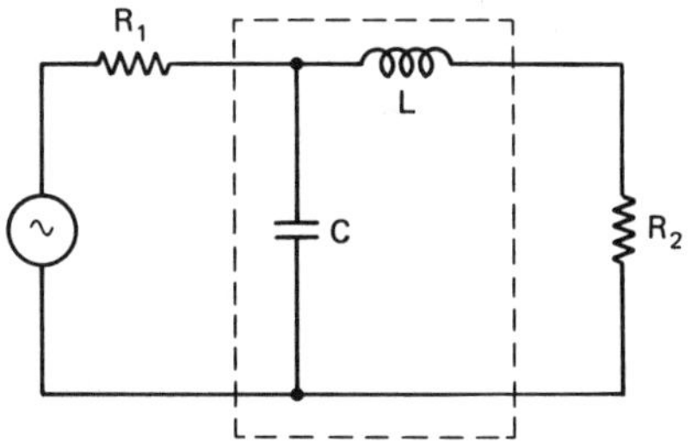

Figure 3–14 Lumped-parameter, step-down, quarter-wave transformer.

We have taken a wave theory point of view in the examples above and have replaced the distributed impedance of a transmission line by a lumped-parameter equivalent. The equivalence is imperfect. The lumped-parameter circuit has a resonant frequency, the transmission line does not. At frequencies above resonance the signal is attenuated about 12 dB/octave; that is, the power to the load is reduced by a factor of 16 with every doubling of the frequency. What we have been calling a quarter-wave transformer also is, from a circuit viewpoint, a two-pole, low-pass filter. In Chapter 4 we discuss lumped-parameter transmission lines as low-pass filters.

Our design technique ensures the operating frequency to be less than the resonant or cutoff frequency, so we need not be concerned. To prove this statement for the step-up case, we combine Eqs. (3–8) and (3–11) and solve for X_L,

$$X_L = \frac{R_1 R_2}{R_2}\sqrt{\frac{R_1}{R_2 - R_1}} = \sqrt{\frac{R_1}{R_2 - R_1}} \qquad (3\text{–}14)$$

At resonance $X_C = X_L$, below resonance $X_C/X_L > 1$. We can find the ratio X_C/X_L by dividing Eq. (3–11) by Eq. (3–14),

$$\frac{X_C}{X_L} = \frac{R_2\sqrt{R_1/(R_2 - R_1)}}{\sqrt{R_1(R_2 - R_1)}} = \frac{R_2}{R_2 - R_1} > 1 \qquad (3\text{–}15)$$

Since $R_2/(R_2 - R_1)$ is always greater than unity, X_C is always greater than X_L; in other words, the operating frequency is always below cutoff and is in the passband.

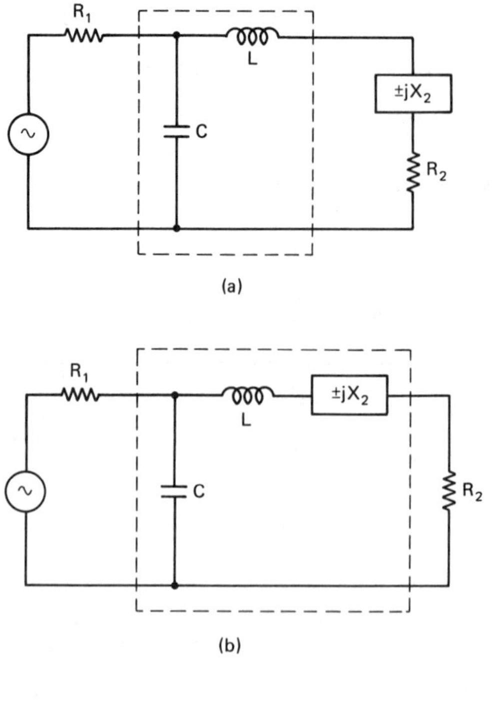

Figure 3–15 Development of a shortened or lengthened quarter-wave transformer for matching to a complex load.

A shortened or lengthened quarter-wave transformer can be used to match to a complex impedance, and the method is demonstrated in Chapter 7. We shall demonstrate it here for the lumped-parameter step-down transformer. We show the transformer with a complex load impedance, $R_2 \pm jX_2$ in Fig. 3–15a. In Fig. 3–15b we show the same circuit rearranged so that the load reactance is included as part of the transformer. Evidently we can effect a match simply by replacing X_L by $X_L \pm X_2$ in Eq. (3–8),

$$X_L \pm X_2 = \frac{R_1 R_2}{\sqrt{X_C}} \tag{3–16}$$

Equation (3–13) remains unchanged.

EXAMPLE 3–10

Design a transformer to match a 50-Ω source to a transistor input impedance of 25 − *j*50 Ω at 100 MHz.

Solution From Eq. (3–13) we have

$$X_C = 50\sqrt{\frac{25}{50 - 25}} = 50\ \Omega$$

and from Eq. (3–16),

$$X_L = \frac{R_1 R_2}{X_C} + X_2 = 50 \times \frac{25}{50} + 50 = 75\ \Omega$$

At 100 MHz the component values are

$$C = \frac{1}{2\pi f X_C} = \frac{1}{2\pi(100 \times 10^6) \times 50} = 31.8\text{ pF}$$

$$L = \frac{X_L}{2\pi f} = \frac{75}{2\pi(100 \times 10^6)} = 0.119\ \mu\text{H}$$

There are many other lumped-parameter techniques for effecting a match between complex impedances. We will not consider them here, since we are primarily interested in the transmission-line techniques used at microwave frequencies. Usually, anything done with wave techniques has its lumped-parameter, low-frequency counterpart.

APPLICATION 3–1: Quarter-Wave Sections

In this application we build and test two quarter-wave sections, one to match a 50-Ω source to a resistive load, another to match a 50-Ω source to a reactive load.

Equipment and Materials

150-MHz 50-Ω signal generator
RF voltmeter
¼-in. copper tubing or rod
polystyrene foam insulation board, or equivalent
assorted connectors, adapters, resistors, capacitors, hand tools, and soldering iron
commercial 300-Ω twin-conductor transmission line (TV twin lead)
caliper micrometer, preferably metric, or wire gauge
Smith chart

PART 1: 50-Ω SOURCE MATCHED TO A RESISTIVE LOAD

Use a quarter-wave section of twin line to match a 50-Ω signal generator to an 1800-Ω load at 150 MHz. The characteristic impedance of the section should be $\sqrt{Z_1Z_2} = \sqrt{50 \times 1800} = 300\ \Omega$. Since the wavelength on twin line is the same as in free space, the required length of the matching section can be calculated with Eq. (1–2). A ⅛-W resistor or a full-wave dipole antenna can be used as a load. The component leads should be as short as possible.

Measure the voltage under load and compare it with the open-circuit voltage. Does it confirm the maximum power transfer theorem? Measure the load voltage at several frequencies above and below the design frequency and show your results in tabular or graphical form in your report. Explain the results in terms of mismatch and reflections.

PART 2: 50-Ω SOURCE MATCHED TO A REACTIVE LOAD

Match a reactive load, a 100-Ω resistor in parallel with a 10-pF capacitor, to a 50-Ω signal generator. Assume the load to be on 300-Ω twin line and the operating frequency to be 150 MHz.

Procedure

Calculate the load admittance at 150 MHz and normalize it to 300 Ω. Plot the normalized admittance on a Smith chart and determine the rotation toward the generator to reach a pure conductance, $g/300 = G = 1/R$. Convert the rotation in wavelengths to distance in centimeters. This is the distance from the load to the start of the quarter-wave matching section. The characteristic impedance of the quarter-wave section must be $\sqrt{R_sR} = \sqrt{50 \times R}$. If your answer is not between 100- and 200-

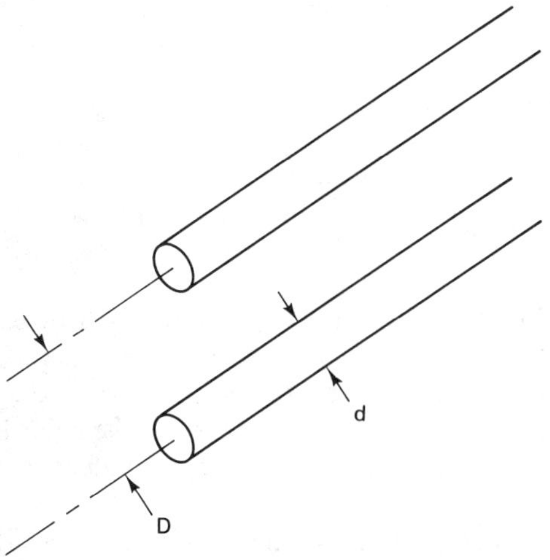

Figure 3–16 Dimensions that determine the characteristic impedance of twin-conductor transmission line.

Ω, check your calculations. Determine the length of the quarter-wave section as you did in Part 1.

You will have to construct a twin-line section with the required characteristic impedance. The equation for twin-line impedance is

$$Z_c = 276 \log \frac{2D}{d} \tag{3–17}$$

where d is the diameter of a conductor and D is the separation of the conductors (Fig. 3–16). Obtain some solid core insulated or bare copper wire or tubing and measure its diameter, d, with a micrometer or wire gauge. By solving Eq. (3–17) for D, find the separation of the conductors that will give the required characteristic impedance,

$$D = \frac{d}{2} \log^{-1} \frac{Z_c}{276} \tag{3–18}$$

EXAMPLE 3–11

Design a twin line of 175-Ω impedance, using copper wire of diameter 0.8 mm.

Solution From Eq. (3–18) we have

$$D = \frac{0.8 \text{ mm}}{2} \log^{-1} \frac{175}{276} = 1.72 \text{ mm}$$

The two conductors must be held in place with a center-to-center separation of 1.72 mm.

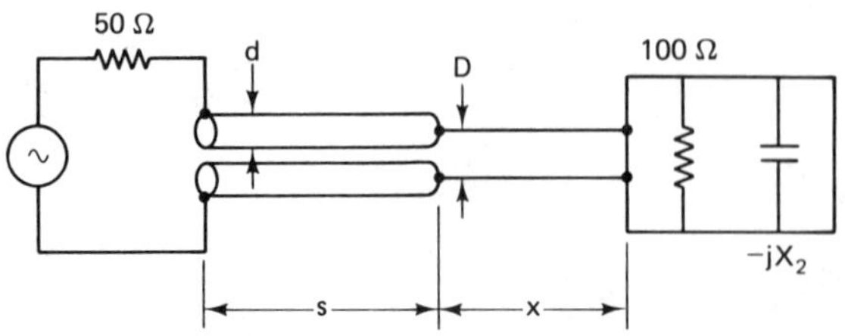

Figure 3–17 Quarter-wave transformer developed in Application 3–1, Part 2.

When you solve Eq. (3–18) for the required twin-line separation, you may find the separation to be impractical unless you use conductors of at least ⅛ to ¼ in. diameter. The line can be constructed by taping the conductors together with the required thickness of paper or other insulation between them, or by cutting two grooves in a block of polyfoam insulation to hold the conductors at the correct separation.

In your report include a sketch of all components and dimensions, similar to that shown in Fig. 3–17; also, include a Smith chart, clearly showing your design calculations. Test your design and report your measurements as you did in Part 1.

SUMMARY

The antireflection layer on a coated lens, and a section of transmission line designed as a quarter-wave transformer, are solutions to the same problem, executed at greatly different frequencies. In either case the purpose is to cause reflections from either side of the layer or transformer to cancel completely, so that all of the incident energy is delivered to the load. Complete cancellation requires that two conditions be satisfied. First, the out-of-phase condition requires that the thickness of the layer or the length of the transformer be a quarter wavelength in the medium. The wavelength in the medium can be found from

$$\lambda = \frac{\lambda_0}{\eta}$$

Second, the magnitudes of the amplitudes of the two reflected waves must be equal, and this will occur if the refractive index of the matching layer is the geometric mean of the refractive indices of the transmission and load media,

$$\eta_2 = (\eta_1\eta_3)^{1/2}$$

or, for a transmission line,

$$Z_2 = (Z_1Z_3)^{1/2}$$

where the Z's are characteristic impedances. The characteristic impedance of a medium or transmission line is defined as

$$Z_c = \frac{E}{H} = \left(\frac{\mu}{\epsilon}\right)^{1/2}$$

$$Z_c = \left(\frac{L}{C}\right)^{1/2}$$

Quarter-wave transformers can only be used to match nonreactive media or resistive sources and loads; however, there are several techniques for matching reactive loads, for example, by combining a stub and a quarter-wave section. A stub is an open or shorted length of auxiliary transmission line that provides a reactance to cancel the reactance at a point on the main line. The design calculations for transformers and stubs are simplified by the use of the Smith chart. The Smith chart is also a visual aid to understanding the wave phenomena that occur on transmission lines.

The wave concepts used in quarter-wave transformer and stub designs have their counterpart in circuit theory and the use of lumped-parameter reactances. The inductive reactance required to match a source resistance R_1 to a load resistance R_2 is

$$X_L = \frac{R_1 R_2}{X_C}$$

where

$$X_C = R_2 \sqrt{\frac{R_1}{R_2 - R_1}} \qquad R_2 > R_1$$

or

$$X_C = R_1 \sqrt{\frac{R_2}{R_1 - R_2}} \qquad R_1 > R_2$$

A match to a reactive load $R_2 \pm jX_2$ can be achieved by using either of the equations above for capacitive reactance and

$$X_L \pm X_2 = \frac{R_1 R_2}{X_C}$$

The fabrication of a quarter-wave section requires a knowledge of how the characteristic impedance of a line is determined by its physical properties; for example, the characteristic impedance of a twin-conductor line is

$$Z_c = 276 \log \frac{2D}{d}$$

The physical properties of various types of transmission lines are considered in more detail in Chapter 4.

PROBLEMS

3–1. Find the wavelength in a glass insulator when the operating frequency is 2 GHz. The relative permittivity (dielectric constant) of the glass is 4.4.

3–2. A camera lens is to be used for infrared photography at a wavelength of 1.5 μm [1 micrometer (micron) = 1 μm = 10^{-6} m). The refractive index of the lens is 2.2. Determine the refractive index and thickness of a quarter-wave coating for the air–glass interface.

3–3. Design a quarter-wave section to match 300-Ω twin line to a 1-kΩ load at 1 MHz. In addition to finding the length and characteristic impedance of the section, determine the separation of the conductors if the section is constructed of 2.07-mm-diameter wires.

3–4. The electric field strength, measured 1 mile from a broadcasting antenna, is 50 mV/m. What is the magnetic field strength?

3–5. Design a stub to match 300-Ω twin line to a 65-Ω load. Start by plotting the normalized load admittance, 4.6, on a Smith chart. Show clearly all your work on the chart and turn it in with your design. Assume an operating frequency of 100 MHz and show distances in centimeters.

3–6. A quarter-wave section is used to match 300-Ω transmission line to load impedance of 600 + j300. What is the distance, in wavelengths, from the load to the transformer? Assume a 300-Ω line between load and transformer. What is the characteristic impedance of the quarter-wave section?

3–7. Determine the lumped-parameter reactances needed to match a 600-Ω source to a 50-Ω load.

3–8. Find the reactances required to match a 50-Ω source to a transistor input impedance of 35 − j20 Ω.

3–9. Twin-conductor line, Z_c = 600 Ω, is to be used to connect a transmitter to a broadcasting antenna. Because of the high voltage, the conductors must be 20 cm apart. What is the diameter of the conductors?

3–10. A 50-Ω source is connected to a 100-Ω load with a 50-Ω transmission line. What would be the percentage increase in power to the load if the 50-Ω line were replaced by a quarter-wave matching section?

4

Transmission Lines

4–1 INTRODUCTION

A transmission line is a means of leading electromagnetic power from one place to another. It can be a conductor or an insulator: a few centimeters of metal between two components in a circuit, or a few kilometers of thin glass fiber between an infrared transmitter and a distant, light-sensitive detector. It can be continuous or can be a periodic distribution of discontinuities: the smooth interior of a conducting cylinder, or a rough exterior of a conducting surface. Its possible forms are extremely varied: a series of lumped-parameter inductors and capacitors can be regarded as a transmission line; free space itself can be regarded as a transmission line.

In this chapter we consider only the more common types of lines: coaxial, twin wire, and microstrip line. Coaxial lines often are used to connect system subcomponents; for example, connecting an oscilloscope to a signal generator is done with coax. Microstrip line is more commonly used for connecting circuit components.

We often choose a transmission line for a particular application because of its characteristic impedance, power- or voltage-handling capability, size, cost, efficiency—where efficiency may be determined by ohmic and radiation losses—or simply on the basis of convenience. For many types of low-loss lines the characteristic impedance is

$$Z_c = \left(\frac{L}{C}\right)^{1/2} \tag{4-1}$$

where L and C are the inductance and capacitance per unit length of line. If, for example, we regard free space as a transmission line, then Eq. (4–1) gives $Z_c = (\mu_0/\epsilon_0)^{1/2} = (2\pi \times 10^{-7}\ \text{H/m})/(8.85 \times 10^{-12}\ \text{F/m}) = 377\ \Omega$.

We can derive Eq. (4–1) from the principle of *equipartition of energy*. In a steady-state system, energy is distributed equally among the various active modes of the system. For a wave propagating on a transmission line, equal amounts of energy are distributed in the electric and magnetic fields, $\frac{1}{2}Cv^2 = \frac{1}{2}Li^2$. From this equation and the definition of impedance as $Z_c = v/i$, we obtain Eq. (4–1).

For some lines, coax and twin wire, for example, finding the inductance and capacitance per unit length is known as a "soluble problem" of electromagnetic theory, and exact formulas are available for the inductance and capacitance. For others, such as microstrip line, we must use empirical formulas. We shall consider the results of soluble problems first.

4–2 COAXIAL LINE

The fundamental propagating mode in coaxial line is transverse electromagnetic, TEM, which has a field configuration like that shown in Fig. 4–1. In the figure, D is the inner diameter of the outer conductor, IDOC, and d is the outer diameter of the inner conductor, ODIC.

The inductance per unit length of a lossless coaxial line is

$$L = \frac{\mu}{2\pi} \ln \frac{D}{d} \tag{4–2}$$

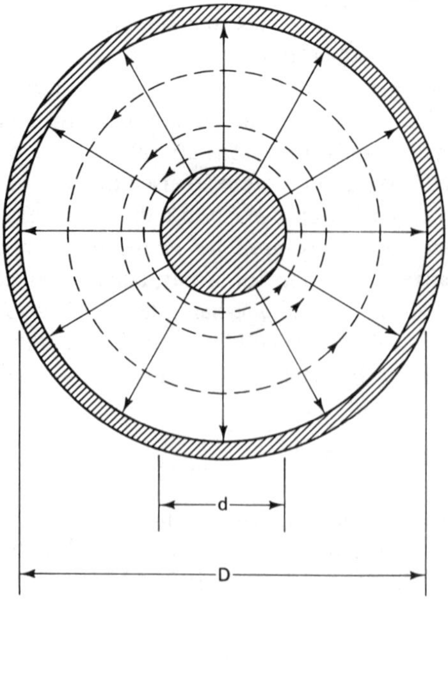

Figure 4–1 Cross-sectional field configuration of the TEM mode in a coaxial line: solid lines, electric field; dashed lines, magnetic field.

and the capacitance per unit length is

$$C = \frac{2\pi\epsilon}{\ln (D/d)} \tag{4–3}$$

We combine these equations with Eq. (4–1) to obtain $Z_c = [(\mu/4\pi^2) \ln^2 (D/d)]^{1/2} = (377/2\pi\epsilon_r^{1/2}) \ln (D/d)$, or

$$Z_c = \frac{60}{\epsilon_r^{1/2}} \ln \frac{D}{d} \tag{4–4}$$

The permeability does not appear in the final equation because we have assumed that $\mu = \mu_0$.

At sufficiently high frequencies, such that the mean circumference of the line is comparable to a wavelength, higher-order modes may be generated. The higher-order modes have longitudinal field components; either the electric or magnetic field has a component in the direction of propagation. The mode with the lowest cutoff frequency has a longitudinal magnetic field component and is designated as TE_{11}. The subscript "11" indicates that both the electric and magnetic fields have nodes which form circumferential standing waves, as shown in cross section in Fig. 4–2. The cutoff wavelength of this mode is approximately $2\pi(d/2 + D/2)/2$, or

$$\lambda_c \approx \frac{\pi(d + D)}{2} \tag{4–5}$$

Since standard lines, connectors, and components are designed for the TEM mode, care must be taken to avoid the higher-order modes. At frequencies higher than S- or X-band it is necessary to use "miniature coax," that is, coaxial lines of small diameter.

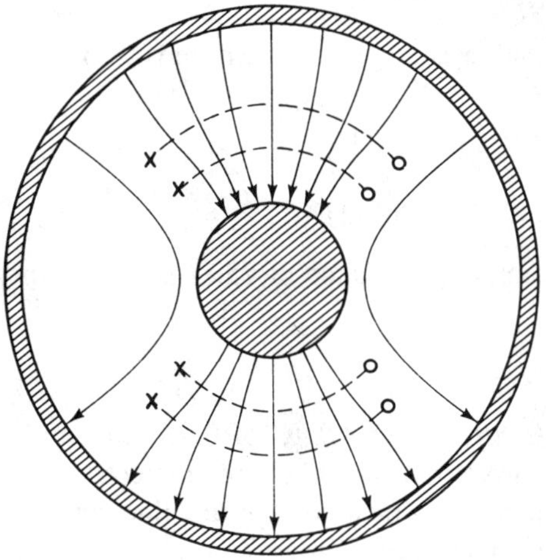

Figure 4–2 TE_{11} mode in a coaxial line.

EXAMPLE 4–1

Calculate the characteristic impedance of a coaxial line with IDOC = 0.285 in., ODIC = 0.080 in., and a relative dielectric constant of $\epsilon_r = 2.2$.

Solution From Eq. (4–4) we have

$$Z_c = \frac{60}{(2.2)^{1/2}} \ln \frac{0.285}{0.080}$$

$$= 51.4\ \Omega$$

EXAMPLE 4–2

A 53.5-Ω coaxial line is to be made with AWG No. 20 copper wire (ODIC = 0.813 mm) and polyethylene insulation ($\epsilon_r = 2.2$). What should be the outer diameter of the insulation?

Solution We solve Eq. (4–4) for D: $\ln (D/d) = Z_c\epsilon_r^{1/2}/60$, or

$$D = d \exp\left(\frac{Z_c\epsilon_r^{1/2}}{60}\right)$$

$$= 0.813 \exp\left(\frac{53.5 \times 2.2^{1/2}}{60}\right)$$

$$= 3.05 \text{ mm}$$

EXAMPLE 4–3

A coaxial directional coupler is contained in a cylindrical can, IDOC = 2.25 in., and within the can there is a ⅝-in.-diameter inner conductor. There are two coaxial connectors at either end of the can. What type of cable should be connected to this device?

Solution Since the insulation is air, $\epsilon_r \approx 1$, we find, from Eq. (4–4),

$$Z_c = 60 \ln \frac{2.25}{0.625}$$

$$= 76.9\ \Omega$$

Any cable with a nominal impedance of 75 Ω is satisfactory: RG−6/U, −12/U, or −11/U.

EXAMPLE 4–4

Standard coaxial cable RG-8/U has an ODIC of 0.285 in., an IDOC of 0.80 in., and its polyethylene insulation has a relative dielectric constant of 2.2. What is the highest frequency at which we can use this cable without losses owing to the generation of higher-order modes?

Solution From Eq. (4-5) we have

$$\lambda_c = \frac{0.080 + 0.285}{2}$$

$$= 0.574 \text{ in., } \quad \text{or} \quad 1.46 \text{ cm}$$

The free-space wavelength is $\lambda_0 = \lambda_c \epsilon_r^{1/2} = 1.46 \times 2.2^{1/2} = 2.16$ cm, and we conclude that the cutoff frequency is

$$f_c = \frac{c}{\lambda_0} = \frac{3 \times 10^{10}}{2.16}$$

$$= 13.8 \text{ GHz}$$

At frequencies greater than 13 GHz we should use a smaller-diameter line.

In Table 4-1 we show the characteristic impedance, the inner and outer diameters, and the breakdown voltage for some of the standard coaxial lines. The power-handling capacity of the lines is usually determined by the heat generated, not the breakdown voltage. Line losses result from radiation, generation of higher-order modes, ohmic heating of the conductors, and dielectric friction or "relaxation losses." Ohmic and relaxation losses generate heat and so limit the power-carrying capability of the cable. These losses increase exponentially with frequency, from about 0.02 dB/ft at 100 MHz to 0.15 dB/ft at 2 GHz. These are typical values but may vary as much as a factor of 10 for different types of standard cable. Power-carrying capability decreases exponentially with increasing frequency from a typical 1 kW at 100 MHz to about 150 W at 2 GHz.

Loss data are usually available from the manufacturer; however, it should be remembered that loss increases with age, with repeated flexing, corrosion, and temperature cycling of the dielectric. There are several different ways of expressing transmission-line loss, and these will be discussed in Chapter 5.

TABLE 4-1

Type	IDOC (mm)	ODIC (mm)	Z_c (Ω)	V_{max} (rms kV)	$\epsilon_r^{1/2}$
RG-58/U	2.95	0.813	53.5	1.9	1.48
RG-8/U	7.24	2.03	52.0	4.0	1.48
RG-17/U	17.27	4.78	52.0	11.0	1.48
RG-55/U	2.95	0.813	53.5	1.9	1.48
RG-5/U	4.70	1.290	53.5	2.0	1.48
RG-9/U	7.11	2.03	51.0	4.0	1.48
RG-59/U	3.71	0.643	73.0	2.3	1.48
RG-11/U	7.24	1.134	75.0	4.0	1.48
RG-6/U	4.70	0.723	76.0	2.7	1.48
RG-62/U, RG-71/U	3.71	0.643	93.0	0.75	1.13
CATV 6/U	4.75	1.063	75.0	—	1.20

In the last column of Table 4–1 we show the square root of the relative permittivities of the dielectric insulations inside the coaxial lines. This is the refractive index and is equal to the ratio of the wavelength in free space to the wavelength in the line, $\epsilon_r^{1/2} = \lambda_0/\lambda = c/v$. The reciprocal of the refractive index is called the *velocity factor*. The product of the refractive index and the physical length of a line is called the *electrical length* of the line,

$$s = \epsilon_r^{1/2} s_0 \tag{4–6}$$

and corresponds to what is called the *optical path length* in physical optics.

EXAMPLE 4–5

A 100-MHz signal is connected through 1.2 m of RG-17/U to channel A of an oscilloscope, and the same signal is transmitted directly to channel B through 1 m of air. What phase difference will be measured between the two channels?

Solution The electrical length of the line to channel A is $s = \epsilon_r^{1/2} s_0 = (2.2)^{1/2} \times 1.2 = 1.78$ m, and the difference in optical path lengths is $\Delta s = 1.78 - 1.00 = 0.780$ m. To find the corresponding phase difference, we first need to determine the operating wavelength: $\lambda = c/f = 3 \times 10^8/10^8 = 3$ m. Since each wavelength corresponds to 360° of phase, we have

$$\Delta\phi = \frac{360\,\Delta s}{\lambda} = \frac{360 \times 0.780}{3}$$

$$= 93.6°$$

Alternative solution The time for the signal to travel through the cable is the electrical length divided by the speed of light, $t_1 = \epsilon_r^{1/2} s_0/c = (2.2)^{1/2} \times 1.2/3 \times 10^8 = 5.933$ ns, and the time for the signal to travel through 1 m of space is $t_2 = 1/c = 1/3 \times 10^8 = 3.333$ ns. Since the difference in travel times is $\Delta t = t_1 - t_2 = 5.933 - 3.333 = 2.60$ ns, and the period is $\tau = 1/f = 1/100 \times 10^6 = 10$ ns, the difference in phase will be the fraction of 360° formed by the ratio of the difference in travel time to the period:

$$\Delta\phi = \frac{\Delta t}{\tau} \times 360° = \frac{2.6}{10} \times 360$$

$$= 93.6°$$

4–3 TWIN-WIRE TRANSMISSION LINE

In its most common form twin-wire line is a ribbon of plastic insulation with stranded copper wires embedded in each edge. Since the middle of the present century it has been widely used to connect rooftop antennas to television receivers and often is called "twin lead." Twisting the line only slightly increases the characteristic impedance and allows a slight tension to be maintained in an installation. Twin wire is also used frequently to connect antennas to low-power transmitters. For transmission of modest power the conductors usually are held in place by periodically spaced insulators, which makes the line look like a ladder.

The advantages of twin wire are its low cost and ease in connecting; its disadvantages are potential shock hazard and radiation losses, especially at bends or in the proximity of material objects. Radiation losses are extreme if the conductor separation is comparable to the operating wavelength.

The propagating mode is TEM, and the field configuration is shown in cross section in Fig. 4–3. For this mode the inductance per unit length is

$$L = \frac{\mu}{\pi} \ln \frac{2D}{d} \tag{4–7}$$

and the capacitance per unit length is

$$\begin{aligned} C &= \frac{\pi \epsilon}{\cosh^{-1}(D/d)} \\ &\approx \frac{\pi \epsilon}{\ln (2D/d)} \qquad D >> d \end{aligned} \tag{4–8}$$

where d is the wire diameter and D is the separation of the wires. With these results and Eq. (4–1) we find the characteristic impedance to be

$$Z_c \approx \frac{120}{\epsilon_r^{1/2}} \ln \frac{2D}{d} \tag{4–9}$$

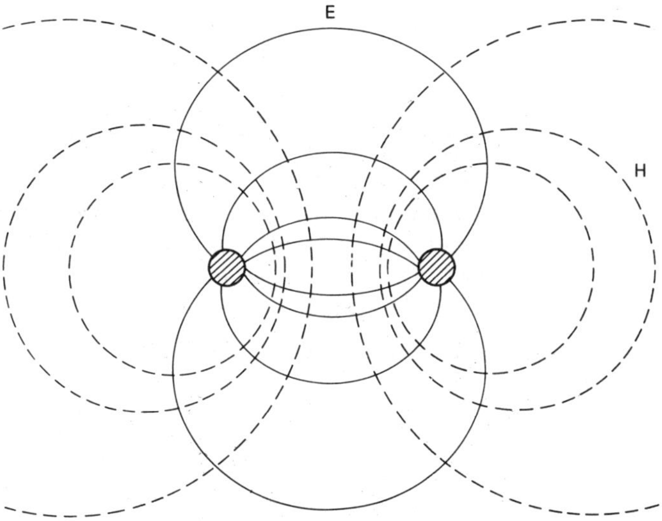

Figure 4–3 Cross-sectional view of the field configuration in twin-wire transmission line.

Usually, only a small amount of insulation is used to separate the wires and most of the electromagnetic field is in the surrounding air; accordingly, we can take $\epsilon_r^{1/2} = 1$, and Eq. (4–9) reduces to

$$Z_c = 120 \ln \frac{2D}{d} \tag{4–10}$$

EXAMPLE 4–6

Calculate the characteristic impedance of twin wire when the center-to-center separation of the wires is ⅜ in. and the wire diameters are 1.56 mm.

Solution Since ⅜ in. is 9.53 mm, we have, from Eq. (4–10),

$$Z_c = 120 \ln \frac{2 \times 9.53}{1.56}$$
$$= 300\ \Omega$$

EXAMPLE 4–7

Twin-wire transmission line with a characteristic impedance of 150 Ω is to be made from wires of diameter 0.3 mm. Find the approximate center-to-center separation of the wires.

Solution We solve Eq. (4–10) for D and find

$$D = \tfrac{1}{2}d \ln^{-1} \frac{Z_c}{120}$$
$$= \tfrac{1}{2} \times 0.3 \exp\left(\frac{150}{120}\right)$$
$$= 0.52 \text{ mm}$$

In Example 4–7 the surface-to-surface separation of the wires is only $D - d = 0.52 - 0.3 = 0.22$ mm, and it is apparent that lower impedances in twin wire are impractical. The limit occurs when $D = d$, and the approximate lowest impedance is $Z_c = 120 \ln (2d/d) = 83\ \Omega$. Coax and microstrip line are better choices if a low characteristic impedance is required.

EXAMPLE 4–8

Calculate the error in characteristic impedance caused by using Eq. (4–10) in Example 4–7.

Solution In Example 4–7 we obtained $D = 0.52$ mm and $d = 0.3$ mm for a characteristic impedance of 150 Ω. To find a more exact value of the impedance, we start with Eqs. (4–7) and (4–8):

$$Z_c = \left(\frac{L}{C}\right)^{1/2} = \frac{377}{\pi(\epsilon_r)^{1/2}}\left(\ln \frac{2D}{d} \cosh^{-1} \frac{D}{d}\right)^2$$

$$= 120\left(\ln \frac{2D}{d} \cosh^{-1} \frac{D}{d}\right)^{1/2} \qquad (4\text{–}11)$$

$$= 120\left(\ln \frac{2 \times 0.52}{0.3} \cosh^{-1} \frac{0.52}{0.3}\right)^{1/2}$$

$$= 143\ \Omega$$

The error is $(150 - 143)/143 = 4.9\%$.

If a scientific calculator is not available that has the hyperbolic functions, the value of the inverse hyperbolic cosine will have to be found from a table of functions. Higher-order modes in twin wire are not usually considered since, when $D \sim \lambda/2$, the line losses are comparable to the radiation from an antenna.

4–4 MICROSTRIP LINE

Microstrip line or ''microstrip'' also is a two-conductor transmission line comprised of a narrow, flat conductor separated from a much larger conductor, or ground plane, by a layer of dielectric insulation. A sketch of microstrip line is shown in Fig. 4–4. It is often constructed from double copper-clad fiberglass or Teflon board. The unwanted copper can be removed by peeling, after cutting the shape of the line with a knife, or the unwanted copper can be removed by etching. Etching kits are available at retail stores and are suitable for making prototypes. Industrial production requires the use of hazardous chemicals which are not advisable for the individual experimenter or for use in the classroom.

Since the electric field is partly in air and partly in the dielectric, the refractive

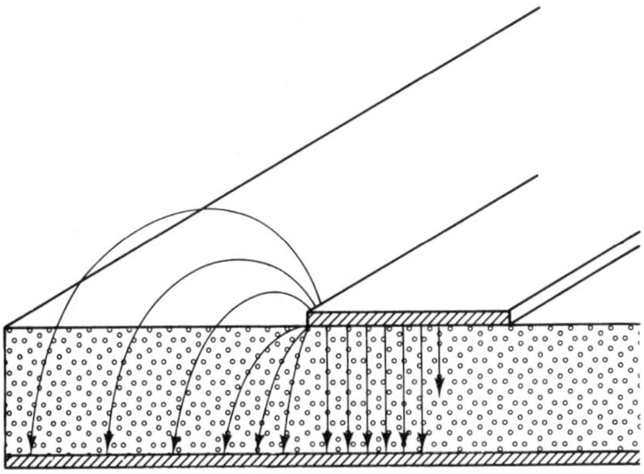

Figure 4–4 Cross-sectional view of the electric field in microstrip line.

index of the dielectric cannot be used as the ratio of the free-space wavelength to the guide wavelength. The effective velocity factor of microstrip can be found from the empirical equation

$$(\epsilon_r')^{-1/2} = \frac{\lambda}{\lambda_0} = \left[1 + 0.63(\epsilon_r - 1)\left(\frac{W}{T}\right)^{0.123}\right]^{-1/2} \tag{4–12}$$

where ϵ_r is the relative permittivity of the dielectric, W is the width of the narrow conductor, and T is the thickness of the dielectric.

EXAMPLE 4–9

Calculate the velocity factor for a microstrip line constructed from Teflon board ($\epsilon_r = 2.6$). The width-to-thickness ratio is 3.

Solution From Eq. (4–12) we have

$$\frac{\lambda}{\lambda_0} = [1 + 0.63(2.6 - 1)(3)^{0.123}]^{-1/2}$$
$$= 0.681$$

On a scientific calculator enter 3 and use the LOG key to obtain 0.4771; multiply the result by 0.123 and then find the inverse log. The result is $(3)^{0.123} = 1.1447$. Multiply this result by $(2.6 - 1) = 1.6$ and by 0.63 to obtain 1.1538. Add 1, take the square root, and invert with the $1/x$ key to obtain the answer, 0.681.

The answer to Example 4–9 can be verified by referring to Fig. 4–5, where we show the velocity factor for various relative permittivities and W/T ratios.

EXAMPLE 4–10

A 5-cm length of microstrip line on a 1/32-in. epoxy board ($\epsilon_r = 4.3$) has a width-to-thickness ratio of 6. What is its electrical length?

Solution We start, as in Example 4–9, by finding the velocity factor:

$$\frac{\lambda}{\lambda_0} = [1 + 0.63(4.3 - 1)(6)^{0.123}]^{-1/2}$$
$$= 0.528$$

The effective refractive index is $(\epsilon_r')^{1/2} = \lambda_0/\lambda = 1/0.528 = 1.895$, and the electrical length is

$$s = (\epsilon_r')^{1/2} s_0 = 1.895 \times 5.0$$
$$= 9.48 \text{ cm}$$

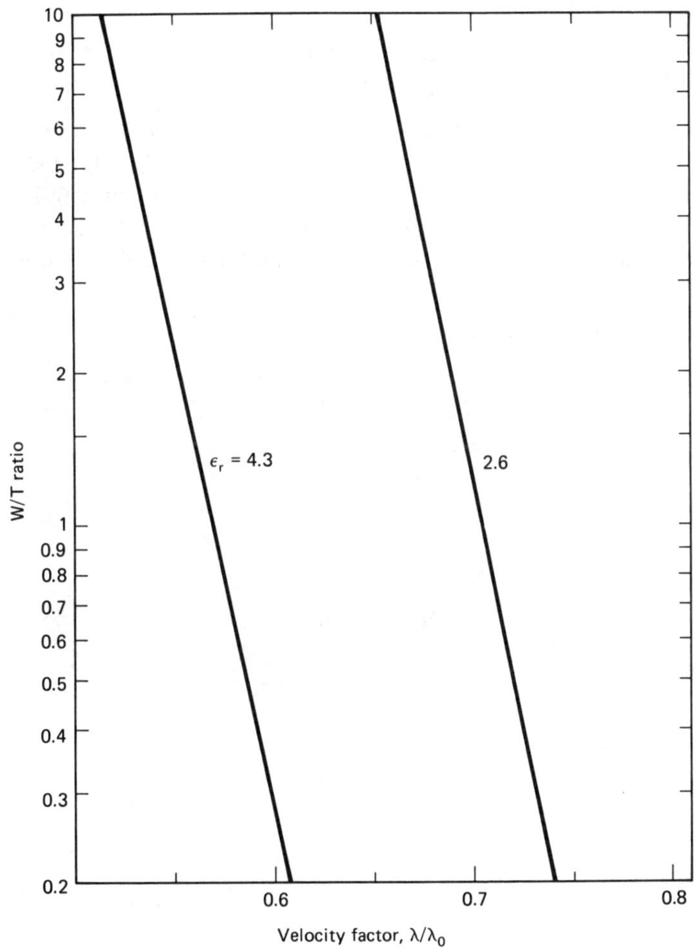

Figure 4–5 Velocity factor in microstrip line for relative permittivities of 2.6 and 4.3. The width-to-thickness ratio is W/T.

The characteristic impedance of microstrip line can be found from the empirical formula

$$Z_c = \frac{377}{\epsilon_r^{1/2}(W/T)[1 + 1.735(W/T)^{-0.836}]} \tag{4–13}$$

EXAMPLE 4–11

Find the characteristic impedance of the microstrip line discussed in Example 4–9.

Solution From Eq. (4–13) we have

$$Z_c = \frac{377}{2.6^{1/2} \times 3 \times (1 + 1.735 \times 3^{-0.836})}$$
$$= 46.0\ \Omega$$

On a scientific calculator, evaluate $3^{-0.836}$ by using the LOG key after entering 3. Multiply the result by -0.836 then use INVERSE LOG to obtain $3^{-0.836} = 0.3991$. Multiply by 1.735 and add 1 to obtain 1.6925. Multiply by 3, again depress the multiplication key, enter 2.6, depress the SQUARE ROOT key and the "=" key; finally, depress $1/x$ key and multiply by 377 to obtain 46.0 Ω.

EXAMPLE 4–12

A microstrip line is to be fabricated from 1/32-in. Teflon board, $\epsilon_r = 2.6$. If a characteristic impedance of 50 Ω is required, what is the width of the strip?

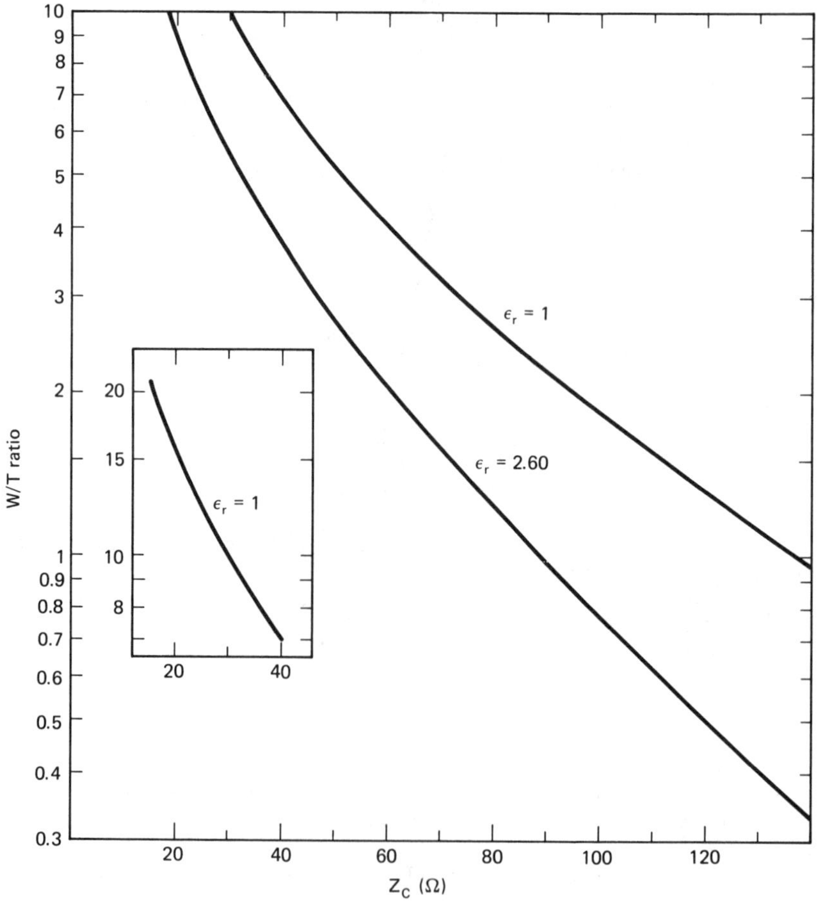

Figure 4–6 Characteristic impedance of microstrip line for insulation permittivities of 1 and 2.6.

Solution Since solving Eq. (4–13) for W/T is a chore, it is convenient to make a graph of the equation, as shown in Fig. 4–6. From the graph we see that $Z_c = 50$ corresponds to $W/T = 2.75$. Since the thickness is $T = \frac{1}{32} \times 25.4 = 0.794$ mm, the required strip width is $W = 2.75 \times 0.794 = 2.18$ mm.

We have observed, at sufficiently high frequencies, a null in the electric field intensity at the center of the strip. We presume that this results from the development of a waveguide mode similar to the TE_{10} mode in rectangular waveguide. Since reflections at open sides are opposite in phase to those in ordinary rectangular, metal waveguide, the transverse standing wave has a null at the center of the strip—rather than a maximum—as shown in Fig. 4–7. The cutoff wavelength of such mode should be approximately

$$\lambda_c = 2W \tag{4–14}$$

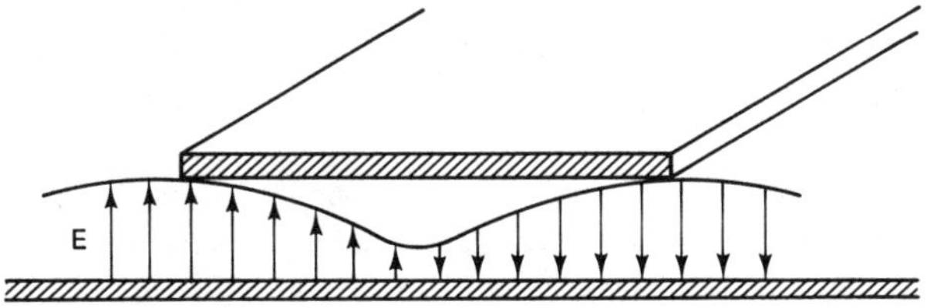

Figure 4–7 Development of a waveguide mode in microstrip line.

EXAMPLE 4–13

A low-impedance microstrip line is made from 0.794-mm Teflon board, $\epsilon_r = 2.6$, and has a conductor width of 7.94 mm. At what frequency may there be higher-order mode difficulties?

Solution The electrical width of the microstrip line is $W = W_0\epsilon_r^{1/2} = 7.94 \times (2.6)^{1/2} = 12.8$ mm, and the cutoff wavelength of the lowest higher-order mode is $\lambda_c = 2W = 2 \times 12.8 = 25.5$ mm. The corresponding cutoff frequency is

$$f_c = \frac{c}{\lambda_c} = \frac{3 \times 10^{11}}{25.6}$$
$$= 11.7 \text{ GHz}$$

We may expect difficulties at even lower frequencies, perhaps 9 or 10 GHz.

4–5 CATV TRANSMISSION-LINE SYSTEM

We think of cable television, CATV, as a coaxial transmission system, at least from the subscriber's viewpoint, but the system may start with several microwave links. Consider, for example, the CATV system in Prescott, a small rural town in the central mountains of Arizona. Signals from satellites, metropolitan areas, and local communities are received at the head end, processed, then transmitted by microwave antennas to the Prescott area for distribution by cable.

By "head end" we mean the initial receiving site and its receiving, processing, and retransmission facilities. The head end for Prescott is on one of the higher mountain peaks in the area, Mt. Francis, at an elevation of 7000 ft and about 2000 ft above the town. The processed signals are retransmitted on microwave links to downtown Prescott, the airport, Groom Creek, and Ponderosa Park (Fig. 4–8). The four receiving sites are geographically dispersed to minimize the length of coaxial trunk lines serving the various population centers. It may seem from the map in Fig. 4–8 that two of the receiving sites are close together, but in this mountainous region there is a large elevation difference separating Groom Creek and Ponderosa Park.

At the head end, signals are received from satellite antennas, from metropolitan Phoenix 100 miles to the south and from Flagstaff 100 miles to the north. As examples of the processing that occurs at the head end we will consider a signal received from Flagstaff and one received by a satellite dish and follow them through the system.

Flagstaff channel 2 (55.25 MHz) is received by one of the end-fire arrays shown in Fig. 4–9. The signal is delivered to a *processor*, where it is heterodyned to another carrier frequency corresponding to a local channel, in this case channel

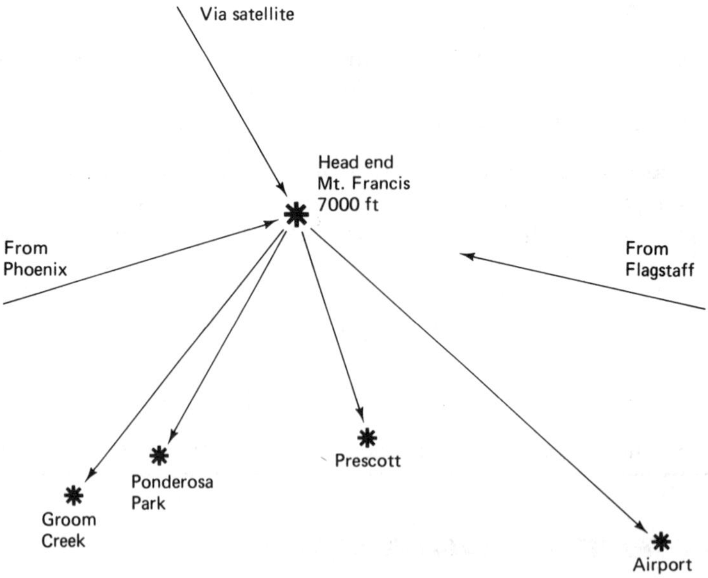

Figure 4–8 At the head end signals are received from various sources, processed, and retransmitted to local stations for distribution by cable.

Figure 4–9 The peak of Mt. Francis is crowded with antennas for CATV, telephone links, and other communications systems. One of these end-fire antennas receives channel 2 from Flagstaff 100 miles to the north, via a repeater on Mingus Mountain 45 miles to the north.

10 (193.25 MHz). What we have described here is only one channel. In the whole system there are an additional 24 channels. The outputs from all 25 processors and modulators are delivered to a *combining network*, also called an amplitude-modulating link (AML), which is an assemblage of directional couplers, pads, amplifiers, and modulators (Fig. 4–10). Here the signals are amplified and equalized and combined as a modulation on a 13.2-GHz carrier. The resultant signal is beamed at 13.2 GHz, by the antennas shown in Fig. 4–11, to the local receiving sites shown in Fig. 4–8.

A satellite signal, for example, 3.76 GHz, is received at a "dish" (paraboloidal reflector antenna, Fig. 4–12) and amplified and down-converted to 750 to 450 MHz in an LNA (low-noise amplifier) and down converter mounted at the dish. The 750 to 450 MHz output is transmitted through 1700 ft of coaxial cable from the dish site up to the retransmission site at the peak of the mountain. Cable loss, at 2.0 dB per 100 ft, is regrettable, but a long line is unavoidable because of the rugged terrain. The cable terminates in a receiver (Fig. 4–13) where the signal is demodulated from the 750- to 450-MHz carrier. If the satellite signal is "scrambled," an additional loop in the receiver is used to unscramble (video decipher, audio decrypt) the signal and the output is ready for further processing before transmission to the local system.

The unscrambled signal is input to a *modulator*, as shown in Fig. 4–13. The modulator strips off the audio and video, generates a new carrier frequency—the frequency of what will be local channel 3—and modulates the new frequency with

Figure 4–10 Head-end AML (amplitude-modulating link) at Mt. Francis. In this combining network the information signals are modulated onto an X-band carrier and carried in waveguide to the antennas shown in Fig. 4–11.

Figure 4–11 At the head end, Mt. Francis, four X-band paraboloidal reflector antennas are used to transmit the modulated 13.2-GHz carrier to local sites for distribution by cable in the town below.

Figure 4–12 Head end, Mt. Francis. Satellite dish antennas, used for reception of satellite transponded downlink signals, are protected by radomes and internal heating. The screen in the background reduces telephone interference.

the audio and video. To summarize, the signal has been unscrambled, demodulated, and remodulated on a new carrier as a local channel. It is now delivered to the combining network, already discussed, and transmitted on a 13.2-GHz carrier to the local distribution sites shown in Fig. 4–8. If necessary, the signal may again be scrambled to prevent local thievery.

At a local receiving site the signal is demodulated, equalized, amplified, and delivered to a coaxial trunk line (Fig. 4–14). Because coaxial-line losses are higher at higher frequencies, the signals must be equalized again at the input of each trunk-line amplifier. Trunk-line amplifiers accept an input of 11 dBmV, are spaced every 1800 ft, and have a gain of 22 dB and a bandwidth of 440 MHz. Within each trunk-line amplifier there is a bridging amplifier which may be used as a directional coupler to a "feeder line." Each feeder line has several "taps," which are directional couplers or "power splitters" that serve one to eight subscribers. Since each power division is a loss of about 3.5 dB, the length of a feeder line is determined by the number of subscribers it serves as well as by line attenuation. A feeder line can be extended with a "line extender" amplifier followed by more taps. Eventually, the line must be terminated in a 75-Ω matched load—we do not want a subscriber's tap point to be at a null in a standing-wave pattern.

The FCC recommends that the signal level to a subscriber be 0 dBmV, but since many subscribers now have two or more TV sets, VCRs, and converters, a

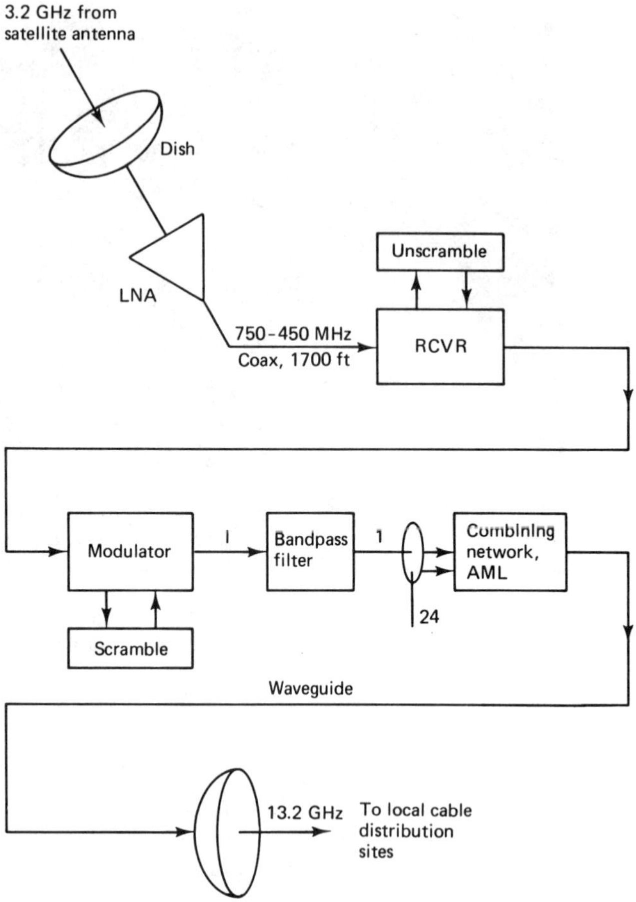

Figure 4–13 Block diagram of the signal-processing system at the head end, Mt. Francis.

more practical specification is +10 dBmV. When we express a signal level in dBmV, we imply a reference to 75 Ω, unless otherwise stated.

> **EXAMPLE 4–14**
>
> Express +30 dBmV in a 75-Ω system as a power.
>
> **Solution** We have +30 dBmV = $10^3 \times 10^{-3} = 1$ V. From the power formula $P = V^2/R$ we have $P = 1^2/75 = 0.0133$ W, or 11.25 dBm.

In addition to the RF system described above, there is also an essential power network. The RF coaxial cables do double duty and serve as ac power conductors.

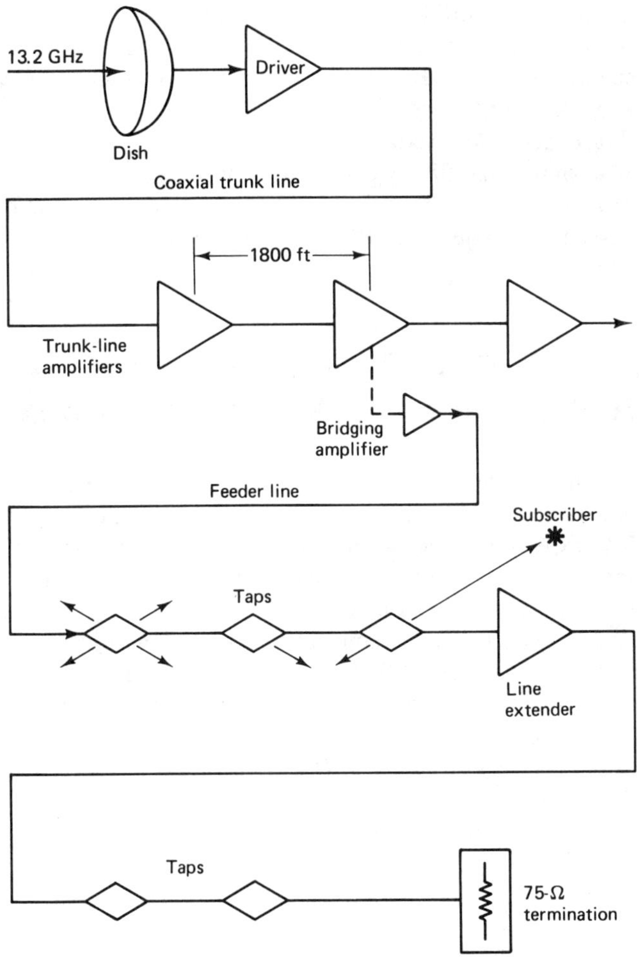

Figure 4–14 Block diagram of a local CATV cable distribution system.

At various points in the distribution network, drop lines are provided by the local power company. The drop lines supply ac power which is transformed to 60 Vac, 12 to 14 A—a power sufficient for five line-amplifiers. The ac power is carried on the RF cables to the line-amplifiers. Each line-amplifier has a rectifier that delivers about 3 A at 24 Vdc.

From the description above we see that a CATV technician needs a basic understanding of electronics from dc to RF; in addition, the technician must be able to troubleshoot, replace, or repair everything from microwave antennas to dc power rectifiers. Since CATV systems are constantly being expanded to accommodate new subscribers, the technician needs to be especially knowledgeable about "powering": the calculation of line attenuation and passive losses. Attenuation in transmission lines is discussed further in Chapter 5; by "passive losses" we mean line losses due to power splitting, couplings and fittings, kinks, turns, and reflections. Finally, the technician needs several mechanical skills, such as soldering, connectoring, and pole climbing.

We expect the coaxial systems eventually to be replaced by fiber optics, but the same knowledge of systems powering, basic electronics, and mechanical skills will be required of the technician. Propagation and attenuation in fiber optic cables is discussed in Chapters 5 and 10. Microwave antennas and propagation are discussed in Chapter 9.

APPLICATION 4–1: *Measurement of Electrical Length and Refractive Index*

Electrical length and refractive index of a coaxial cable can be measured by determining the number of standing waves on an open- or short-circuited length of line. The experimental arrangement is shown in Fig. 4–15. We have indicated three standing waves, or three half-wavelengths on the line, but we shall assume that the number, n_1, initially is unknown.

We assume that the frequency, f_1, has been adjusted so that a standing-wave null is detected on the VSWR meter; thus n_1 is an integer equal to the number of half-wavelengths, $\lambda_1/2$, on the line. We now increase the frequency until another null is detected, and there are $n_2 = n_1 + 1$ half-wavelengths on the line at frequency f_2. We now have $n_1(\lambda_1/2) = n_2(\lambda_2/2) = (n_1 + 1)(\lambda_2/2)$. For the case $n_1 = 3$, we would have $n_2 = 4$, as indicated by the broken waveform shown in Fig. 4–15. Rearranging the equations, we have $(n_1 + 1)/n_1 = \lambda_1/\lambda_2 = f_2/f_1$, and solving for n_1 we obtain

$$n_1 = \frac{f_1}{f_2 - f_1} \tag{4–15}$$

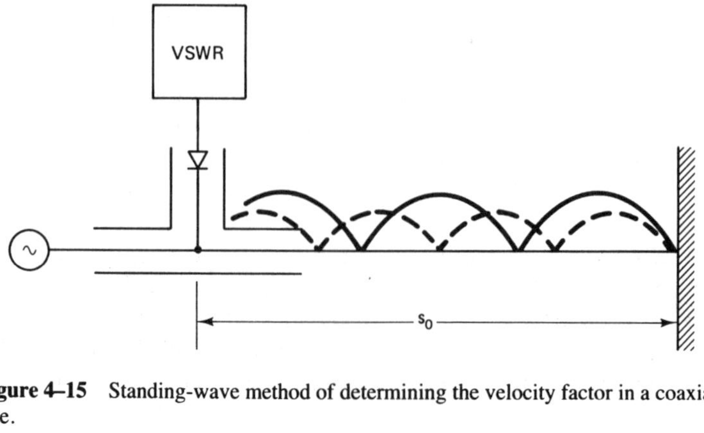

Figure 4–15 Standing-wave method of determining the velocity factor in a coaxial line.

Now that we have determined n_1, we can find the electrical length, $s = n_1 (\lambda_0/2) = (f_1/(f_2 - f_1))/(c/2f_1)$, or

$$s = \frac{\frac{1}{2}c}{f_2 - f_1} \tag{4–16}$$

We can find the refractive index from Eqs. (4–6) and (4–16), $\sqrt{\epsilon_r} = s/s_0$,

$$\sqrt{\epsilon_r} = \frac{c}{2s_0(f_2 - f_1)} \tag{4–17}$$

EXAMPLE 4–15

In performing the experiment above we find a null at 100 MHz and an additional null at 1.333×10^8 Hz. The measured line length is $s_0 = 3.0$ m. Find the electrical length of the line and the refractive index.

Solution From Eq. (4–16) we have

$$s = \frac{3 \times 10^8}{2[(1.333 \times 10^8) - (1.0 \times 10^8)]} = 4.5 \text{ m}$$

and the refractive index is

$$\sqrt{\epsilon_r} = \frac{s}{s_0} = \frac{4.5}{3} = 1.5$$

Check: From Eq. (4–17) we have

$$\sqrt{\epsilon_r} = \frac{3 \times 10^8}{2 \times 3[(1.333 \times 10^8) - (1.0 \times 10^8)]} = 1.5$$

As an alternative we may leave the end of the line open. When the frequency is adjusted to f_1 we have N_1 quarter-wavelengths on the line, and when the frequency is increased to N_2 for the next null, we have $N_2 = N_1 + 2$ quarter-wavelengths. Proceeding as before, we find that Eq. (4–15) is changed to

$$N_1 = \frac{2f_1}{f_2 - f_1} \tag{4–18}$$

and the electrical length becomes $s = N_1(\lambda_0/4)$ or

$$s = \frac{\frac{1}{2}c}{f_2 - f_1}$$

Since this is identical to Eq. (4–16), we see that the equation for $\sqrt{\epsilon_r}$ [Eq. (4–17)] is also unchanged.

Equipment and Materials

flexible coaxial cable, 1 to 8 m long
coaxial tee
modulated signal generator, 10 MHz to 1 GHz or higher
coaxially mounted crystal detector
caliper micrometer
tape measure
SWR meter

Procedure

Choose an operating frequency in the suggested range and determine the electrical length, refractive index, and velocity factor for the cable. Repeat, using two other frequencies; also, repeat, using the open-circuit technique. Measure the ODIC and the diameter of the insulation (IDOC) of the cable, and use Eq. (4–4) to determine the characteristic impedance.

In your report, present the data and results in tabular form; also, compare your results with the manufacturer's specifications and compute the percentage difference between the two. In discussing your results, state why you believe your values are less or more reliable than the manufacturer's.

APPLICATION 4–2: Measurement of Electrical Length and Refractive Index by the Time-Domain Method

The electrical length of a line can be determined by measuring the time required for a pulse to travel the length of the line; or, if the line is short- or open-circuited at the far end, by measuring the time required for the pulse to travel twice the length of the line.

EXAMPLE 4–16

The time required for a pulse to travel down a line, be reflected, and travel back is measured on an oscilloscope display as indicated in Fig. 4–16. If the measured time is $t = 0.1\ \mu s$ and $s_0 = 10$ m, determine the electrical length, refractive index, and velocity factor of the line.

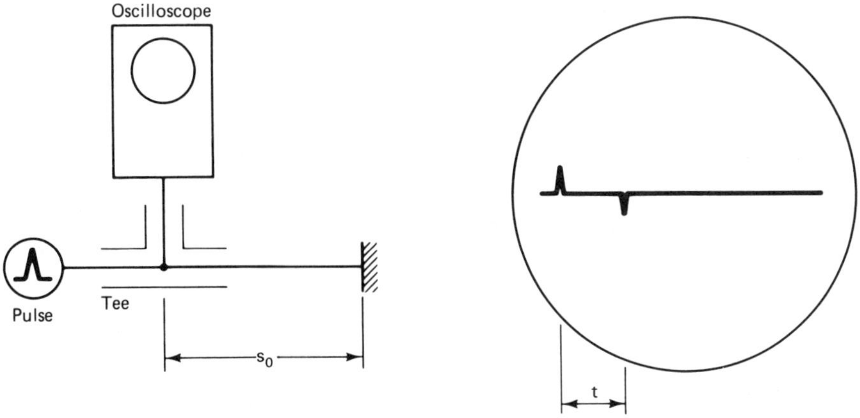

Figure 4–16 Time-domain method of determining the velocity factor in a coaxial line.

We start by finding the propagation speed on the line:

$$v = \frac{2s_0}{t} = \frac{2 \times 10}{0.1 \times 10^{-6}}$$
$$= 2 \times 10^8 \text{ m/s}$$

The refractive index is

$$\epsilon_r^{1/2} = \frac{c}{v} = \frac{3 \times 10^8}{2 \times 10^8}$$
$$= 1.5$$

and the velocity factor is

$$\epsilon_r^{-1/2} = \frac{1}{1.5}$$
$$= 0.667$$

If a suitable pulse generator is not available, the experiment can be done with a bounceless switch and a dc source of voltage. When the switch is turned on, the oscilloscope trace starts its sweep from the trigger pulse input. After reflection the advancing wavefront reaches the oscilloscope input and cancels the original input. The resultant pulse length displayed on the oscilloscope is equal to the total travel time. A bounceless switch can be made from an ordinary switch and a pair of 7400 NAND gates, as shown in Fig. 4–17.

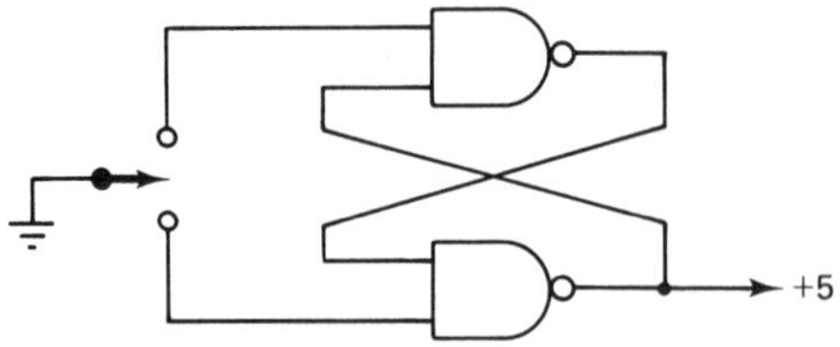

Figure 4–17 Bounceless switch constructed from two 7400 NAND gates.

Equipment and Materials

bounceless switch
oscilloscope
coaxial line, 10 m
coaxial tee

Procedure

Set up the apparatus as shown in Fig. 4–16; use a pulse generator or build the bounceless switch shown in Fig. 4–17. Measure the travel time and determine the electrical length, refractive index, and velocity factor of the cable. Repeat for several trials and average the results. Measure the ODIC and IDOC of the cable, determine the characteristic impedance, and compare your results with the manufacturer's specifications. If multiple reflections are annoying in the course of measurements, reduce them by including a series resistance between the pulse source and the line.

APPLICATION 4–3: Connectoring Coaxial Cable

The coaxial cables and fittings developed for the cable television industry (CATV) are less expensive and can be assembled more quickly and easily than the standard types developed to army/navy specifications. CATV coaxial components—tees, terminations, adapters, boosters—are available and also inexpensive. We suggest their use by the individual experimenter and in the classroom.

PART 1: CATV COAXIAL CABLE

Equipment and Materials

CATV coaxial cable, doubly shielded RG-6/U, 1.5- to 2-m lengths
type F connectors (Fig. 4–18)
crimping tool
electrician's knife
scissors

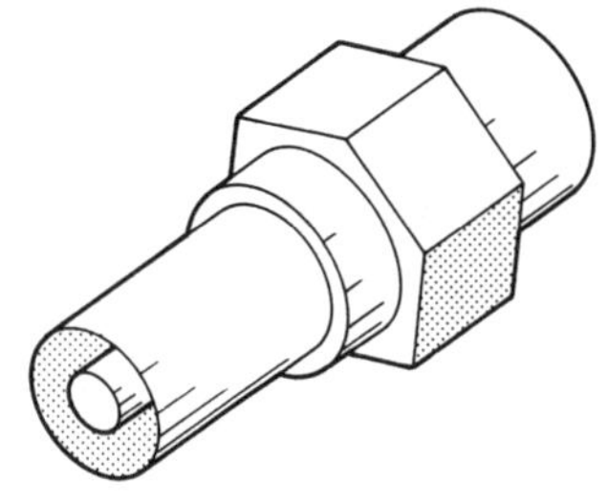

Figure 4–18 Type F CATV cable fitting.

Procedure

To connector the cable, start by removing ⅜ in. of outer cover insulation to expose the outer braid. Straighten the outer braid and fold it back over the outer insulation (refer to Fig. 4–19).

Remove the outer foil and straighten and fold back the inner braid; some of

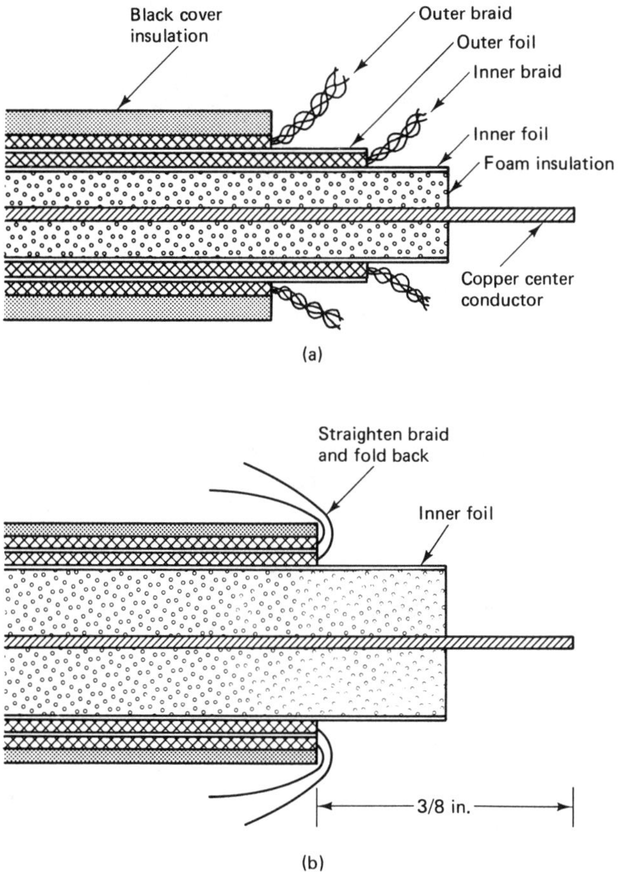

Figure 4–19 Longitudinal sections of CATV RG-6/U cable and its preparation for connecting.

the strands of the braid will be longer than 3/8 in.; fan out the braid and with a pair of scissors trim it to 1/4 to 5/16 in. and fold back.

The inner foil is now exposed; at the midpoint of the inner foil cut through the inner foil and inner insulation and remove, taking care not to cut or nick the inner conductor. Inspect: no foil or braid should touch or come close to the center conductor. Push the F fitting over the end of the foil-covered inner insulation; the inner insulation should be flush with the face of the inner brass tube. The strands of braid should not extend beyond the back of the fitting; the end of the center conductor should be almost flush with the face of the fitting. If the assembly is not satisfactory, correct it. When the inspection is satisfactory, use the crimping tool to crimp the shank of the connector onto the outer insulation of the cable.

Test the cable for electrical continuity of inner and outer conductors and for short circuits. Test for noise by flexing the cable when it is used to connect a weak signal from a signal generator to an oscilloscope, with the oscilloscope gain set at its most sensitive level.

PART 2: COMMERCIALLY CONNECTORED COAXIAL CABLE

Equipment and Materials

1.5- to 2-m lengths of commercially connectored coaxial cable, for example, RG-58/U with UG-88/U-type BNC connectors
compatible new connector kit
electrician's knife
soldering iron and solder
set of thin end wrenches

Procedure

Carefully dissemble the connectored cable, using the end wrenches on the flats of the connector to unscrew the parts and a soldering iron to unsolder the bayonet pin and body of the plug if it is soldered to the braid as in the four solder holes of the PL-259. As the connector is dissembled, make a sketch, similar to that in Fig. 4–19b, showing all dimensions, cable number, connector type, and location of components, for reassembly. After the sketch is complete, form a fresh end on the cable by removing a few inches, and connector the cable.

Connector the other end of the cable with a new connector, using your sketch as a guide. Test the cable electrically, as in Part 1; in addition, test the connector for mechanical strength. A BNC connector, for example, should withstand enough tension to activate the spring loading without pulling off the end of the cable. Common failings are: a short circuit owing to pieces of braid in the connector; burred or rounded flats owing to use of the wrong size of end wrench; and an oval-shaped body owing to excessive clamping or wrenching.

SUMMARY

Of the many types of transmission lines, three often-used types are coaxial line, twin conductor or twin line, and microstrip line. Theoretical equations for the inductance and capacitance per unit length are available for coax and for twin line, and from these we can derive equations for the characteristic impedances:

$$Z_c = \frac{60}{\epsilon_r^{1/2}} \ln \frac{D}{d}$$

$$Z_c = \frac{120}{\epsilon_r^{1/2}} \ln \frac{2D}{d} \approx 120 \ln \frac{2D}{d}$$

When using commercial coaxial cables to connect components or subsystems it is important to match the characteristic impedance of the cable to the input and output resistances of sources and loads. If the lines are long, we should remember that the lines are lossy, especially at high frequencies. Loss increases exponentially with frequency. The power-handling capability of coaxial cables is limited by heat generated in the cable and by high-voltage breakdown. Another limitation at very high frequencies is the possibility of generating modes other than the TEM. The "cutoff wavelength" is the wavelength at which the first higher mode can be generated,

$$\lambda_c \approx \frac{\pi(d + D)}{2}$$

The wavelength in coaxial cable is the same as the wavelength in the dielectric insulation,

$$\lambda = \frac{\lambda_0}{\epsilon_r^{1/2}} = \frac{\lambda_0}{\eta}$$

Since the wavelength in the line is shorter than the wavelength in air, the electrical length of the line is greater than its physical length,

$$s = \eta s_0$$

Electrical length is also called optical path length.

One way to learn to connector coax is to take apart and copy a commercially connectored cable. Twin-conductor line is inexpensive and easy to fabricate. Because it is unshielded it has the disadvantages of possible radiation loss and shock hazard when used with transmitting gear; conversely, when used with receiving equipment, it may be noisey. The wavelength in twin line is approximately the same as the wavelength in air.

Microstrip line is often used to connect components in devices operated at UHF/microwave frequencies. It is easy to fabricate and is suitable for the circuit board techniques used in mass-produced devices. There are no exact equations for the characteristic impedance and wavelength in microstrip line, but there are many empirical equations. Two of these are

$$\frac{\lambda}{\lambda_0} = \left[1 + 0.63(\epsilon_r - 1)\left(\frac{W}{T}\right)^{0.123}\right]^{-1/2}$$

$$Z_c = \frac{377}{[\epsilon_r^{1/2}(W/T)][1 + 1.735(W/T)^{-0.836}]}$$

Other types of often-used transmission lines are waveguide (Chapter 6) and fiber optic cable (Chapter 10). Surface-wave guiding is discussed briefly in Chapter 9.

PROBLEMS

4–1. A coaxial cable has an IDOC of 0.300 in. and an ODIC of 0.060 in. and a polyethylene dielectric with a relative dielectric constant of 2.25. What is the characteristic impedance of the line? What is the wavelength in the line if the operating frequency is 150 MHz?

4–2. A rigid coaxial line is made of aluminum tubing, inside diameter 3.81 cm. The inner conductor is 0.61-cm-diameter tubing supported on polystyrene foam, $\epsilon_r \approx 1.0$. At what frequency would you expect higher-order mode difficulties?

4–3. A twin-wire line must have a 5-in. conductor spacing to avoid electrical breakdown. If the characteristic impedance is to be 300 Ω, what must be the diameter of the conductors?

4–4. A transmission line, 30 m long, is to have a characteristic impedance of 3000 Ω. Is twin-wire line practical? Back up your answer with calculations.

4–5. What is the inductance per unit length of the transmission line described in Problem 4–1?

4–6. If the IDOC is 10 mm, the ODIC 1.2 mm, and the characteristic impedance 90 Ω, what is the relative permittivity of the insulation?

4–7. A coaxial line used for pulse transmission has water for a dielectric, $\epsilon_r = 80$, an IDOC of 8.2 cm, and an ODIC of 2.54 cm. What is the characteristic impedance?

4–8. If the line in Problem 4–7 is 3.5 m long, what is the total capacitance? What is the electrical length? The velocity factor?

4–9. Common 120-V lamp cord is made of two conductors embedded in rubber insulation. The conductors are each 1.6 mm in diameter and are separated 5.3 mm, center to center. What is the characteristic impedance?

4–10. If the capacitance per unit length is 260 pF m^{-1} and the inductance 30 μH m^{-1}, what is the characteristic impedance of the transmission line?

4–11. Twin wire, $D = 1.2$ cm and $d = 1.0$ mm, is immersed in oil, dielectric constant 3.5. What is the characteristic impedance of the immersed line? The characteristic impedance in air? If the line enters the oil at normal incidence, what is the magnitude of the reflection coefficient at the air–oil interface?

5

Transmission-Line Theory

5–1 INTRODUCTION

We have seen that the characteristic impedance of a transmission line is determined by the ratio of its inductance and capacitance per unit length, which can be represented schematically as shown in Fig. 5–1. Here the inductance and capacitance are continuously distributed and are measured by an arbitrary unit of length; however, if we consider them to be discrete or lumped parameters we can build the line from coils and capacitors, as suggested by the schematic diagram.

Just because we can use the same picture to represent different things is not, of course, assurance that the different things will have the same behavior. To decide the matter, let us try to form an intuitive idea of how energy moves down a transmission line. For the first capacitor to be charged, charge must flow through the first inductor. A magnetic field is associated with a current, and the field that fills the space around the inductor is an energy density $\frac{1}{2}Li^2/d$, where d is the *cell length* shown in Fig. 5–1. Since an *ac* signal or wave is incident on the line, the magnetic field eventually collapses and the first capacitor is charged to an energy density $\frac{1}{2}Cv^2/d$. The voltage on the first capacitor now is impressed across the second coil, charge starts to flow through the second coil, and the process continues down the line. The energy that was in the inductor must equal the energy now in the capacitor. Equating the two and solving for $v/i = Z_c$, we again obtain

$$Z_c = \left(\frac{L}{C}\right)^{1/2} \qquad \lambda >> d \qquad (4\text{–}1)$$

The cell length d seems inconsequential since it cancels in the ratio L/C; however, if we calculate the propagation speed on the line, the result is slightly different

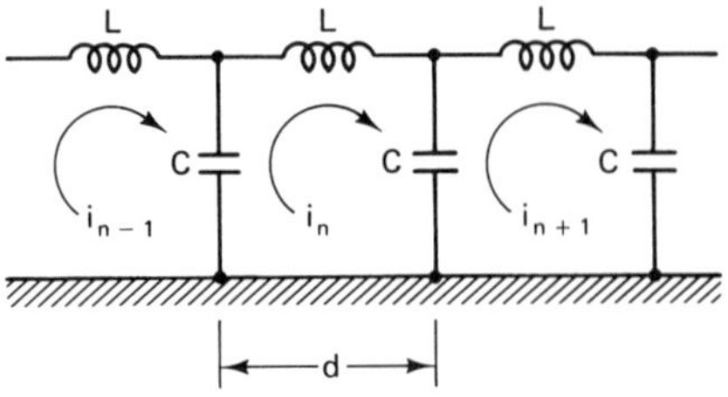

Figure 5–1 Lumped-parameter transmission line. The lattice constant, *d*, is the physical length of a cell. The symbols *L* and *C* can represent lumped parameters or distributed inductance and capacitance per unit length.

from what we obtained earlier. Equation (1–7) for the wave velocity now becomes $v = [(L/d)(C/d)]^{-1/2}$, or

$$v = \frac{d}{(LC)^{1/2}} \qquad \lambda >> d \tag{5–1}$$

The difference seems trivial: for the distributed line we use inductance and capacitance per unit length; for the lumped-parameter line we use inductance and capacitance divided by *d*.

The difference between an arbitrary unit of length and a cell length is physically significant. There is a resonant effect when $\lambda = 2d$, which causes the lumped-parameter line to act as a low-pass filter, and transmission is strongly attenuated at higher frequencies. We can consider continuous and lumped lines to be equivalent only at frequencies well below the resonance; we have to add the condition $\lambda >> d$ to Eqs. (4–1) and (5–1). Our intuitive thinking went awry when we assumed that all of the energy always moves in the forward direction; as the frequency approaches resonance, some of the energy starts to flow back toward the source. If a line is very long, or infinite, the attenuation at resonance is total. If a line is short and there are only a few cells, it is called a *multipole filter*, and the number of cells determines the number of poles.

A linear arrangement of cells, as shown in Fig. 5–1, also can be called a one-dimensional *lattice*. Crystalline substances, such as common salt or NaCl, silicon, and germanium, are examples of three-dimensional lattices. Usually an atom or molecule is contained in each cell. By observing the transmission or reflection of electromagnetic waves from such lattices it is possible to determine the cell spacings or *lattice constants* as well as the structure of crystalline substances. Such transmissions or reflections at or near resonant frequencies are called *Bragg scattering*, after W. H. Bragg and his son, W. L. Bragg, who used x-rays to investigate the structure of crystals. Since the lattice constants are very small for crystals, resonant effects occur for very small wavelengths corresponding to frequencies in the x-ray spectrum of electromagnetic waves.

The study of waves in lattices is a branch of physics called "propagation in periodic structures," and the study of waves in a linear lattice of only a few cells is a part of network analysis called "filter theory." Here we shall be interested primarily in the long linear lattice as a model for transmission lines or media, and we shall use the lattice model to determine the effects on waveforms or "information" that is transmitted through transmission lines.

5–2 LUMPED-PARAMETER TRANSMISSION LINES

We shall use mesh analysis to find the cutoff frequency of a lumped-parameter transmission line. As we show in Fig. 5–1, i_n is the mesh current in the nth cell, i_{n-1} and i_{n+1} the currents in the preceding and following cells. By Kirchhoff's voltage law, the sum of the voltages in a closed circuit must be zero:

$$ji_n\omega L - \frac{ji_n}{\omega C} - \frac{ji_n}{\omega C} + \frac{ji_{n-1}}{\omega C} + \frac{ji_{n+1}}{\omega C} = 0 \qquad (5\text{–}2)$$

where the first three terms are the voltage drops across the inductor and capacitors in the nth cell, and the last two terms are the reverse drops owing to the currents in the neighboring cells.

If the wavelength is large compared to the lattice constant, $\lambda >> d$, the currents in the neighboring cells will be equal to that in the nth cell and in phase; thus Eq. (5–2) reduces to $i_n\omega L = 0$. The voltage drops across the inductors are zero, and there is no attenuation on the line. If, however, the frequency is increased until the currents in neighboring cells are equal to, but out of phase with, the current in the nth cell, Eq. (5–2) becomes $i_n\omega_c L = i_n 4/\omega_c C$, or

$$f_c = [\pi(LC)^{1/2}]^{-1} \qquad (5\text{–}3)$$

and we see that the voltage drop across the inductor, $i_n\omega_c L$, is a maximum. The signal is rapidly attenuated along the line, and for this reason we call f_c the cutoff frequency. We shall see later that the corresponding cutoff wavelength is

$$\lambda_c = 2d \qquad (5\text{–}4)$$

This result is known as the *Bragg condition* and has a simple interpretation. Since each reflected wave from a cell travels one wavelength farther than a wave reflected from a previous cell, all the reflected waves add in phase and produce a maximum total reflected amplitude. This is one of the effects used in crystallography to determine lattice constants.

EXAMPLE 5–1

A lumped-parameter transmission line is made of 5-mH inductors and 0.01-μF capacitors spaced 12.5 cm apart. Find the cutoff frequency and cutoff wavelength.

Solution From Eqs. (5–3) and (5–4) we have

$$f_c = \{\pi[(5 \times 10^{-3}) \times (0.01 \times 10^{-6})]^{1/2}\}^{-1}$$
$$= 45.0 \text{ kHz}$$
$$\lambda_c = 2 \times 12.5 = 25.0 \text{ cm}$$

EXAMPLE 5–2

Find the characteristic impedance and wave velocity on the transmission line of Example 5–1 when the operating frequency is much less than 45 kHz.

Solution Since $f << 45$ kHz or $\lambda >> d$, we can use Eqs. (4–1) and (5–1):

$$Z_c = \left(\frac{L}{C}\right)^{1/2} = \left(\frac{5 \times 10^{-3}}{0.01 \times 10^{-6}}\right)^{1/2}$$

$$= 707\ \Omega$$

$$v = \frac{d}{(LC)^{1/2}} = \frac{12.5}{[(5 \times 10^{-3}) \times (0.01 \times 10^{-6})]^{1/2}}$$

$$= 1.77 \times 10^6 \text{ cm/s}$$

In Example 5–2 we see that the wave speed is quite slow compared to the speed of light in free space; indeed, the refractive index of the line is $\epsilon_r^{1/2} = c/v = (3 \times 10^{10})/(1.77 \times 10^6) = 16{,}900$. Since the wave speed is so small, lumped-parameter lines can be used to obtain large phase delays and often are called *delay lines*.

EXAMPLE 5–3

If the delay line in Examples 5–1 and 5–2 is 1 m long and the operating frequency is 5 kHz, find the phase difference between a wave that has propagated on the line and one that has propagated the same distance in free space.

Solution The travel times for the two waves are

$$t_1 = \frac{s}{v} = \frac{100 \text{ cm}}{1.77 \times 10^6 \text{ cm/s}}$$

$$= 56.50\ \mu\text{s}$$

$$t_2 = \frac{s}{c} = \frac{100 \text{ cm}}{3 \times 10^{10} \text{ cm/s}}$$

$$= 3.33 \text{ ns}$$

The difference in times is $t_1 - t_2 \approx t_1 = 56.50\ \mu$s. The operating period is $\tau = 1/f = 1/5 \times 10^3 = 200\ \mu$s. Since each period corresponds to 360° of phase angle, the phase difference is $360°(t_1/\tau) = 360°(56.50/200) = 101.7°$.

We have considered the easier, extreme cases in which either $\lambda >> d$ or $\lambda = 2d$. To consider a more general case in which wavelength is unrestricted, we shall assume that the phase and amplitude of the mesh currents vary sinusoidally along the line. At the capacitors, where two mesh currents are added together, we shall assume the currents are 180° out of phase but have different magnitudes; thus the mesh equation becomes

$$i_0\left(\omega L \sin \phi_n - 2\frac{\sin \phi_n}{\omega C} + \frac{\sin \phi_{n-1}}{\omega C} + \frac{\sin \phi_{n+1}}{\omega C}\right) = 0$$

Let x be the location of the nth cell, so that the phase at the nth cell is $2\pi x/\lambda$; then the mesh equation becomes

$$\omega^2 LC \sin 2\pi x/\lambda = 2 \sin 2\pi x/\lambda - \sin 2\pi(x - d)/\lambda - \sin 2\pi(x + d)/\lambda$$

By expanding the last two terms we find

$$\begin{aligned}\omega^2 LC \sin 2\pi x/\lambda = 2 \sin 2\pi x/\lambda &- \sin 2\pi x/\lambda \cos 2\pi d/\lambda + \sin 2\pi d/\lambda \cos 2\pi x/\lambda \\ &- \sin 2\pi x/\lambda \cos 2\pi d/\lambda - \sin 2\pi d/\lambda \cos 2\pi x/\lambda\end{aligned}$$

or

$$\omega^2 LC = 2\left(1 - \cos\frac{2\pi d}{\lambda}\right)$$

Finally, by expanding the cosine term we have

$$\begin{aligned}\omega^2 LC &= 2\left(1 - \cos^2\frac{\pi d}{\lambda} + \sin^2\frac{\pi d}{\lambda}\right) \\ &= 4 \sin^2\frac{\pi d}{\lambda}\end{aligned}$$

or

$$f = [\pi(LC)^{1/2}]^{-1} \sin\frac{\pi d}{\lambda} \qquad (5\text{–}5)$$

With Eq. (5–1) and a rearrangement of terms in Eq. (5–5), we obtain the *Baden-Powell velocity equation*: $v = \lambda f = (\lambda/\pi d)[d/(LC)^{1/2}] \sin \pi d/\lambda$, or

$$v = v_0 \frac{\sin \pi d/\lambda}{\pi d/\lambda} \qquad (5\text{–}6)$$

where we have used v_0 to represent the extreme low-frequency velocity $d/(LC)^{1/2}$. This equation was obtained by Baden-Powell in the nineteenth century as a result of an analysis of mechanical wave propagation on a line of weights interconnected with springs.

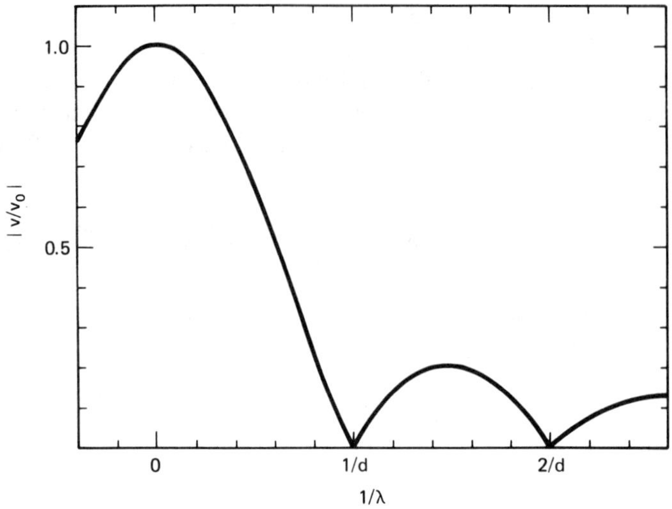

Figure 5–2 Baden-Powell relation between velocity and wavelength on a lumped-parameter transmission line.

EXAMPLE 5–4

Calculate the ratio v/v_0 and the velocity factor for the transmission line of Example 5–1 when the operating wavelength is $\lambda = 30$ cm.

Solution From Eq. (5–6) we have

$$\frac{v}{v_0} = \frac{\sin(\pi \times 12.5/30)}{\pi \times 12.5/30}$$

$$= 0.738$$

In Example 5–2 we found $v_0 = 1.77 \times 10^6$ cm/s; thus the velocity is $v = (1.77 \times 10^6) \times 0.738 = 1.306 \times 10^6$ cm/s, and the velocity factor is

$$\epsilon_r^{-1/2} = \frac{v}{c} = \frac{1.306 \times 10^6}{3 \times 10^{10}}$$

$$= 4.35 \times 10^{-5}$$

By repeating the calculations in Example 5–4 we can make a graph of the magnitude of the Baden-Powell function, as shown in Fig. 5–2. We shall see this function again, in Chapter 9, as the diffraction pattern of a slit or the radiation pattern of a slot antenna.

We can find the Bragg condition as a special case of the Baden-Powell equation. From Eqs. (5–3) and (5–6) we have

$$\lambda_c = \frac{v}{f_c}$$

$$= \frac{d}{(LC)^{1/2}} \frac{\sin(\pi d/\lambda_c)}{\pi d/\lambda_c} [\pi(LC)^{1/2}] = \lambda_c \sin\frac{\pi d}{\lambda_c}$$

or

$$\sin\frac{\pi d}{\lambda_c} = 1$$

This last result is satisfied when the argument of the sine is $\pi/2$: $\pi d/\lambda_c = \pi/2$, or $\lambda_c = 2d$, which is the Bragg condition, Eq. (5–4).

To find the characteristic impedance when the wavelength is comparable to the lattice constant, we recall that impedance is the reciprocal of the refractivity, $z_c = 1/\eta$, and from Eq. (5–6) we have $z_c = v/c = (v_0/c)(\sin \pi d/\lambda)/(\pi d/\lambda)$, or

$$Z_c = Z_0 \frac{\sin \pi d/\lambda}{\pi d/\lambda} \tag{5–7}$$

where $Z_0 = (L/C)^{1/2}$.

EXAMPLE 5–5

A delay line is made of 1000-pF capacitors and 200-μH inductors spaced at intervals of 8 cm. Find the characteristic impedance when $\lambda >> d$ and when the operating frequency is 0.5 MHz.

Solution At low frequencies we have

$$Z_0 = \left(\frac{L}{C}\right)^{1/2} = \left(\frac{200 \times 10^{-6}}{1000 \times 10^{-12}}\right)^{1/2}$$

$$= 447\ \Omega$$

We shall solve the second part of the problem by first finding an expression for the wavelength on the line, then we can use Eq. (5–7) to find the impedance. From Eq. (5–6) we have

$$v = \lambda f = v_0 \frac{\sin \pi d/\lambda}{\pi d/\lambda} \text{ from which we find}$$

$$\sin\frac{\pi d}{\lambda} = \frac{\pi d f}{v_0} = \pi f(LC)^{1/2} \qquad \frac{\pi d}{\lambda} = \sin^{-1}[\pi f(LC)^{1/2}]$$

or

$$\lambda = \frac{\pi d}{\sin^{-1}[\pi f(LC)^{1/2}]}$$

$$= \frac{\pi 8}{\sin^{-1}\{\pi \times (0.5 \times 10^6)[(200 \times 10^{-6}) \times (1000 \times 10^{-12})]^{1/2}\}} \quad (5\text{–}8)$$

$= 32.3$ cm

From Eq. (5–7) we now have

$$Z_c = \frac{447 \sin(\pi 8/32.3)}{\pi 8/32.3}$$

$$= 403.2\ \Omega$$

5–3 DELAY-LINE DISPERSION

A waveform will propagate down a transmission line without distortion, provided that the transmission-line characteristics do not change with frequency or wavelength. It is evident that the characteristic impedance, wave speed, and attenuation of a delay-line change markedly as the operating frequency approaches cutoff, and we can expect pulse distortion because of dispersion. The term "dispersion" is from optics and is exemplified by the fragmentation of white light into a rainbow of color by a glass prism. The colors are the various frequency components of the original white light, violet being the highest frequency visible to the eye and red the lowest. Since the refractive index of glass increases with frequency, the higher frequencies are refracted or "bent" more than the lower, and the white light is "dispersed" into its spectrum of frequency components.

If a pulse or waveform propagates on a delay line, its higher-frequency components travel more slowly than the lower, and the waveform is spread out or dispersed into its various frequency components. In addition, some of the higher-frequency components of the pulse may be beyond the cutoff frequency and be strongly attenuated; thus the received pulse will be missing some of its original components and be a distorted form of the original.

To study this further we shall have to understand how a waveform can be formed from sine waves of various frequencies, and this requires some familiarity with Fourier's theorem.

5–4 THE BEGINNINGS OF FOURIER ANALYSIS

The eighteenth-century mathematician James Bernoulli discovered that the curves traced by waveforms on a string could be represented by a series of sine waves. His mathematical proof lacked the rigor to be acceptable to many of the mathematicians of that time, and it was not until the following century that the mathematician Dirichlet established a proof and defined the class of curves that could be represented by a series of sinusoidal waves. A French engineer, Fourier, applied the theory to physical problems and established methods for finding the series that would represent

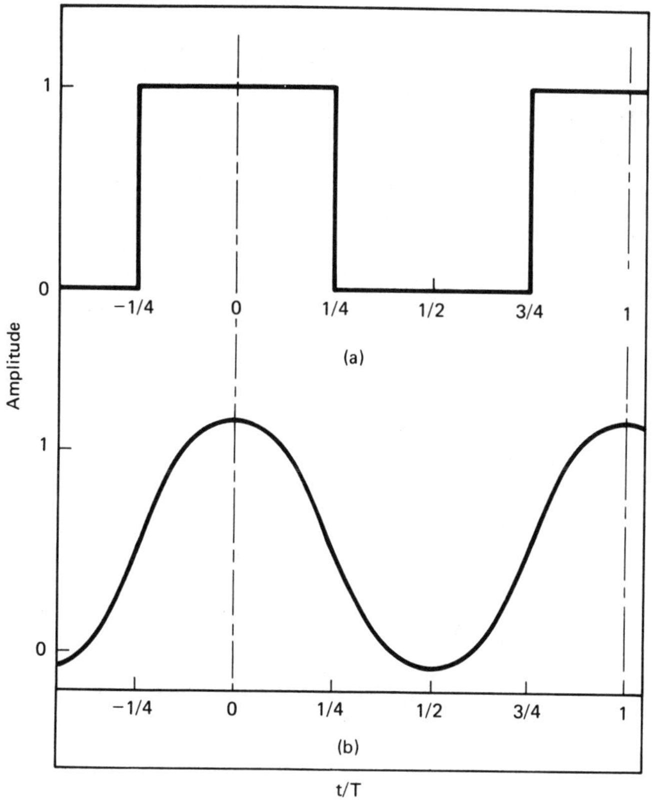

Figure 5–3 Rectangular wavetrain and its fundamental Fourier component.

a given curve or function, and today the theory is known by his name. As an example of the technique of "harmonic analysis" we shall use a Fourier series representation of a train of rectangular pulses to find the effect of transmission through a delay line.

Let the pulse-train of a signal source or function generator be a series of rectangular voltage pulses of unit amplitude and period T (Fig. 5–3), and let the symbol $f(t)$—read as a "function of the time, t"—represent the pulse-train. The Fourier series representation of this function is

$$f(t) = \frac{1}{2} + \frac{2}{\pi}\left(\cos \omega t - \frac{1}{3}\cos 3\,\omega t + \frac{1}{5}\cos 5\,\omega t - \frac{1}{7}\cos 7\,\omega t + \cdots\right) \qquad (5\text{–}9)$$

where $\omega = 2\pi/T$.

The accuracy of the representation increases with the number of terms included in the series. For example, if only the first harmonic or "fundamental" is included, we have

$$f(t) \approx \frac{1}{2} + \frac{2}{\pi} \cos \omega t$$

as shown in Fig. 5–3. If the third harmonic is included, we have

$$f(t) \approx \frac{1}{2} + \frac{2}{\pi} \left(\cos \omega t - \frac{1}{3} \cos 3\omega t \right)$$

and in Fig. 5–4 we show each term and the sum. If the fifth harmonic is included, we obtain a sum that begins to look somewhat like the rectangular pulse (Fig. 5–5). In Fig. 5–6 we show a tracing of an oscilloscope photograph of a pulse comprising all harmonics up to the thirteenth. We notice that the overshoot on the leading

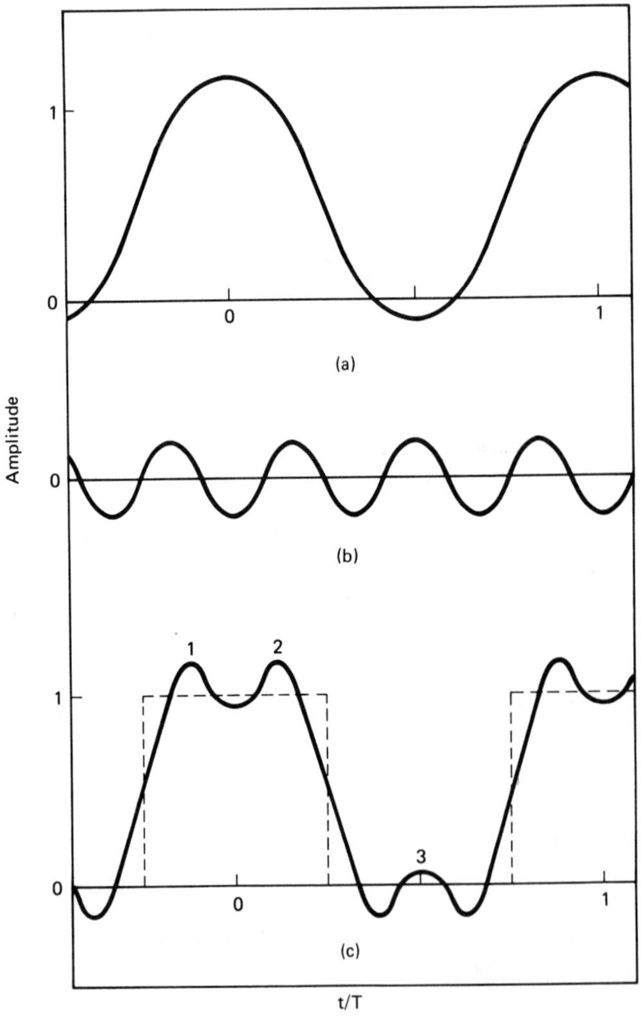

Figure 5–4 Fourier components of a rectangular wavetrain: (a) fundamental; (b) third harmonic; (c) sum of the first and third harmonics.

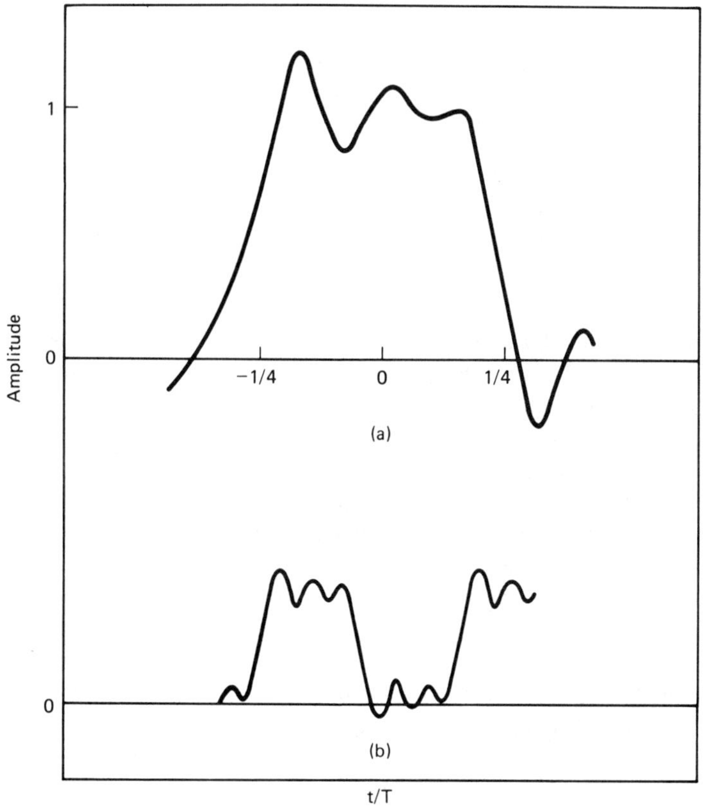

Figure 5–5 (a) Waveform obtained by graphically adding the first, third, and fifth Fourier components of a rectangular wavetrain; (b) tracing of an oscilloscope photograph of the sum of the first, third, and fifth harmonics of a rectangular wavetrain at the output of a lumped-parameter transmission line.

edge remains about 9% of the pulse height, no matter how many harmonics are included; however, the width of the overshoot becomes narrower as the number of harmonics increases. The overshoot is called *Gibbs phenomenon* and ordinarily appears on both edges of the pulse. When there is velocity dispersion the overshoot on the trailing edge may be suppressed.

We have explored the result of building up a pulse with an increasing number of harmonics, but we have not considered dispersion. The oscilloscope photographs in Figs. 5–5 and 5–6 show rectangular voltage pulses that were detected at the end of a lumped-parameter transmission line. Since the operating frequencies were just below the cutoff frequency, there was considerable dispersion. Referring to Eq. (5–6), or Fig. 5–2, we see that the wave velocity decreases as the wavelength decreases; thus the higher-frequency components are delayed more because they travel more slowly.

To illustrate the dispersive effect we delayed the fifth harmonic $\frac{1}{20}$ of a cycle

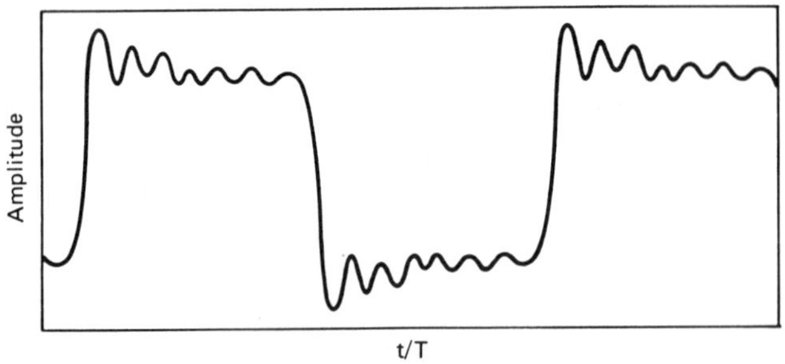

Figure 5–6 Tracing of an oscilloscope photograph of a rectangular pulse formed from the sum of all odd harmonics from 1 to 13.

before graphically adding it to the fundamental and third harmonics; the result (Fig. 5–5a) resembles the observed pulse. The resemblance would be better if we also had delayed the third harmonic slightly. The technique of making the observations electronically will be discussed in an application at the end of this chapter.

We can conclude that when a rectangular pulse, or any waveform comprised of many harmonics, is transmitted through a delay line, all those harmonics above the cutoff frequency will be removed and the received pulse will be distorted. In addition, the slower propagation speed of the higher harmonics will give additional distortion, adding asymmetry to the pulse and, ultimately, spread it out or disperse it.

To sum up, Fourier analysis can be used to determine the effect of transmission through a lossy, dispersive medium. We first resolve the initial waveform into its Fourier components, we next compute the attenuation and time delay of each component as it propagates through the medium, and finally, we add all the altered components together to see what the received pulse will look like.

5–5 TRANSMISSION-LINE ATTENUATION

We have assumed a transmission-line model (Fig. 5–1) that is lossless and infinitely long; accordingly, the attenuation beyond cutoff is absolute, all the energy incident on the line is reflected. In practice, a line is finite, perhaps short, and we need a way to estimate the attenuation in a short line.

We recall that the voltage drop across a series inductor increases with frequency, since its *ac* resistance increases with frequency, $X_L = 2\pi fL$; also, the reactance of a shunt capacitor decreases with frequency, $X_C = 1/2\pi fC$. If the operating frequency is doubled, the series reactance doubles, the shunt reactance is halved, and the signal amplitude is reduced fourfold. This means that we can estimate the reduction in voltage by reducing it by one-half for each inductor and capacitor in the line. For example, if a delay line is made of five inductors and six capacitors, the voltage attenuation is $1/2^{11} = -33.1$ dB for each octave change in frequency beyond cutoff. Since power is proportional to the square of the voltage, the power attenuation is -66.2 dB.

EXAMPLE 5–6

A delay line is made of four capacitors and four inductors. Estimate the power attenuation at a frequency four times the cutoff frequency.

Solution The attenuation per octave is 2^{-8} or -24.1 dB in voltage. An operating frequency that is 4 times cutoff is 2 octaves beyond cutoff; therefore, the power attenuation is $2 \times 2 \times (-24.1) = -96.4$ dB.

On a scientific calculator, enter 2 and find LOG 2 = 0.30103. Multiply by -8 to obtain the voltage attenuation in bels and by 10 to obtain -24.1 dB. Double this once to obtain the power attenuation per octave and double it again to find the power attenuation for two octaves, -96.4 dB.

EXAMPLE 5–7

Design a 600-Ω low-pass filter with a cutoff frequency of 250 MHz and a power attenuation of at least 20 dB at 400 MHz.

Solution We find the required inductor and capacitor sizes by first solving Eq. (4–1) for $(L)^{1/2}$,

$$(L)^{1/2} = Z_c(C)^{1/2}$$

and we then eliminate $(L)^{1/2}$ from Eq. (5–3), $f_c = [\pi(LC)^{1/2}]^{-1} = 1/\pi CZ_c$, or

$$\begin{aligned} C &= \frac{1}{f_c \pi Z_c} \\ &= \frac{1}{(250 \times 10^6) \times \pi \times 600} \qquad (5\text{–}10) \\ &= 2.12 \text{ pF} \end{aligned}$$

With this value of the capacitance we return to Eq. (4–1) to obtain the inductance,

$$\begin{aligned} L &= Z_c^2 C = 600^2 \times 2.12 \times 10^{-12} \\ &= 0.764\ \mu\text{H} \end{aligned}$$

Check:

$$\begin{aligned} Z_c &= \left(\frac{L}{C}\right)^{1/2} = \left(\frac{0.764 \times 10^{-6}}{2.12 \times 10^{-12}}\right)^{1/2} \\ &= 600\ \Omega \end{aligned}$$

To find the required number of cells we plot unit voltage amplitude on semilog graph paper (point A in Fig. 5–7). A power attenuation of 20 dB at 400 MHz is a voltage attenuation of 10 dB or 1/10 (point B in the figure). We draw a straight line through points A and B and extend it to point C, where it intersects the 500-MHz ordinate at 1/45. We see that a doubling of the frequency, from 250 to 500 MHz, implies a corresponding voltage attenuation of 1/45. Since each element of the filter halves the voltage we require n elements, where

$$\frac{1}{2^n} = \frac{1}{45}$$

$$-n \log 2 = -1.653$$

$$n = 5.49$$

We need six elements, or three cells.

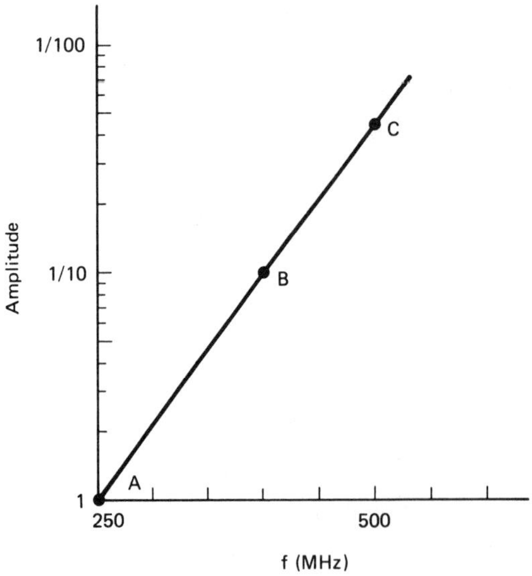

Figure 5–7 Graphical method of finding attenuation, Example 5–7.

We show the measured attenuation of an 11-element delay line in Fig. 5–8. This five-cell line was constructed with five 10-mH inductors, four 0.022-μF capacitors, and an 0.01-μF capacitor at each end. From Eq. (5–3) we expect the cutoff frequency to be 21.7 kHz; the observed cutoff, to the −1.5-dB level (0.35 V) is 22.5 kHz. The amplitude beyond cutoff initially drops more abruptly than expected; however, it ultimately drops about 33.1 dB/octave frequency change, as indicated by the dashed tangent line in the figure. Since there are 11 elements in the line we should expect an attenuation of $10 \log 2^{11} = 33.1$ dB/octave.

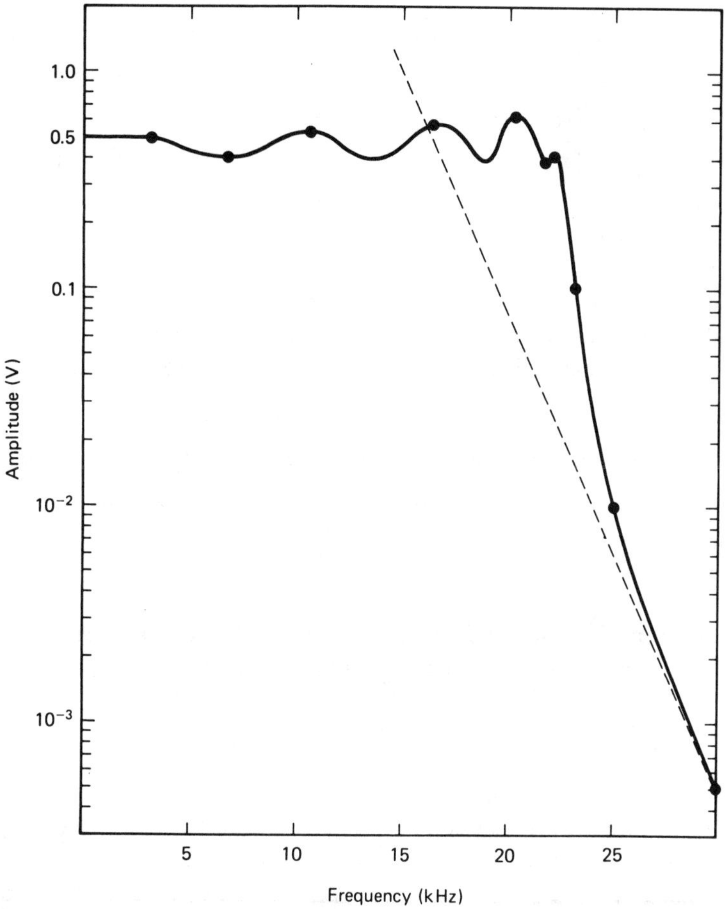

Figure 5–8 Wave propagation through a lumped-parameter transmission line, amplitude versus frequency.

5–6 TRANSMISSION-LINE LOSS

We have been discussing attenuation on a lossless line. Since our model comprises only reactances, the attenuation must result from reflections. We can include dissipative losses in our transmission-line model by adding series and shunt resistances (Fig. 5–9). The series resistance represents ohmic losses in the conductors, per unit length. The shunt conductance primarily is relaxation loss in the dielectric insulation. "Relaxation time" is a phrase used to describe the length of time required for a substance to return to its original state after being stressed. The greater the internal friction, the longer it takes a substance to return to its original form. The retardation due to internal friction is called *hysteresis*. In *ac* circuits, especially at

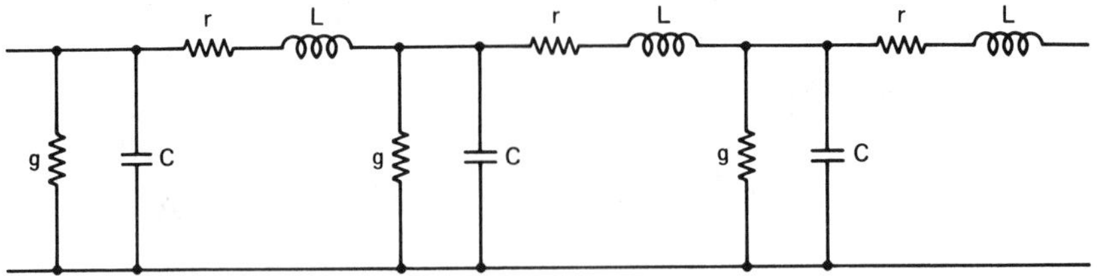

Figure 5–9 Lumped-parameter model of a lossy transmission line.

very high frequencies, the dielectric is alternately stressed in every cycle, and appreciable energy loss can occur because of internal friction.

Referring to Fig. 5–9, we see that the voltage drop across a cell is

$$\Delta v = \Delta i(r + jX_L) \tag{5–11}$$

and the change in shunt current from one side of a cell to the next is

$$\Delta i = v_n(g + jb_C) - v_{n+1}(g + jb_C)$$

or

$$\Delta i = \Delta v(g + jb_C) \tag{5–12}$$

and from the ratio of Eqs. (5–11) and (5–12) we find $(\Delta v/\Delta i)^2 = (r + jX_L)/(g + jb_C)$, which gives a characteristic impedance of

$$Z_c = \left(\frac{r + jX_L}{g + jb_C}\right)^{1/2} \tag{5–13}$$

At UHF and microwave frequencies the change in characteristic impedance due to conductor losses is negligible, even in special high-attenuation lines; however, at low frequencies the change can be appreciable.

EXAMPLE 5–8

A special coaxial cable (RG-21/U) is made with an inner conductor of resistance wire, $r = 0.76\ \Omega$/m, has an inductance of 0.267 μH/m and a nominal characteristic impedance of 53 Ω. Find the characteristic impedance at 200 kHz. Neglect dielectric loss.

Solution With $g = 0$, we can reduce Eq. (5–13) to

$$Z_c = \left(\frac{r + j\omega L}{j\omega C}\right)^{1/2} = \left(\frac{L}{C}\right)^{1/2}\left(1 - \frac{jr}{\omega L}\right)^{1/2}$$

$$= 53\left\{1 - j\frac{0.76}{2 \times \pi \times (2 \times 10^5) \times (0.267 \times 10^{-6})}\right\}^{1/2}$$

$$= 53(1 - j2.26)^{1/2} = 53(2.47\angle{-66.1}^\circ)^{1/2}$$
$$= 83.3\angle{-33.1}^\circ\ \Omega$$

Although the change in characteristic impedance due to lossiness is negligible at higher frequencies, the change in attenuation is not; it increases exponentially with frequency and line length. Just as the energy stored in a capacitor or inductor decreases exponentially with time if the element is connected to a resistor, so does the energy content of a wave decrease exponentially in the time required for it to propagate a distance x along a line or through a medium. At optical frequencies the relation between power and distance is called *Lambert's law*,

$$\frac{I}{I_0} = \frac{P}{P_0} = \exp(-2\alpha x) \tag{5–14}$$

where the left side is the ratio of intensity or power to initial intensity or power, and 2α is the *absorption coefficient*. If the law is expressed in amplitudes, we have

$$\frac{a}{a_0} = \exp(-\alpha x) \tag{5–15}$$

and α is called the *attenuation constant* or *attenuation factor*, measured in nepers/meter.

EXAMPLE 5–9

A coaxial line has an attenuation constant of 0.02 Np/m. Find the power loss in decibels for 10 m of line.

Solution From Eq. (5–15) the voltage attenuation is

$$\frac{v}{v_0} = \exp(-0.02 \times 10) = \exp(-0.2\ \text{Np})$$

$$= 0.8187$$

and the power loss is

$$\frac{P}{P_0} = \left(\frac{v}{v_0}\right)^2 = 0.8187^2 = 0.6703$$

or

$$10 \log \frac{P}{P_0} = 10 \log (0.6703)$$

$$= 1.7372\ \text{dB}$$

On examining Example 5–9 we see, since $\ln e^{-0.2}\text{Np} = -0.2$ Np, that 0.2 Np corresponds to a power loss of 1.7372 dB; thus we have 1.7372/0.2 Np = 8.686 dB/Np. To convert nepers to decibels all we have to do is multiply by 8.686.

EXAMPLE 5–10

Find the power loss in decibels for a 20-m line if $\alpha = 0.015$ Np/m.

Solution The loss is

$$\begin{aligned} -\alpha x &= 0.015 \times 20 = 0.3 \text{ Np} \\ &= 0.3 \times 8.686 \\ &= 2.61 \text{ dB} \end{aligned}$$

We have distinguished between 2α and α by calling them the absorption coefficient and the attenuation factor. In practice, no such distinction is made; this is a consequence of overlapping fields of study. In optics, Lambert's law is written with α in the exponential, not 2α as we wrote it. The difference between units in optics and electronics is hidden in the use of 8.686 to convert nepers to decibels. The conversion should be 4.343. The extra factor of 2 corresponds to the difference between α and 2α. In recent years some writers have resolved the difference by distinguishing between *optical power* and *electrical power*, and the relation between the two is

$$\text{dB (electrical)} = 2 \text{ dB (optical)}$$

We have a model for a transmission line, as shown in Fig. 5–9, and another model for transmission loss, as expressed by Eq. (5–14) or (5–15). Our task now is to put the two together so that we can express loss directly in terms of the lumped-parameter equivalents of the transmission line.

We start by introducing the *wave number*, $n = 1/\lambda$, and the *angular wave number*, $k = 1/\lambda\!\!\!{}^-$. In optical spectroscopy the wavelengths are so small that it is easier to determine wavelength by counting the number of standing waves in a unit length rather than trying to measure a single wavelength directly. Thus the term "wave number" is used as much as wavelength. The term has been adopted in electronics but usually is called the *phase constant*, $\beta = 1/\lambda\!\!\!{}^-$. The reason it is called the phase constant comes from the convenience of multiplying the phase constant by a distance to find the phase change in a wave that travels that distance:

$$\phi = \beta x = \frac{2\pi x}{\lambda} \qquad \text{rad}$$

We can relate the phase constant to our model of a lossless line through the velocity equations $v = \lambda\!\!\!{}^-\omega$ and $v = (LC)^{-1/2}$, and we have $\beta = 1/\lambda\!\!\!{}^- = \omega/v$, or

$$\beta = \omega(LC)^{1/2} \tag{5–16}$$

where L and C are inductance and capacitance per unit length. The right-hand side of this equation can be written as $(\omega L \omega C)^{1/2} = (x_L b_C)^{1/2}$, that is, as the square root of the product of the series reactance and shunt susceptance. For a lossy line we replace x_L and b_C by the series impedance and shunt admittance, z_s and y_p, and we obtain

$$\beta = [(r + j\omega L)(g + j\omega C)]^{1/2}$$

Evidently, the phase constant now is a complex number, so we give it a new name, the *propagation constant*, and a new symbol, γ, and Eq. (5–16) becomes

$$\begin{aligned}\gamma &= [(r + j\omega L)(g + j\omega C)]^{1/2} \\ &= [rg - \omega^2 LC + j(g\omega L + r\omega C)]^{1/2} \\ &= (u + jv)^{1/2}\end{aligned} \tag{5–17}$$

Now that we have defined the propagation constant we can state our objective more plainly. We shall separate the propagation constant into its real and imaginary parts and identify the real part as the absorption coefficient in Lambert's law. Thus loss is expressed in terms of the physical parameters of the transmission line or medium, and we shall find the results have wide application, from low frequencies to microwave and optical.

Separation of the propagation constant into real and imaginary parts requires the extraction of roots of a complex function. This requires some mathematical manipulation, and the reader already familiar with the algebra of complex variables may wish to go directly to Eq. (5–21). To save writing we have abbreviated the real and imaginary terms under the radical with the symbols u and v.

The root of a complex number is found by transforming to polar form, taking the square root of the magnitude, and dividing the polar angle by 2. We justify the last step by recalling that in multiplying complex numbers the angles are added, as though they were exponentials, $\exp(\theta/2) \times \exp(\theta/2) = \exp(\theta)$.

EXAMPLE 5–11

Find the square root of $-5 + j12$.

Solution We transform to polar form and obtain $13\angle 112.6^\circ$. The square root is

$$\begin{aligned}(13)^{1/2}\angle 112.6/2 &= \pm 3.606\angle 56.3^\circ \\ &= \pm(2 + j3)\end{aligned}$$

Check:

$$\begin{aligned}(2 + j3)(2 + j3) &= 4 - 9 + j(6 + 6) \\ &= -5 + j12\end{aligned}$$

To carry out the procedure above algebraically, we square Eq. (5–17), $\gamma^2 = u + jv$, and write the result in polar form,

$$\begin{aligned}\gamma^2 &= (u^2 + v^2)^{1/2}\angle \tan^{-1}\frac{v}{u} \\ &= (u^2 + v^2)^{1/2}\angle\theta\end{aligned} \tag{5–18}$$

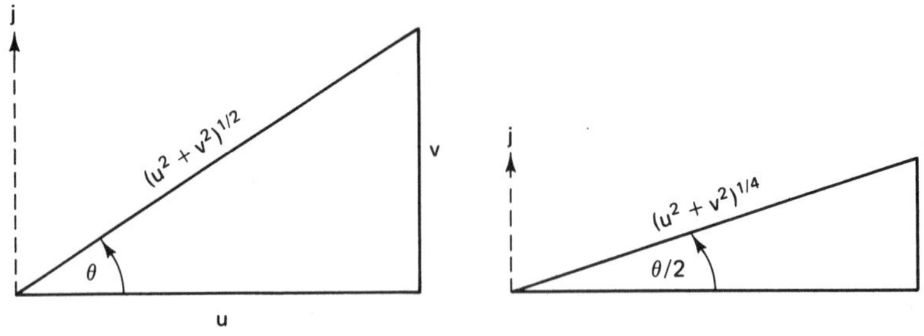

Figure 5–10 Geometrical representation of the extraction of the square root of a complex function.

where, for brevity, $\theta = \tan^{-1} v/u$. We now find the root of the equation in the usual manner,

$$\gamma = (u^2 + v^2)^{1/4} \angle \theta/2$$

and by referring to Fig. 5–10, we see that γ can be separated into its real and imaginary parts as

$$\gamma = (u^2 + v^2)^{1/4}\left(\cos\frac{\theta}{2} + j\sin\frac{\theta}{2}\right) \tag{5–19}$$

EXAMPLE 5–12

Use Eq. (5–19) to find the square root of $-5 + j12$.

Solution We have $u = -5$, $v = 12$, and from Fig. 5–10,

$$\cos\theta = \frac{u}{(u^2 + v^2)^{1/2}} = \frac{-5}{(5^2 + 12^2)^{1/2}}$$
$$= -0.385$$

$$\theta = \cos^{-1}(-0.385) = 1.966 \text{ rad} \qquad \text{and} \qquad \frac{\theta}{2} = 0.983 \text{ rad}$$

From Eq. (5–19) we now have

$$\begin{aligned}(-5 + j12)^{1/2} &= (5^2 + 12^2)^{1/4}(\cos 0.983 + j\sin 0.983)\\ &= (169)^{1/4}(0.555 + j0.832)\\ &= \pm(2 + j3)\end{aligned}$$

in agreement with Example 5–11.

EXAMPLE 5–13

Use Eq. (5–19) to evaluate $1^{1/4}$.

Solution We have $1^{1/4} = (1^{1/2})^{1/2} = (+1)^{1/2}$, $(-1)^{1/2} = +1, -1$, and $(-1)^{1/2}$. We shall use Eq. (5–19) to find $(-1)^{1/2}$:

$$\begin{aligned}(-1)^{1/2} &= (1\angle 180^\circ)^{1/2}\\ &= \pm 1(\cos 90^\circ + j\sin 90^\circ)\\ &= \pm j\end{aligned}$$

and the four answers are ± 1 and $\pm j$.

We now need only to substitute into Eq. (5–19) for the terms we abbreviated as u, v, and θ; we have

$$\begin{aligned}\gamma &= [(rg - \omega^2 LC)^2 + (g\omega L + r\omega C)^2]^{1/4}\left(\cos\frac{\theta}{2} + j\sin\frac{\theta}{2}\right)\\ &= (r^2g^2 + \omega^4L^2C^2 + g^2\omega^2L^2 + r^2\omega^2C^2)^{1/4}\left(\cos\frac{\theta}{2} + j\sin\frac{\theta}{2}\right)\\ &= \omega(LC)^{1/2}\left[1 + \frac{g^2}{\omega^2C^2} + \frac{r^2}{\omega^2L^2}\left(1 + \frac{g^2}{\omega^2C^2}\right)\right]^{1/4}\left(\cos\frac{\theta}{2} + j\sin\frac{\theta}{2}\right)\\ &= j\beta_0\left[\left(1 + \frac{g^2}{\omega^2C^2}\right)\left(1 + \frac{r^2}{\omega^2L^2}\right)\right]^{1/4}\left(\cos\frac{\theta}{2} + j\sin\frac{\theta}{2}\right)\end{aligned}$$

or

$$\begin{aligned}\gamma &= \beta_0\left[\left(1 + \frac{g^2}{\omega^2C^2}\right)\left(1 + \frac{r^2}{\omega^2L^2}\right)\right]^{1/4}\left(-\sin\frac{\theta}{2} + j\cos\frac{\theta}{2}\right) \qquad (5\text{–}20)\\ &= -\alpha + j\beta\end{aligned}$$

where

$$\alpha = \beta_0\left[\left(1 + \frac{g^2}{\omega^2C^2}\right)\left(1 + \frac{r^2}{\omega^2L^2}\right)\right]^{1/4}\sin\frac{\theta}{2} \qquad (5\text{–}21)$$

$$\beta = \beta_0\left[\left(1 + \frac{g^2}{\omega^2C^2}\right)\left(1 + \frac{r^2}{\omega^2L^2}\right)\right]^{1/4}\cos\frac{\theta}{2} \qquad (5\text{–}22)$$

$$\beta_0 = \omega(LC)^{1/2} \quad \text{or} \quad \omega(\mu\epsilon)^{1/2} \qquad (5\text{–}23)$$

and

$$\theta = \tan^{-1}\frac{v}{u} = \tan^{-1}\frac{g\omega L + r\omega C}{rg - \omega^2 LC} \qquad (5\text{–}24)$$

as shown in Fig. 5–10.

In extracting the fourth root of unity we had four choices, ± 1 and $\pm j$, and we have chosen $\pm j$ to conform to the convention of expressing the attenuation factor α as real and the phase constant β as imaginary.

Although we are primarily interested in microwave and infrared frequencies, we shall first consider some simpler examples of loss at low frequencies.

5–7 LOW-FREQUENCY LOSS

At low frequencies the absorption in transmission lines is mainly ohmic loss in the conductors. We can ignore the dielectric loss, $g = 0$, and simplify Eq. (5–21) to

$$\alpha = \beta_0\left(1 + \frac{r^2}{\omega^2 L^2}\right)^{1/4} \sin\left(\frac{1}{2}\tan^{-1}\frac{r}{\omega L}\right) \tag{5–25}$$

EXAMPLE 5–14

Calculate the attenuation constant at 200 kHz for the special resistive coaxial cable described in Example 5–8, RG-21/U.

Solution Our data are

$$r = 0.76\ \Omega/\text{m}$$
$$L = 0.267\ \mu\text{H/m}$$
$$C = 95.1\ \text{pF/m}$$

The argument of the tangent function is

$$\frac{r}{\omega L} = \frac{0.76}{2 \times \pi \times (2 \times 10^5) \times (0.267 \times 10^{-6})}$$
$$= 2.265$$

and we have

$$\tfrac{1}{2}\tan^{-1}\frac{r}{\omega L} = \tfrac{1}{2}\tan^{-1} 2.265$$
$$= 0.578 \text{ rad}$$

From Eq. (5–25) we have

$$\alpha = \omega(LC)^{1/2}\left(1 + \frac{r^2}{\omega^2 L^2}\right)^{1/4} \sin 0.578$$
$$= 2 \times \pi \times (2 \times 10^5)[(0.267 \times 10^{-6}) \times (95.1 \times 10^{-12})]^{1/2} \times [1 + (2.265)^2]^{1/4} \times 0.546$$
$$= 5.44 \times 10^{-3}\ \text{Np/m}$$

$$= (5.44 \times 10^{-3}) \times 8.686 \text{ dB/m}$$
$$= 0.0472 \text{ dB/m}$$
$$= 1.44 \text{ dB/100 ft}$$

5–8 SKIN EFFECT AND LOSS IN A CONDUCTING MEDIUM

The conductivities of a few common metals are shown in Table 5–1. We see that at high frequencies we have $g/\omega C >> 1$ and $r/\omega L << 1$, and we can reduce Eq. (5–21) to

$$\alpha = \omega(\mu\epsilon)^{1/2}\left(\frac{g}{\omega\epsilon}\right)^{1/2} \sin\frac{\theta}{2}$$

where

$$\theta = \tan^{-1}\frac{g}{\omega C}$$

TABLE 5–1 CONDUCTIVITY OF VARIOUS COMMON METALS

Metal	DC Conductivity (S/m)
Al	3.54×10^7
Cu	5.65×10^7
Lead	4.6×10^6
Nichrome	1.0×10^6
Ag	6.13×10^7
Tin	8.7×10^6

The argument of the sine function is so large that the only sensible thing is to replace $\sin \theta/2$ by its root-mean-square average value, $(2)^{-1/2}$, and the loss equation becomes

$$\alpha = \left(\frac{\omega\mu g}{2}\right)^{1/2} \text{ Np/m} \tag{5–26}$$

The distance a wave penetrates into a conducting medium before decreasing to $1/e = 0.368$ or 36.8% of its initial amplitude is called the *skin depth*, δ_s. From Lambert's law we have $a/a_0 = \exp(-\alpha\delta_s) = e^{-1}$, or $\delta_s = 1/\alpha$, and from Eq. (5–26) we have

$$\delta_s = \left(\frac{2}{\omega\mu g}\right)^{1/2} \tag{5–27}$$

EXAMPLE 5–15

Calculate the skin depth in aluminum when the operating frequency is 100 MHz.

Solution From Table 5–1 we have $g = 3.54 \times 10^7$ S/m. The permeability of nonmagnetic substances is approximately $4\ \pi \times 10^{-7}$ H/m, and from Eq. (5–27) we have

$$\delta_s = \left[\frac{2}{(2 \times \pi \times 10^8) \times (4 \times \pi \times 10^{-7}) \times (3.54 \times 10^7)}\right]^{1/2}$$

$$= 8.46 \times 10^{-6}\ \text{m}$$

EXAMPLE 5–16

Repeat Example 5–15 for copper.

Solution We have

$$\delta_s = \left[\frac{2}{(2 \times \pi \times 10^8) \times (4 \times \pi \times 10^{-7}) \times (5.65 \times 10^7)}\right]^{1/2}$$

$$= 6.70 \times 10^{-6}\ \text{m}$$

When working at optical frequencies we find it convenient to measure the skin effect by the *extinction coefficient*, $K_0 = \lambda_0\alpha/2$, or

$$K_0 = \frac{\lambda_0}{2}\left(\frac{\omega\mu g}{2}\right)^{1/2} \qquad (5\text{–}28)$$

We see that the intensity or power carried by a wave falls to exp $(-2K_0)$ times its initial value when the wave has penetrated a distance of a free-space radian wavelength, $X = \lambda_0$: exp $(-\alpha x)$ = exp $(-2K_0)$.

EXAMPLE 5–17

Calculate the extinction coefficient for radiation at normal incidence on a front-silvered mirror. The frequency is that of the yellow line of the sodium spectrum. At this frequency the conductivity of silver is 3×10^6 S/m and the refractive index is 0.177.

Solution The yellow line in the spectrum of the element sodium has a wavelength of 5893 A; thus the operating frequency is

$$f = \frac{c}{\lambda} = \frac{3 \times 10^8}{5893 \times 10^{-10}}$$

$$= 5.09 \times 10^5\ \text{GHz}$$

From Eq. (5–28) we have

$$K_0 = \frac{\lambda_0}{2}\left(\frac{\omega\mu g}{2}\right)^{1/2}$$

$$= \frac{5893 \times 10^{-10}}{4\pi}(½ \times 2 \times \pi \times 5.09 \times 10^{14} \times 4 \times \pi \times 10^{-7} \times 3 \times 10^{6})^{1/2}$$

$$= 3.64$$

In one free-space radian wavelength [$\lambda_0 = (5893 \times 10^{-10})/2\pi = 9.38 \times 10^{-8}$ m], the amplitude falls to $e^{-3.64} = 0.0263$ times its initial value.

The *absorption index* is measured by the wavelength in the medium, rather than a free-space wavelength,

$$K = K_0\epsilon_r^{-1/2} \tag{5–29}$$

EXAMPLE 5–18

Calculate the absorption index for the problem given in Example 5–17.

Solution From the answer to Example 5–17 and Eq. (5–29) we have

$$K = \frac{3.64}{0.177}$$

$$= 20.6$$

In metallic reflection the amplitude reflection coefficient is

$$r = \frac{\eta(1 - jK) - 1}{\eta(1 - jK) + 1} \tag{5–30}$$

where $\eta = \epsilon_r^{1/2}$ is the refractive index and K the absorption index. For $K = 0$ the expression reduces to Eq. (2–8). The power reflection coefficient is $\Gamma = rr^*$ and is called the *reflecting power*,

$$\Gamma = \frac{(\eta - 1)^2 + K_0^2}{(\eta + 1)^2 + K_0^2} \tag{5–31}$$

EXAMPLE 5–19

Calculate the reflecting power of silver under the conditions described in Examples 5–17 and 5–18.

Solution From Eq. (5–31) we have

$$\Gamma = \frac{(0.177 - 1)^2 + 3.64^2}{(0.177 + 1)^2 + 3.64^2}$$

$$= 0.952 \text{ or } 95.2\%$$

5–9 HIGH-FREQUENCY LOSS IN DIELECTRICS

For dielectric media at almost all operating frequencies the shunt conductivity is small, $g^2/\omega^2C^2 << 1$, there are no series conductors, $r = 0$, and our model simplifies to that shown in Fig. 5–11. There is no series conduction current, but there is a shunt conduction current and series and shunt displacement currents. Equation (5–21) is much simplified:

$$\alpha = \beta_0\left(1 + \frac{g^2}{\omega^2C^2}\right)^{1/4} \sin\frac{\theta}{2}$$

or

$$\alpha \approx \beta_0 \sin\frac{\theta}{2} \tag{5–32}$$

where

$$\theta = \tan^{-1}\frac{g}{\omega C} \tag{5–33}$$

The angle θ is called the *loss angle* and usually is represented by the symbol δ,

$$\delta = \tan^{-1}\frac{g}{\omega C} \tag{5–34}$$

The *loss tangent* is

$$\tan\delta = \frac{g}{\omega C} \tag{5–35}$$

Dielectric loss sometimes is included in the definition of permittivity. The permittivity is assumed to be complex,

$$\epsilon_r = \epsilon_r' + j\epsilon_r'' \tag{5–36}$$

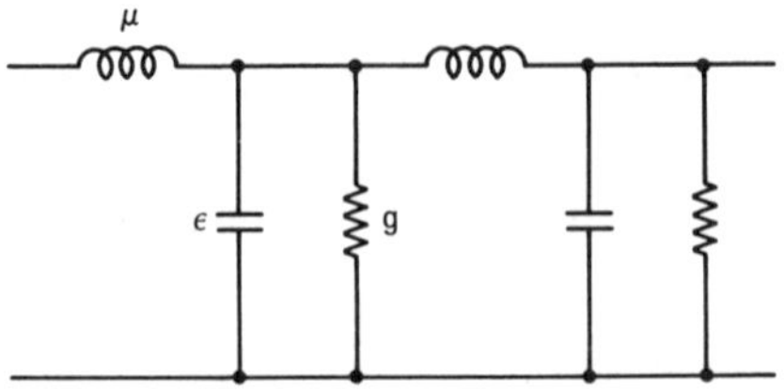

Figure 5–11 Simplified model of dielectric transmission lines or media.

where ϵ_r' is the relative permittivity, which we have been writing as ϵ_r, and ϵ_r'' is called the *loss factor*. With this convention the loss tangent is

$$\tan\delta = \frac{\epsilon_r''}{\epsilon_r'} \tag{5–37}$$

as shown in Fig. 5–12. For whatever convention is used, the loss tangent always is the ratio of loss current to displacement current. In circuit theory the *power factor* is $\cos\phi$, where ϕ is the angle between the loss current and the total current (Fig. 5–12). Since δ is the complement of ϕ, the *dielectric power factor* is $\sin\delta$. As in circuit theory, the quality is defined as the ratio of stored to dissipated power per radian cycle:

$$Q_d = \frac{vi\,(\text{displacement})}{vi\,(\text{loss})}$$

or

$$Q_d = \frac{1}{\tan\delta} = \cot\delta \tag{5–38}$$

where Q_d is the *dielectric quality*. There are several other loss definitions in this subject, where engineering and optics overlap, and before using data from a handbook it is advisable to check the author's terminology.

We continue with Eq. (5–32). For small loss angles we can approximate the sine or tangent function by the angle only, $\sin\theta/2 = \sin(\frac{1}{2}\tan^{-1} g/\omega C) \approx g/$

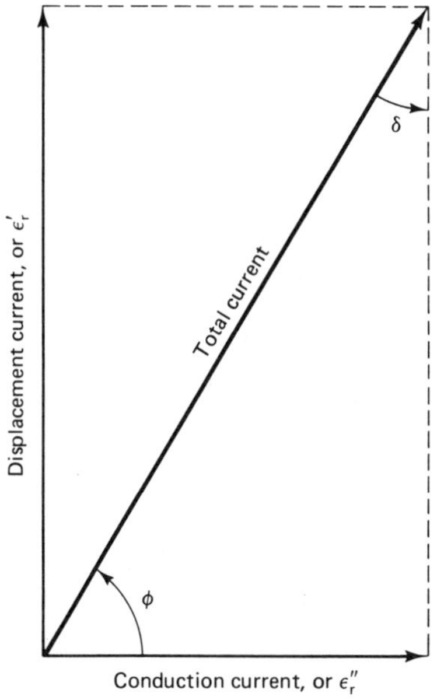

Figure 5–12 Geometrical representation of the relationship between power factor, $\cos\phi$, loss angle, δ, and the conductive and reactive currents.

$2\omega C$, and the loss coefficient equation simplifies to $\alpha \approx \beta_0 g/2\omega C = \omega(LC)^{1/2}g/2\omega C = \frac{1}{2}(L/C)^{1/2}g$, or

$$\alpha = \tfrac{1}{2}Z_c g \tag{5–39}$$

EXAMPLE 5–20

At an operating frequency in the infrared at $\lambda = 1.6\mu\text{m}$, the glass in a fiber optic cable has a resistivity of 110 MΩ-cm and a refractivity of 1.5. Find the attenuation coefficient in dB/km.

Solution We calculate Z_c by recalling that $Z_c = Z_0/\epsilon_r^{1/2} = 377/1.5 = 251$ Ω. Since $g = 1/\rho$, we have, from Eq. (5–39),

$$\alpha = \tfrac{1}{2} \times \frac{251}{110 \times 10^6}$$

$$= 1.141 \times 10^{-6} \text{ Np/cm}$$
$$= 1.141 \times 10^{-4} \text{ Np/m}$$
$$= 1.141 \times 10^{-4} \times 8.686 \text{ dB/m}$$
$$= 0.991 \text{ dB/km}$$

EXAMPLE 5–21

A fiber optic cable is made of glass that has a relative permittivity of 4.28 and a resistivity of 50 MΩ-cm. What is the absorptive loss in a 1500-m length of line?

Solution As in Example 5–20, we first find the characteristic impedance, $Z_c = 377/(4.28)^{1/2} = 182$ Ω, and again from Eq. (5–39) we have

$$\alpha = \tfrac{1}{2} \times \frac{182}{50 \times 10^6}$$

$$= 1.822 \times 10^{-6} \text{ Np/cm}$$
$$= 1.583 \times 10^{-3} \text{ dB/m}$$

The absorptive loss in a 1500-m line is

$$\alpha x = 1.583 \times 10^{-3} \times 1.5 \times 10^3$$
$$= 2.37 \text{ dB}$$

In Examples 5–20 and 5–21 we have calculated only the absorptive loss. In addition, there are losses due to bending, connecting, and Rayleigh scattering. These are discussed in Chapter 10.

5–10 COAXIAL CABLE LOSS AT HIGH FREQUENCIES

Now that we have discussed skin depth we can turn to the calculation of loss in coaxial cables at higher frequencies. Since the electromagnetic field can penetrate conductors not much farther than a skin depth, we conclude that only the cross-sectional area of the skin depth is available for conduction current; further, we assume that the shunt loss in the dielectric is negligible, $g = 0$. Equation (5–21) now simplifies to

$$\alpha = \beta_0\left(1 - \frac{r^2}{\omega^2L^2}\right)^{1/4} \sin\left(\tfrac{1}{2}\tan^{-1}\frac{r}{\omega L}\right)$$

At sufficiently high frequencies we also have $(r/\omega L)^2 << 1$, and the equation simplifies further to

$$\alpha = \frac{\beta_0 r}{2\omega L} \qquad (5\text{–}40)$$

where r is the series resistance/meter based on a skin-depth cross section.

The skin-depth cross section of the inner conductor of the cable can be calculated as the difference between the areas of circles of diameters d and $d - 2\delta_s$, the shaded region shown in Fig. 5–13. The area of the shaded region is $A = \pi d^2/4 - (\pi/4)(d - 2\delta_s)^2 = (\pi d^2/4)[1 - (1 - 2\delta_s/d)^2]$, or

$$A \approx \pi d\delta_s \qquad (5\text{–}41)$$

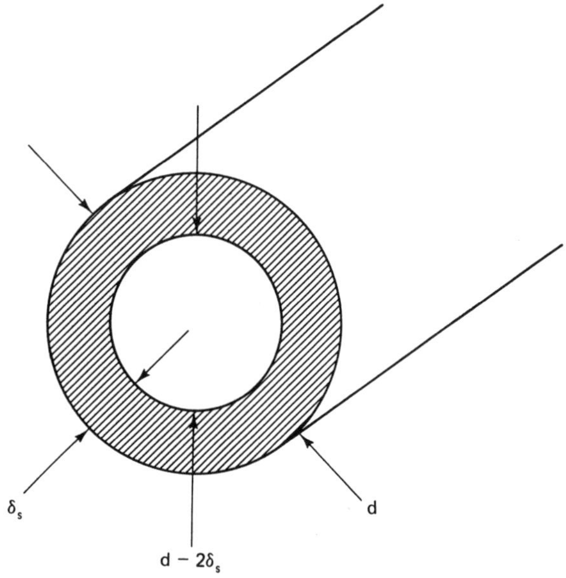

Figure 5–13 Cross section of the inner conductor of a coaxial cable. d, Diameter of the conductor; δ_s, skin depth of the conducting layer at a high frequency.

Since resistance is inversely proportional to cross-sectional area, the high-frequency resistance is greater than the *dc*. The *dc* conductivities of a few common metals are listed in Table 5–1.

EXAMPLE 5–22

Compute the attenuation factor for an RG-8/U CATV cable at an operating frequency of 100 MHz. Assume that the outer braid is covered on both sides with aluminum foil.

Solution From Tables 4–1 and 5–1 and from Example 5–16 we gather the following data:

$$\text{ODIC} = d = 2.03 \text{ mm}$$
$$\text{IDOC} = 7.24 \text{ mm}$$
$$\epsilon_r^{1/2} = 1.48$$
$$L = 2.62 \times 10^{-7} \text{ H/m}$$
$$g^{-1} = \rho(\text{copper}) = 1.77 \times 10^{-8}\ \Omega\text{-m}$$
$$\rho(\text{aluminum}) = 2.82 \times 10^{-8}\ \Omega\text{-m}$$
$$\delta_s(\text{copper}) = 0.0067 \text{ mm}$$

In addition, we need the skin depth in aluminum:

$$\delta_s = \left(\frac{2}{\omega\mu g}\right)^{1/2} = \left[\frac{2}{(2 \times \pi \times 10^8) \times (4 \times \pi \times 10^{-7}) \times (3.55 \times 10^7)}\right]^{1/2}$$
$$= 0.00845 \text{ mm}$$

We first calculate the high-frequency resistance/meter from the ratio of resistivity to cross-sectional area,

$$r = \frac{\rho}{\pi d \delta_s} \tag{5–42}$$

$$= \frac{1.77 \times 10^{-8}}{\pi \times (2.03 \times 10^{-3}) \times (6.7 \times 10^{-6})}$$

$$= 0.414\ \Omega/\text{m}$$

Before calculating the absorption coefficient from Eq. (5–40) we express β_0 in terms that match the data: $\alpha = \beta_0 r/2\omega L = (r\omega/v)/(2\omega L)$, or

$$\alpha = \frac{\epsilon_r^{1/2} r}{2cL} \tag{5–43}$$

$$= \frac{1.48 \times 0.415}{2 \times (3 \times 10^8) \times (2.62 \times 10^{-7})}$$

$$= 3.90\ 10^{-3}\ \text{Np/m}$$
$$= 0.0339\ \text{dB/m}$$
$$= 1.035\ \text{dB/100 ft}$$

We next must repeat the calculation for the aluminum outer conductor:

$$r = \frac{2.82 \times 10^{-8}}{\pi \times (7.24 \times 10^{-3}) \times (8.46 \times 10^{-6})}$$
$$= 0.1467\ \Omega/\text{m}$$

$$\alpha = \frac{1.48 \times 0.1467}{2 \times (3 \times 10^{8}) \times (2.62 \times 10^{-7})}$$
$$= 1.381 \times 10^{-3}\ \text{Np/m}$$
$$= 0.366\ \text{dB/100 ft}$$

The total absorption is the sum, 1.401 dB/100 ft.

The calculations in Example 5–22 are appropriate for CATV-type cable which has the braid sandwiched between aluminum foil. It is more difficult to estimate the absorptive loss in flexible cables that have only braid for an outer shield. An oxide layer on the strands tends to insulate the strands from one another so that the braided length of current path is longer than the length of the line; in addition, the effective skin depth is reduced. The loss in an old cable that has not been moved often can be reduced by repeatedly flexing it. For braid, the loss may be two or three times that of a solid outer conductor. If, in Example 5–22, the outer shield were braid only, the additional loss could perhaps be 1 dB per 100 ft, for a total absorption of 2.04 dB per 100 ft.

The skin effect is sometimes described as a relocation of charge carriers by magnetic fields at the center of the conductor. It is difficult to see how a relocation of charge carriers could increase resistance, since the same number of carriers are available; further, the magnetic field at the center of the conductor is zero.

5–11 TRANSMISSION LINES AS CIRCUIT ELEMENTS

We have expressed transmission-line theory in terms of equivalent lumped parameters: inductance, capacitance, and resistance per unit length of line. We can reverse our point of view and regard a section of distributed-parameter transmission line as equivalent to a lumped parameter, a passive circuit element.

Consider a transmission line of characteristic impedance Z_c to which we wish to add an inductance or capacitance at point A (Fig. 5–14). We add an auxiliary short-circuited transmission line of length D and characteristic impedance Z_s. The auxiliary line is called a *stub*. The use of stubs for matching transmission lines and

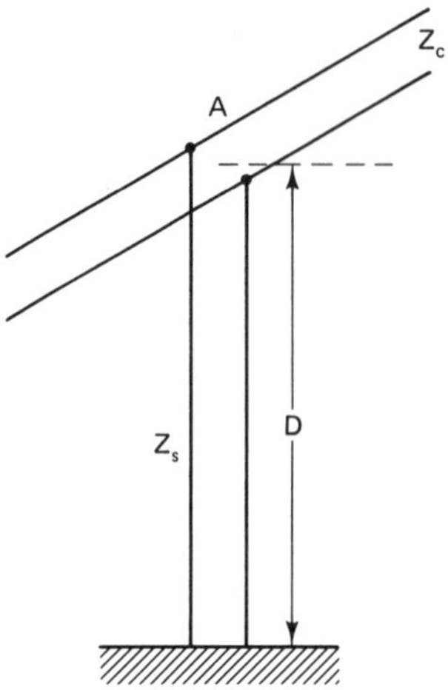

Figure 5–14 Use of a stub as a circuit element.

loads has been discussed in Section 3–6. Here we consider the stub as a circuit element.

To find the additional impedance or admittance contributed by the stub at point A we use a Smith chart, starting at the short circuit, point S in Fig. 5–15. We rotate toward the generator a relative distance of $d = D/\lambda$ to find the impedance at point A.

EXAMPLE 5–23

If the relative length of the stub is $d = D/\lambda_s = 0.125$ and $Z_s = 50\ \Omega$, find the impedance contributed to the transmission line at point A.

Solution We start at the short circuit, $z = 0$, at point S on the Smith chart (Fig. 5–15). We rotate 0.125 wavelength toward the generator to point T on the chart. We now have $z = 0 + j$ or $Z = z_s Z_c = j50\ \Omega$, and we see that we have added an inductance of $j50\ \Omega$ in parallel with Z_c at point A; that is, we have added $-j0.02$ S of susceptance.

In Example 5–23 we see that we could have added a shunt capacitive reactance 50 Ω if we had used an open-circuited stub. The starting point would have been at $z = \infty$, point U, on the chart shown in Fig. 5–15.

By selecting various values of Z_s and D, we can obtain a wide range of equivalent lumped-circuit parameters added in parallel to the transmission-line characteristic impedance at point A in Fig. 5–14. An example of special significance is the quarter-wavelength stub. If we rotate a quarter-wavelength from a short circuit on the Smith chart, we find an infinite impedance, point S to point U in Fig. 5–15. A short-circuited, quarter-wavelength stub has an infinite input impedance, but if the operating frequency is varied, the impedance will decrease, inductively or

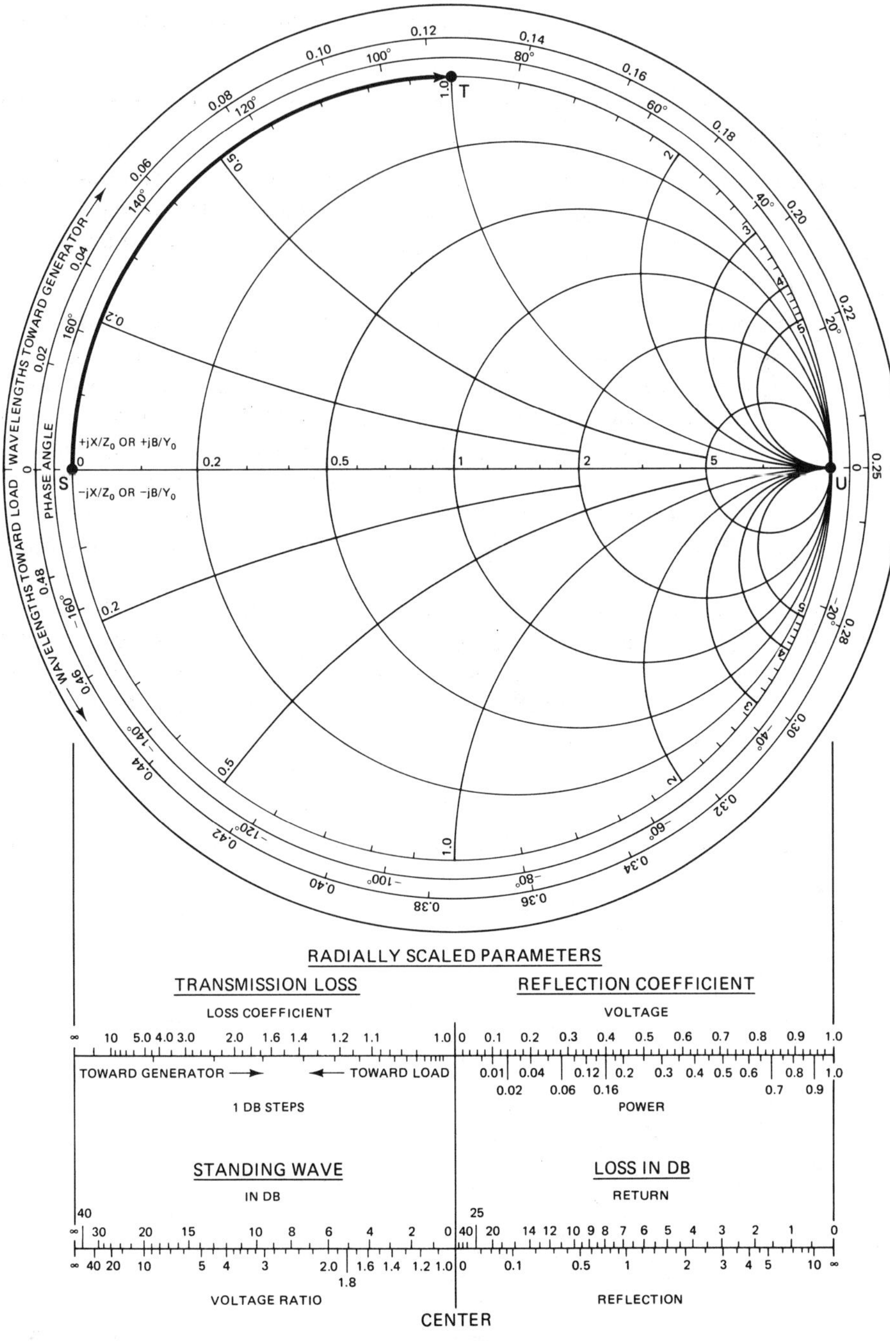

Figure 5–15 Smith chart representation of the use of a stub transmission line as a lumped-parameter circuit element.

capacitively. In other words, a quarter-wave stub has the characteristics of a parallel resonant circuit.

EXAMPLE 5–24

A quarter-wave stub of characteristic impedance 100 Ω is terminated in a 10-Ω load (Fig. 5–16). Draw a graph of the magnitude and phase of the input impedance of the stub as the frequency is varied and determine the quality of the resonance.

Solution The normalized load impedance is 10/100 = 0.1, and the corresponding VSWR circle is shown in Fig. 5–17. At frequency $f = \omega/\omega_0 = 1$, the stub is a quarter-wavelength long and the normalized input impedance is $z = 10\angle 0°$, as shown by the intersection of the VSWR circle and the horizontal axis of a Smith chart. At a lower frequency, say $f = 0.85$, the wavelength increases to 0.25/0.85 = 0.294, and the electrical length of the line is reduced to $0.25 \times 0.85 = 0.213\lambda$. The electrical distance to the load has been reduced to 0.213λ and the input impedance now is $1.2 + j3.1$, as shown at point *A* in Fig. 5–17.

By transforming from rectangular to polar form, we obtain $z = 3.32\angle 68.8°$, and this is plotted as points *A* in Fig. 5–18. After repeating the calculations for several values of f we obtain the curves shown in Fig. 5–18.

From the amplitude curve we find the bandwidth at the $-(2)^{-1/2}$ points to be $\Delta f = 0.116$, and from the formula $Q = f/\Delta f$ we have

$$Q = \frac{1}{0.116} = 8.6$$

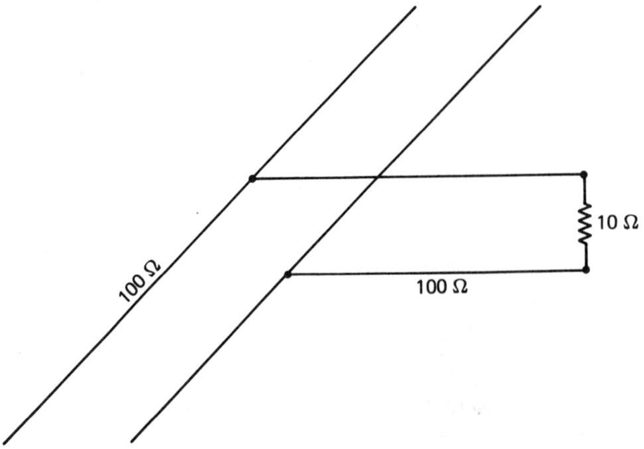

Figure 5–16 Schematic representation of the use of a stub as a resonant circuit, Example 5–24.

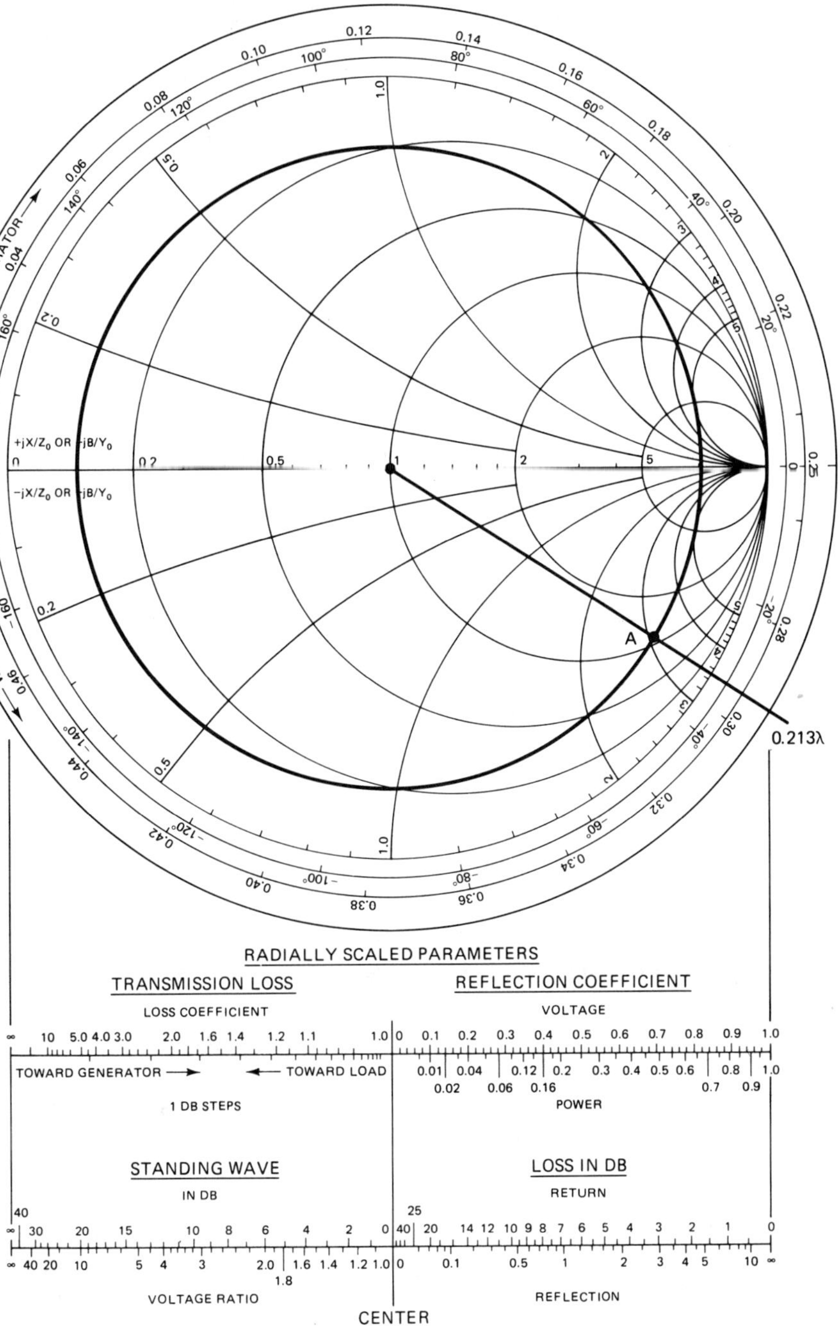

Figure 5–17 Smith chart representation of a stub used as a resonant circuit, Example 5–24.

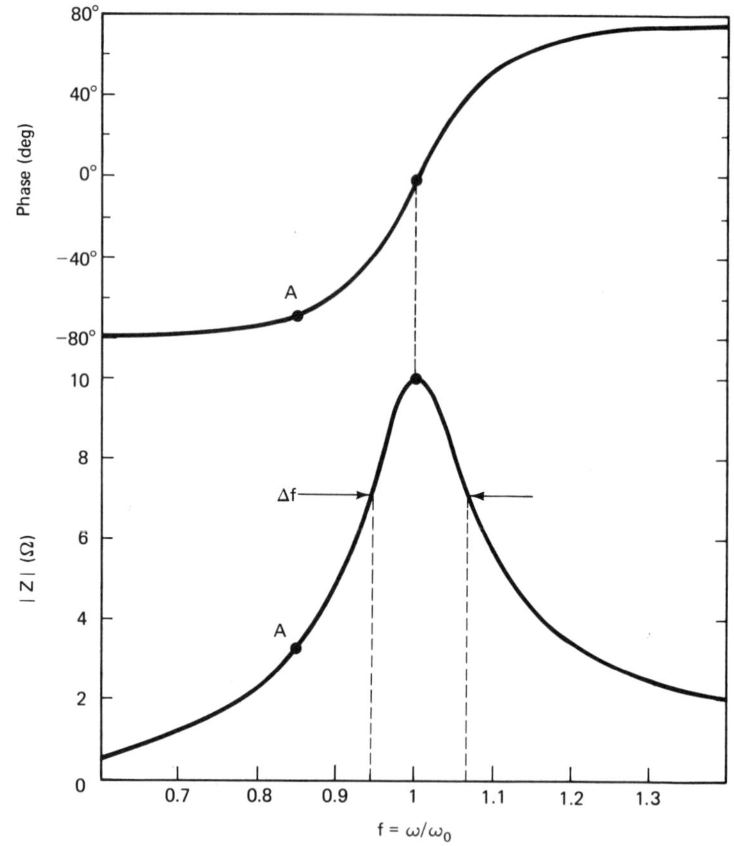

Figure 5–18 Amplitude and phase response of a stub used as a parallel resonant circuit.

5–12 DEVICE EMBEDDED IN TRANSMISSION LINES

When the input and output of a device are connected to transmission lines, the device is said to be *embedded* in transmission lines. If the device is being tested, for example, the measurement of the gain of an amplifier, we refer to it as the device under test or DUT. We shall consider first the simple case where the device is passive.

The reflection coefficient at the input interface of a passive device is given by Eq. (2–11). We can find the transmission coefficient from conservation of energy. Let Z_1 be the characteristic impedance of the first medium or transmission line and Z_2 the impedance of the second medium or load. We assume that the incident amplitude is unity and that r_{12} and t_{12} are the reflection and transmission coefficients, as shown in Fig. 5–19. From conservation of energy we have $1^2/Z_1 = r_{12}^2/Z_1 + t_{12}^2/Z_2$ or

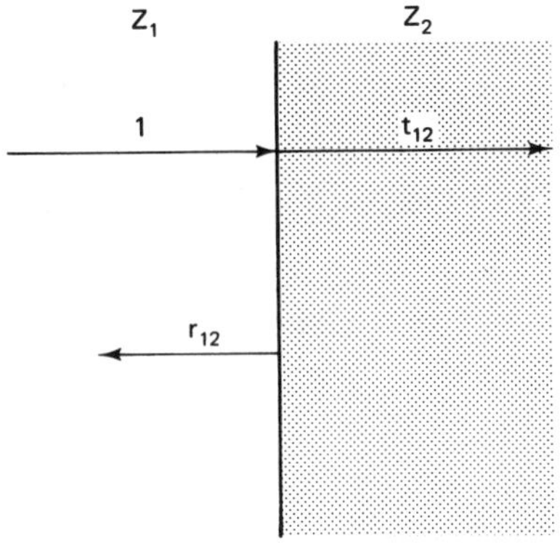

Figure 5–19 Reflection and transmission coefficients at an interface between two media.

$$1 = r_{12}^2 + \frac{t_{12}^2}{z_2} \tag{5–44}$$

and from Eq. (2–11),

$$t_{12}^2 = z_2\left[1 - \frac{(z_2 - 1)^2}{(z_2 + 1)^2}\right]$$

or

$$t_{12} = \frac{2z_2}{z_2 + 1} \tag{5–45}$$

Until now we have used only reflection coefficients. Since the incident and reflected amplitudes were in the same medium, it was not necessary to normalize to the impedance of the medium; however, when a wave propagates from one medium into another, it is necessary to normalize to the impedance of the second medium since intensity or power is proportional to a^2/z. This means that if we use normalized impedances, as in Eq. (5–44), only the transmission coefficients need be normalized by $(z_c)^{1/2}$; for example, t_{12} becomes $t_{12}/(z_2)^{1/2}$ when normalized.

Some workers define all coefficients as not normalized, and the appropriate impedances must be written explicitly; others define coefficients as though the normalization impedances were included, which makes equations containing reflection and transmission coefficients appear simpler. We shall not use either convention exclusively, but will choose whichever is convenient for the problem at hand.

EXAMPLE 5–25

Find the reflection and transmission coefficients for a 10-Ω resistor embedded in 50-Ω transmission lines.

Solution We refer to Fig. 5–20. The load impedance is R and Z_c in parallel, $RZ_c/(R + Z_c)$, and the normalized impedance is $Z_L/Z_c = R/(R + Z_c)$. The reflection coefficient is

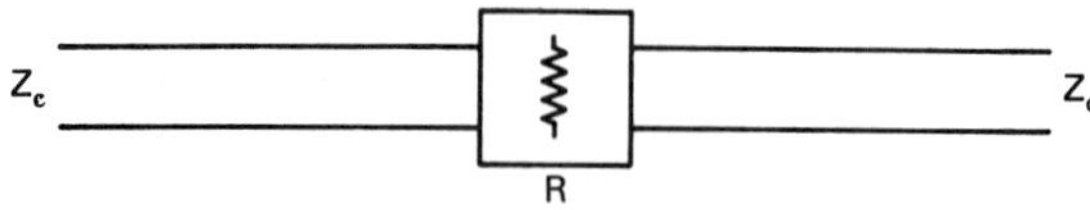

Figure 5–20 The DUT is a resistor embedded in transmission lines of characteristic impedance Z_c, Example 5–25.

$$r = \frac{z-1}{z+1} = \frac{R/(R+Z_c)-1}{R/(R+Z_c)+1}$$

$$= \frac{-Z_c}{2R+Z_c}$$

$$= \frac{-50}{2 \times 10 + 50}$$

$$= -0.714$$

and the unnormalized transmission coefficient is

$$t = \frac{2z}{z+1} = \frac{2R/(R+Z_c)}{R/(R+Z_c)+1}$$

$$= \frac{2R}{2R+Z_c}$$

$$= \frac{2 \times 10}{2 \times 10 + 50}$$

$$= 0.286$$

Check:

The load impedance is $Z_L = 10 \times 50/(10+50) = 8.33\ \Omega$. Conservation of energy requires that the reflected and transmitted energies equal the incident energy,

$$\frac{1^2}{Z_c} = \frac{r^2}{Z_c} + \frac{t^2}{Z_L}$$

$$\frac{1}{50} = \frac{0.714^2}{50} + \frac{0.286^2}{8.33}$$

$$0.02 = 0.0200$$

EXAMPLE 5–26

Repeat Example 5–25 when the DUT is a capacitor having an admittance of 50 mS.

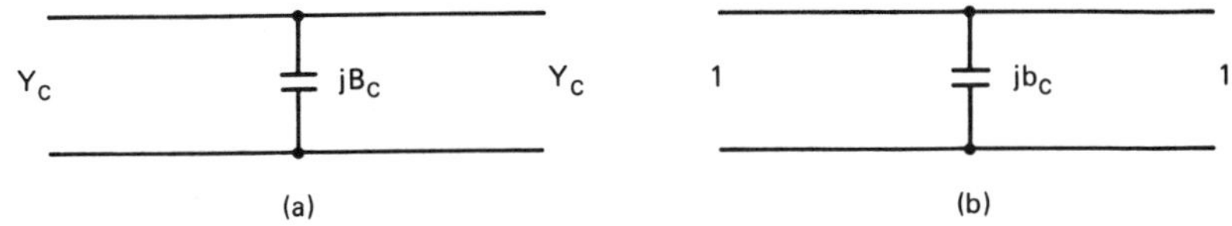

Figure 5–21 (a) The DUT is a capacitor embedded in transmission lines of characteristic admittance Y_c; (b) same problem normalized to Y_c.

Solution We refer to Fig. 5–21, where jb_C is the admittance normalized to $1/50 = 0.02$ S; that is, $jb_C = j50$ mS/20 mS $= j2.5$. The load admittance is $y_L = 1 + jb_C = 1 + j2.5$; the reflection coefficient is

$$r = \frac{1 - y_L}{1 + y_L} = \frac{1 - 1 - j2.5}{1 + 1 + j2.5}$$

$$= \frac{2.5\underline{/-90^\circ}}{3.20\underline{/51.3^\circ}}$$

$$= 0.781\underline{/-141.3^\circ}$$

and the transmission coefficient is

$$t = \frac{2}{1 + y_L} = \frac{2}{1 + 1 + j2.5}$$

$$= \frac{2}{3.20\underline{/51.3^\circ}}$$

$$= 0.625\underline{/-51.3^\circ}$$

Check: On a Smith chart plot the load admittance $1 + j2.5$ and find $r = 0.79\underline{/-141^\circ}$, as shown in Fig. 5–22.

In Examples 5–25 and 5–26 we have assumed that the transmission lines either are infinite or terminated in matching resistances at source and load, so that there are no reflections other than those at the DUT; however, if the DUT is itself of finite size, comparable to a wavelength, there will be multiple reflections within the device itself. Consider, for example, a DUT that is a section of transmission line of length d, or a slab of dielectric of length d, embedded between two others, as shown in Fig. 5–23. The amplitudes of the various multireflected and transmitted rays are shown on the figure. Our problem is to sum all the amplitudes to find the total reflected and transmitted amplitudes. The problem is not as difficult as first appears if we use the binomial theorem in the form

$$\frac{1}{1 - x} = 1 + x + x^2 + x^3 + \cdots$$

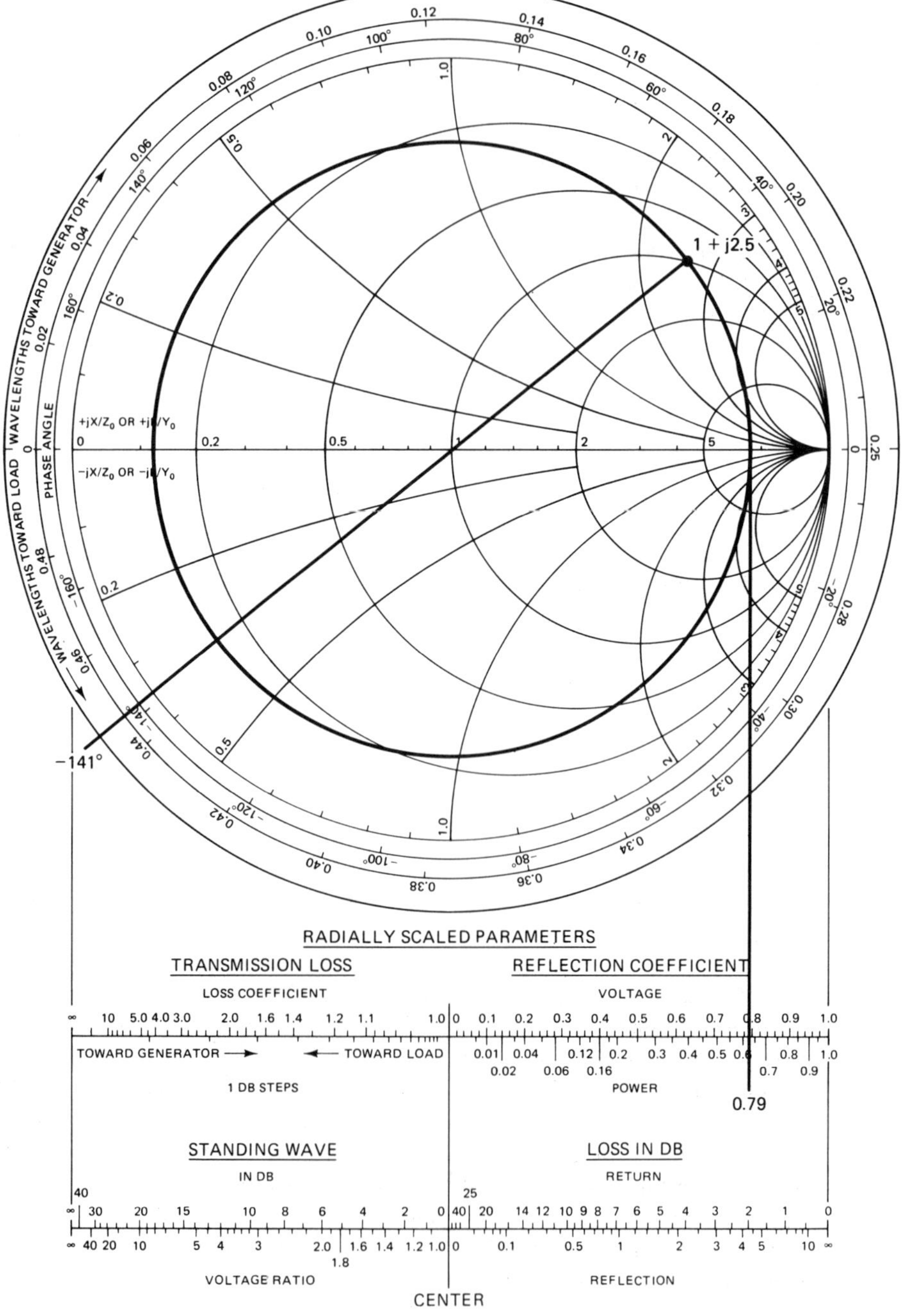

Figure 5–22 Smith chart solution to Example 5–26.

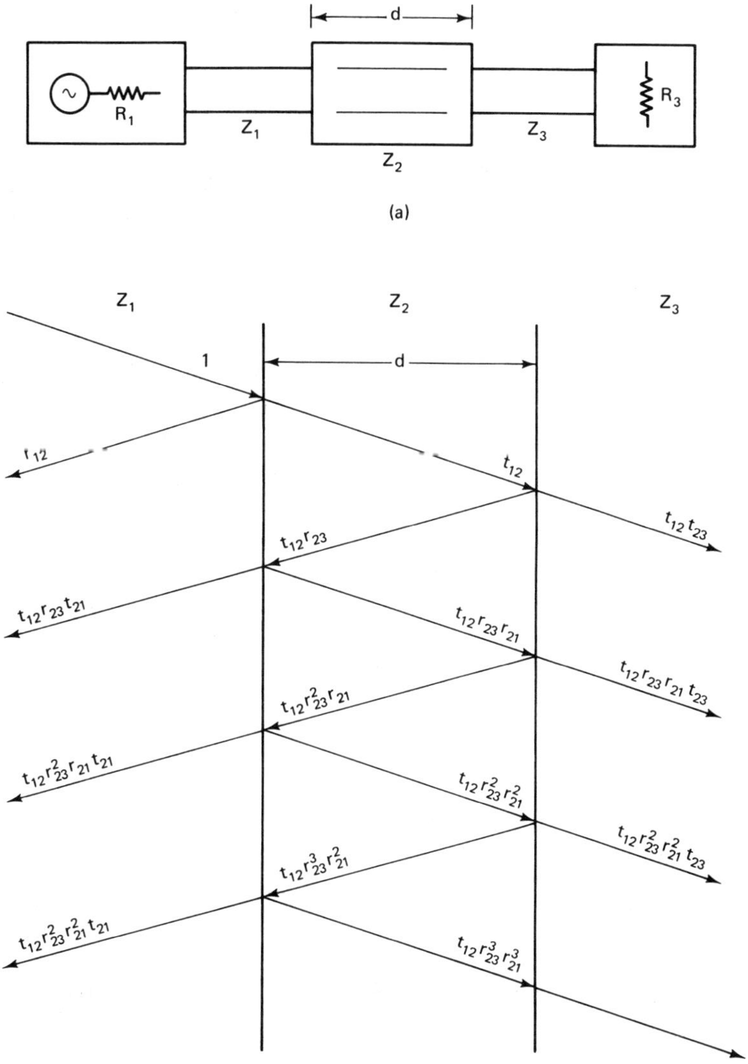

Figure 5–23 (a) The DUT is itself a length of transmission line; (b) amplitudes of the multireflected waves transmitted, reflected, and stored.

where x can be real or complex.

The sum of all the reflected rays is $R = r_{12} + t_{12}t_{21}r_{23}\,[1 + r_{23}r_{21} + (r_{23}r_{21})^2 + \cdots]$ or, by the binomial theorem,

$$R = r_{12} + \frac{t_{12}t_{21}r_{23}}{1 - r_{23}r_{21}} \tag{5–46}$$

The sum of all the transmitted wave amplitudes is $T = t_{12}t_{23}[1 + r_{23}r_{21} + (r_{23}r_{21})^2 + \cdots]$, or

$$T = \frac{t_{12}t_{23}}{1 - r_{23}r_{21}} \tag{5–47}$$

In evaluating the expressions above we must remember that the reflection and transmission coefficients include not only the phase change on reflection or transmission but also include the phase delay due to the propagation path through the device. This means that the interior coefficients must include a phase delay of $\beta = 2\pi d/\lambda$:

$$\begin{aligned} t_{12} &= |t_{12}| \underline{/\psi_{12} + \beta} \\ r_{23} &= |r_{23}| \underline{/\psi_{23} + \beta} \\ r_{21} &= |r_{21}| \underline{/\psi_{21} + \beta} \end{aligned}$$

where d is the electrical length of the device. The exterior coefficients include only the phase change on reflection or transmission:

$$\begin{aligned} t_{23} &= |t_{23}| \underline{/\psi_{23}} \\ t_{21} &= |t_{21}| \underline{/\psi_{12}} \end{aligned}$$

With these provisions our results are general and applicable to complex media of finite size and complex sources and terminations.

Before we consider examples of the results above, we shall calculate the amplitude of the internal reflections, which is indicative of the quantity of energy stored within the device. We shall see that a passive device can have the characteristics of a resonant circuit, and by calculating the stored energy we can find the Q of the device.

The sum of the amplitudes of all the waves moving to the left in Fig. 5–23 is $S_L = t_{12}r_{23}(1 + r_{23}r_{21} + (r_{23}r_{21})^2 + \cdots)$ or

$$S_L = \frac{t_{12}r_{23}}{1 - r_{23}r_{21}} \tag{5–48}$$

and the sum of the amplitudes of the internal waves moving to the right is $S_R = t_{12}[1 + r_{23}r_{21} + (r_{23}r_{21})^2 + \cdots]$, or

$$S_R = \frac{t_{12}}{1 - r_{23}r_{21}} \tag{5–49}$$

With the results above we are now able to calculate the Q of the DUT. Energy stored in a half-wavelength is proportional to the voltage or amplitude squared of the internal waves, divided by the (real) impedance,

$$U_2 = \frac{Z_1}{Z_2} \frac{|t_{12}|^2|r_{23}|^2 + |t_{12}|^2}{|1 - r_{23}r_{21}|^2} \tag{5–50}$$

from Eqs. (5–48) and (5–49). The energy lost to the right side of the DUT is, from Eq. (5–47),

$$U_3 = \frac{Z_1|t_{12}|^2|t_{23}|^2}{Z_3|1 - r_{23}r_{21}|^2} \qquad (5-51a)$$

and the energy lost from the left side of the device is given by the second term of Eq. (5–46):

$$U_1 = \frac{|t_{12}|^2|t_{21}|^2|r_{23}|^2}{|1 - r_{23}r_{21}|^2} \qquad (5\text{–}51b)$$

The quality of the DUT is the ratio of the energy stored to the energy lost per radian cycle,

$$Q = \frac{2\pi}{Z_2} \frac{(Z_1/Z_2)(|t_{12}|^2|r_{23}|^2 + |t_{12}|^2)}{(Z_1/Z_3)(|t_{12}|^2|t_{23}|^2 + |t_{12}|^2|t_{21}|^2|r_{23}|^2)}$$

or

$$Q = \mathrm{Re}\,\frac{2\pi Z_1 Z_3}{Z_2} \frac{1 + |r_{23}|^2}{Z_1|t_{23}|^2 + Z_3|t_{21}|^2|r_{23}|^2} \qquad (5\text{–}52)$$

EXAMPLE 5–27

A half-wavelength transmission line with characteristic impedance of 50 Ω is embedded between a 300-Ω source and an 8.33-Ω termination (Fig. 5–24). Calculate the quality of the resonance.

Solution We start by calculating the required reflection and transmission coefficients:

$$|r_{23}| = \left|\frac{8.33 - 50}{8.33 + 50}\right| = |-0.714| \qquad |r_{23}|^2 = 0.510$$

$$|t_{23}| = \left|\frac{2 \times 8.33}{50 + 8.33}\right| = |0.286| \qquad |t_{23}|^2 = 0.0816$$

$$|t_{21}| = \left|\frac{2 \times 300}{300 + 50}\right| = |1.714| \qquad |t_{21}|^2 = 2.939$$

From Eq. (5–52) we have

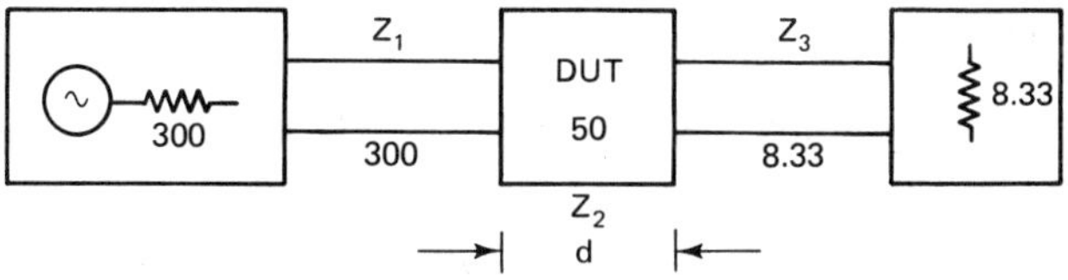

Figure 5–24 In this example the DUT is a 50-Ω transmission line of length d embedded between 300- and 8.33-Ω lines.

$$Q = \frac{2 \times \pi \times 300 \times 8.33}{50}\left(\frac{1 + 0.510}{300 \times 0.0816 \times 8.33 \times 2.939 \times 0.510}\right)$$

$$= 12.82$$

The half-wave mismatched section of transmission line has the characteristic of a parallel-resonant circuit with a Q of 12.82.

We now can calculate more accurately the quality of the stub in Example 5–24.

EXAMPLE 5–28

Find the Q of a quarter-wave stub of characteristic impedance 100 Ω; the stub is terminated in a 10-Ω load. The characteristic impedance of the main line also is 100 Ω, as shown in Fig. 5–16.

Solution We again start by calculating the required reflection and transmission coefficients:

$$|r_{23}| = \left|\frac{10 - 100}{10 + 100}\right| = 0.818 \qquad |r_{23}|^2 = 0.669$$

$$|t_{23}| = \left|\frac{2 \times 10}{100 + 10}\right| = 0.1818 \qquad |t_{23}|^2 = 0.03306$$

$$|t_{21}| = \left|\frac{2 \times 100}{100 + 100}\right| = 1 \qquad |t_{21}|^2 = 1$$

and from Eq. (5–52) we have

$$Q = \frac{2 \times \pi \times 100 \times 10}{100}\left(\frac{1 + 0.669}{10 \times 0.669 + 100 \times 0.03306}\right)(0.900)^2$$

$$= 8.51$$

where 0.900 is the mean peak voltage on the line.

By the graphical method of Example 5–24, in which multiple reflections were ignored, we obtained the less accurate result $Q = 8.6$.

EXAMPLE 5–29

Find the reflection coefficient of the DUT described in Example 5–27 and Fig. 5–24.

Solution Since we need to evaluate Eq. (5–46), we first calculate the reflection and transmission coefficients at the interfaces. The propagation delay through the device is $\beta = 2\pi d/\lambda = \pi$. We have

$$r_{12} = \frac{50 - 300}{50 + 300} = -0.714$$

$$r_{23} = \frac{8.33 - 50}{8.33 + 50} \underline{/\pi} = 0.714$$

$$r_{21} = \frac{300 - 50}{300 + 50} \underline{/\pi} = -0.714$$

$$t_{12} = \frac{2 \times 50}{50 + 300} \underline{/\pi} = -0.286$$

$$t_{21} = \frac{2 \times 300}{300 + 50} = 1.714$$

and Eq. (5–46) becomes

$$R = -0.714 + \frac{0.286 \times 1.714 \times (-0.714)}{1 - 0.714 \times (-0.714)}$$

$$= -0.946$$

As a final example we consider the DUT in Fig. 5–23 to be a quarter-wave transformer or antireflection layer. Since $R = 0$, Eq. (5–46) becomes

$$r_{12} = \frac{-t_{12}t_{21}r_{23}}{1 - r_{23}r_{21}}$$

Since we must take into account the propagation delays, the equation becomes

$$r_{12} = \frac{-t_{12}\underline{/\pi/2}\, t_{21}r_{23}\underline{/\pi/2}}{1 - r_{23}\underline{/\pi/2}\, r_{21}\underline{/\pi/2}}$$

or

$$r_{12} = \frac{t_{12}t_{21}r_{23}}{1 + r_{23}r_{21}}$$

and we see that either r_{12} and r_{23} are both negative or both positive, conditions that are satisfied if $Z_1 > Z_2 > Z_3$ or $Z_1 < Z_2 < Z_3$.

If this is to be a matching layer, no reflection occurs and the transmitted amplitude should be unity. Equation (5–47) becomes, after normalization,

$$1 = \frac{t_{21}/z_2 + t_{23}/z_3}{1 - r_{23}r_{21}}$$

where $z_2 = Z_2/Z_1$ and $z_3 = Z_3/Z_2$. We now substitute from Eqs. (5–45) and (2–11) to obtain

$$1 = \frac{[(Z_1/Z_2)^{1/2}2Z_2/(Z_2 + Z_1)][(Z_2/Z_3)^{1/2}2Z_3/(Z_3 + Z_2)\angle \pi/2]}{1 - \frac{Z_3 - Z_2}{Z_3 + Z_2}\angle \pi/2 \; \frac{Z_1 - Z_2}{Z_1 + Z_2}\angle \pi/2}$$

or

$$Z_2^2 - 2Z_2(Z_1Z_3)^{1/2} + Z_1Z_3 = 0$$
$$Z_2 = (Z_1Z_3)^{1/2} \tag{5–53}$$

We obtained this quarter-wave transformer condition more simply in Eq. (3–4). We have repeated it here to illustrate the use of normalized reflection and transmission coefficients.

In the preceding examples we see that the inclusion of multiple reflections increases the accuracy of our analyses; however, in some cases it only verifies answers obtained more simply. In general, multiple reflections usually decrease the bandwidth of a device by increasing its Q. In many cases, an analysis of multiple reflections is equivalent to a consideration of mutual impedances.

5–13 SCATTERING PARAMETERS

We have been using a reflection and transmission coefficient notation, r_{12} or t_{12}, in which the first subscript indicates the medium or transmission line of origin and the second subscript shows the medium or load from which a wave is reflected or into which the wave is transmitted. In another frequently used convention, only the resultant reflection and transmission coefficients of a DUT are considered (Fig. 5–25). The DUT is embedded in identical transmission lines that are matched to source and load, so that there are no reflections, single or multiple, between source or load and the DUT. The industry standard is to use 50- or 75-Ω lines, source, and termination. There may be, of course, multiple reflections in the DUT, but these are included in the definitions of S_{11} and S_{21} as the *resultant* reflection and transmission coefficients, which now are called *scattering parameters*.

The name "scattering parameters" comes from particle physics. Physicists have investigated the structure of matter by accelerating subatomic particles and causing them to collide with target atoms or nuclei. The technique requires the measurement of the angle and momentum or energy of the scattered projectile and the recoil of the target. When part of the theory of scattering was adopted into transmission-line theory, the name was adopted also. Since one-dimensional transmission lines allow only two directions, there are only forward scattering and back scattering, which correspond to transmission and reflection.

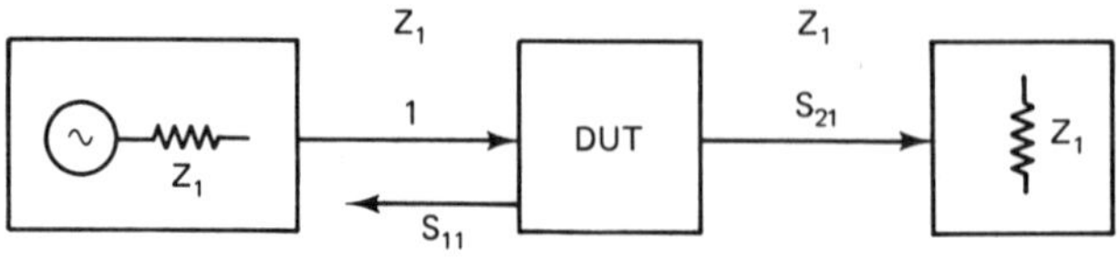

Figure 5–25 Scattering parameters S_{11} and S_{21} for a DUT embedded in transmission lines.

If we interchange source and load in Fig. 5–25, we obtain the reverse coefficients, S_{22} and S_{12}. The measurement or calculation of the four scattering parameters determines the properties of the DUT.

EXAMPLE 5–30

Find the scattering parameters for a 10-Ω resistor embedded in 50-Ω transmission lines.

Solution From Example 5–25 we have

$$S_{11} = S_{22} = -0.714$$
$$S_{21} = S_{12} = 0.286$$

Check: For conservation of energy we require

$$1 = S_{11}^2 + S_{21}^2 \frac{Z_1}{Z_2}$$
$$= (-0.714)^2 + (0.286)^2 \frac{50}{8.33}$$
$$= 1$$

Note: If we follow the convention of normalizing the scattering parameters,

$$S_{11} = \frac{S_{11}}{(Z_1)^{1/2}}$$
$$S_{21} = \frac{S_{21}}{(Z_2)^{1/2}}$$
$$1 = \frac{1}{(Z_1)^{1/2}}$$

the conservation equation becomes simpler,

$$1 = S_{11}^2 + S_{21}^2$$

but there is an increased chance of error of oversight.

The subscripts of S-parameters have no obvious connection with the designations of the source or load. They are in matrix notation, which will be discussed presently, but first we shall consider some slightly more complicated examples.

EXAMPLE 5–31

Find the S-parameters of the DUT shown in Fig. 5–24.

Solution From Example 5–29 we have

$$S_{11} = S_{22} = R = -0.946$$

From Eq. (5–47) we have

$$S_{21} = T = \frac{t_{21}t_{23}}{1 - r_{23}r_{21}}$$

where $t_{23} = 2Z_3/(Z_3 + Z_2) = 2 \times 8.33/(8.33 + 50) = 0.286$, and the remaining coefficients were evaluated in Example 5–28. We have

$$S_{12} = S_{21} = \frac{(-0.286)(0.286)}{1 - 0.714 \times (-0.714)}$$

$$= -0.0542$$

Check: For conservation of energy we require

$$1 = S_{11}^2 + S_{21}^2 \frac{Z_1}{Z_3}$$

$$= (-0.946)^2 + (-0.0542)^2 \times \frac{300}{8.33}$$

$$= 1.00$$

In Examples 5–30 and 5–31 we have used unnormalized S-parameters. Many manufacturers now use normalized parameters for describing active devices; therefore, it is necessary to use normalized parameters in reading their catalogs. The value of S-parameters for describing active devices lies in the ease with which they can be measured. This means that with the knowledge of the values of the S-parameters we can design a circuit around the device without having to analyze or understand what is happening inside the device. To make the measurements, an active device, a transistor, for example, is embedded in transmission lines (Fig. 5–26). If a_1, a_2 and b_1, b_2 are the incident and reflected wave amplitudes, the scattering parameters can be measured as

$$S_{11} = \left.\frac{b_1}{a_1}\right|_{a_2=0} \qquad S_{21} = \left.\frac{b_2}{a_1}\right|_{a_2=0}$$
$$S_{12} = \left.\frac{b_1}{a_2}\right|_{a_1=0} \qquad S_{22} = \left.\frac{b_2}{a_2}\right|_{a_1=0} \tag{5–54}$$

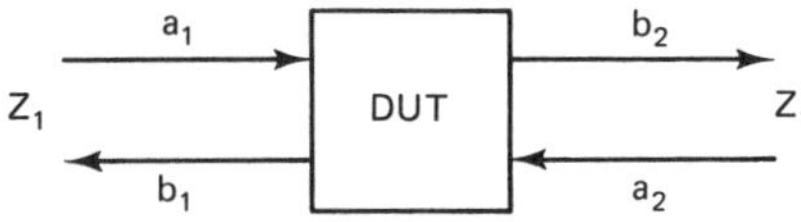

Figure 5–26 To determine *S*-parameters, the DUT is embedded in identical transmission lines matched to source and load. For purpose of illustration, in the discussions following Examples 5–30 and 5–31 in the text, the medium on the right has an impedance Z_2.

The measurement conditions $a_1 = 0$ and $a_2 = 0$ are achieved by matching source and load to the lines, which eliminates reflections between the DUT and the ends of the lines. The parameters are normalized through the wave amplitudes:

$$a_1 = \frac{v_{i1}}{(Z_1)^{1/2}} \qquad b_1 = \frac{v_{r1}}{(Z_1)^{1/2}}$$

$$a_2 = \frac{v_{i2}}{(Z_2)^{1/2}} \qquad b_2 = \frac{v_{r2}}{(Z_2)^{1/2}}$$

where v_{i1}, v_{r1} and v_{i2}, v_{r2} are the voltage amplitudes at the first and second interfaces. It is customary to call the interfaces the entrance and exit *ports*. We see, as in Section 5–12, that the reflection parameters are unchanged, but the transmission parameters become

$$S_{21} = \frac{b_2}{a_1}\left(\frac{Z_1}{Z_2}\right)^{1/2} = \frac{b_2}{a_1}\left(\frac{1}{z_2}\right)^{1/2}$$

$$S_{12} = \frac{b_1}{a_2}\left(\frac{Z_2}{Z_1}\right)^{1/2} = \frac{b_1}{a_2}(z_2)^{1/2}$$

EXAMPLE 5–32

Repeat Example 5–31 using normalized scattering parameters.

Solution As before, we have $S_{11} = S_{22} = -0.946$, but for the transmission coefficients we have

$$S_{21} = S_{12} = \frac{t_{12}(Z_1/Z_2)^{1/2}t_{23}(Z_2/Z_3)}{1 - r_{23}r_{21}}$$

$$= \frac{t_{12}t_{23}(1/z_3)^{1/2}}{1 - r_{23}r_{21}}$$

$$= \frac{-0.286 \times 0.286(300/8.33)^{1/2}}{1 - 0.714 \times (-0.714)}$$

$$= -0.325$$

Check: For conservation of energy we require

$$\begin{aligned} 1 &= S_{11}^2 + S_{21}^2 \\ &= 0.946^2 + 0.325^2 \\ &= 1.00 \end{aligned}$$

We see in Fig. 5–26 that the amplitudes transmitted through a DUT can be expressed in terms of the incident amplitudes:

$$\begin{aligned} b_1 &= S_{11}a_1 + S_{12}a_2 \\ b_2 &= S_{21}a_1 + S_{22}a_2 \end{aligned} \tag{5–55}$$

We can write Eq. (5–55) in matrix form as

$$\begin{pmatrix} b_1 \\ b_2 \end{pmatrix} = \begin{pmatrix} S_{11} & S_{12} \\ S_{21} & S_{22} \end{pmatrix} \begin{pmatrix} a_1 \\ a_2 \end{pmatrix} \tag{5–56}$$

and we see that the scattering parameters are in standard matrix notation. An introduction to matrix algebra is given in Section 5–15.

5–14 TRANSDUCER GAIN

In the examples we have considered so far, the DUTs were enmeshed in transmission lines that were matched to the source and load. Although this is standard procedure for measuring scattering parameters, it is not necessarily an ideal operating condition or a condition that occurs in applications. We know, for example, that maximum power is transferred from source to DUT if the source impedance and the DUT impedance are complex conjugates, a condition that does not necessarily require flat matching. For most applications we shall have to consider the more general case shown in Fig. 5–27.

Since the source and load are not matched to the transmission line we have to consider three possible resonances or sets of multiple reflections: one between the source and DUT; another between the DUT and the load; and an overall set of reflections between the source and load, through the DUT.

The sum of all the wave amplitudes transferred into the load, as shown in Fig. 5–27, is

$$t_1S_{21}t_2[1 + r_LS_{22} + (r_LS_{22})^2 + \cdots] = \frac{t_1S_{21}t_2}{1 - r_LS_{22}} = t_1S_{21}T_2$$

where

$$T_2 = \frac{t_2}{1 - r_LS_{22}} \tag{5–57}$$

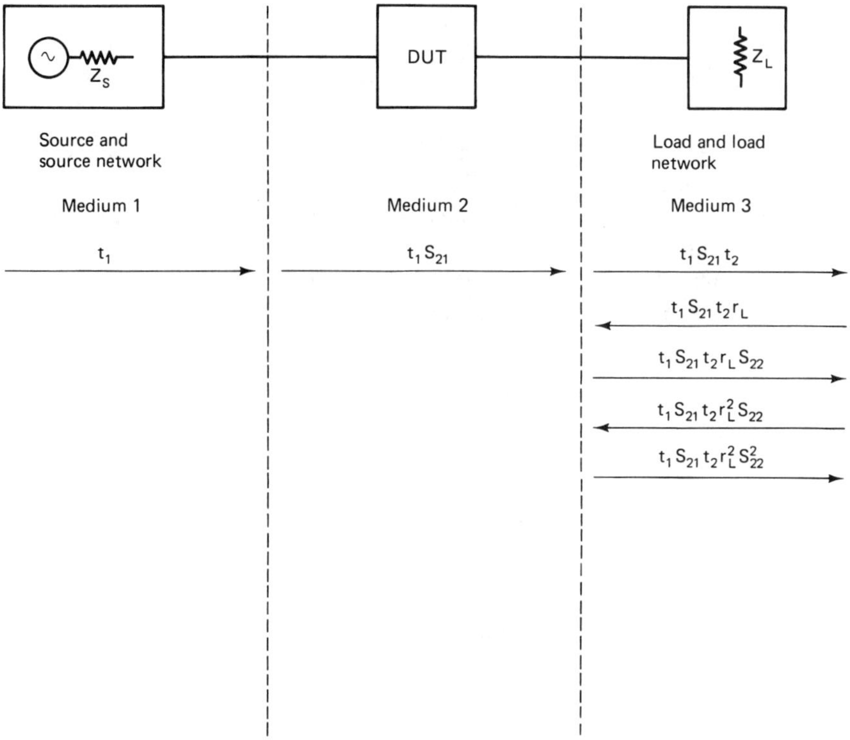

Figure 5–27 General case in which source, DUT, and load have different impedances. In the figure we show only the multiple reflections between the DUT and the load.

and r_L is the reflection coefficient at the load interface. Similarly, we find the sum of all the amplitudes reflected from the load.

$$t_1S_{21}t_2r_L[1 + r_LS_{22} + (r_LS_{22})^2 + \cdots] = \frac{t_1S_{21}t_2r_L}{1 - r_LS_{22}} = t_1S_{21}t_2R_L$$

where

$$R_L = \frac{r_L}{1 - r_LS_{22}} \qquad (5\text{–}58)$$

From a similar consideration of the multiple reflections between the DUT and the source, we find

$$T_1 = \frac{t_1}{1 - r_sS_{11}} \qquad (5\text{–}59)$$

and

$$R_S = \frac{r_S}{1 - r_S S_{11}} \tag{5–60}$$

We can now consider the higher-order set of reflections between the load and source. In general, the DUT is not *unilateral*; that is, it is not a one-way street. Because of the feedback represented by S_{12} the device may act like a two-way street: just as waves were led across or "transduced" from source to load, so may multireflected waves be led back to the source.

The higher-order set of transduced waves are shown in Fig. 5–28. We see that the sum of all the wave amplitudes delivered to the load now is

$$\begin{aligned} A_v &= T_1 S_{21} T_2 [1 + S_{21} S_{12} R_L R_S + (S_{21} S_{12} R_L R_S)^2 + \cdots] \\ &= \frac{T_1 S_{21} T_2}{1 - S_{21} S_{12} R_L R_S} \end{aligned}$$

or, with Eqs. (5–57) to (5–60),

$$A_v = \frac{t_1 S_{21} t_2}{(1 - r_S S_{11})(1 - r_L S_{22})\left[1 - \dfrac{S_{21} S_{12} r_L r_S}{(1 - r_S S_{11})(1 - r_L S_{22})}\right]}$$

or

$$A_v = \frac{t_1 S_{21} t_2}{(1 - r_S S_{11})(1 - r_L S_{22}) - S_{21} S_{12} r_L r_S} \tag{5–61}$$

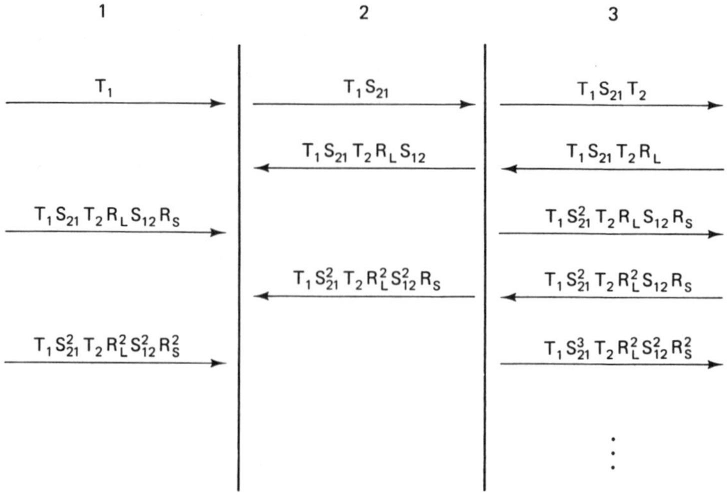

Figure 5–28 Transducer action in terms of multiple-reflected waves.

Since the parameter S_{21} is a forward transfer ratio, we can regard A_v as a transducer voltage gain. We obtain the *transducer gain* by squaring the modulus of the voltage gain, $G = |A_v|^2$. With the use of the conservation equations $|t_1|^2 = 1 - |r_S|^2$ and $|t_2|^2 = 1 - |r_L|^2$, we find

$$G = \frac{(1 - |r_S|^2)|S_{21}|^2(1 - |r_L|^2)}{|(1 - r_S S_{11})(1 - r_L S_{22}) - S_{21}S_{12}r_L r_S|^2} \tag{5–62}$$

For some devices, transistors for example, the reverse transfer parameter S_{12} may be quite small; that is, the device is nearly unilateral. This means that the last term in the denominator of Eq. (5–62) can be neglected, and we have

$$G = \frac{(1 - |r_S|^2)|S_{21}|^2(1 - |r_L|^2)}{|1 - r_S S_{11}|^2|1 - r_L S_{22}|^2} \tag{5–63}$$

We shall use Eq. (5–63) in the design of a transistor amplifier in Chapter 7.

For complicated network problems it is convenient to use flowcharts to keep track of all the possible sets of multiple reflections. A flowchart for the transducer gain given by Eq. (5–63) is shown in Fig. 5–29. The first loop symbolizes the effect of multiple reflections between the source and DUT interfaces, the loop on the right side represents multiple reflections between the load and DUT interfaces, and the loop $a_1b_2a_2b_1a_1$ represents the ''transduced'' reflections between source and load. A set of algorithms has been devised for finding flowchart equivalents of networks.

Multiple reflections can be regarded as a form of feedback. The theory of nuclear scattering usually was addressed as a feedback problem rather than one of multiple reflections. We shall, in Chapter 8, present multiple reflections as a feedback

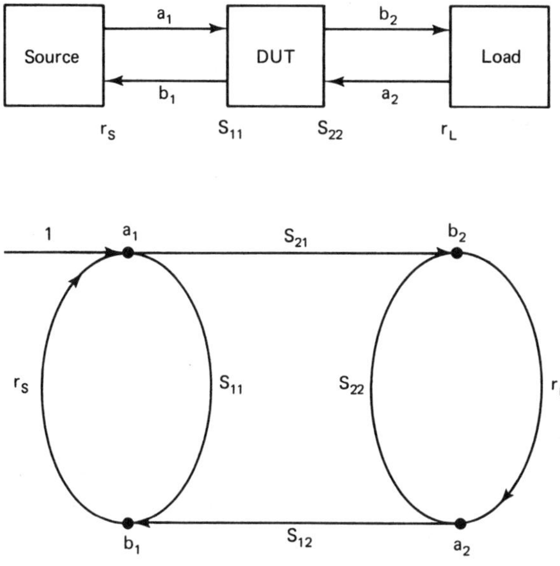

Figure 5–29 Flowchart equivalent of a network.

problem, which may be easier for readers who are more familiar with circuit theory than with wave theory.

5–15 ELEMENTS OF OPERATOR ALGEBRA

Scattering parameters are presented in matrix notation for convenience in using them as elements of matrices. Since matrix algebra is not usually presented as something useful in electronics applications, we shall include in this chapter an elementary introduction to the subject.

An *operator* does something to an *operand*; for example, the multiplication $2 \times 3 = 6$ may be written

$$2 \circ 3 = 6$$

Here the operator 2 doubles the operand 3 to yield the result 6. We use "$\circ$" to stand for "operates on."

As another example consider the operation "sine phi." Recall from the definition of the sine function that if phi is the phase angle of a point on a circle, measured from the horizontal axis, sin ϕ is the ratio of the height of the point to the radius of the circle, as shown in Fig. 5–30.

EXAMPLE 5–33

$$\sin \circ 30° = 0.5$$

Enter 30° on a calculator and depress the SIN key to obtain 0.5. The height of the phase point is one-half the radius of the circle.

Some operations have corresponding inverse operations. If you pick up a book, you can lay it down. Inverse operators are usually indicated by a superscript "-1".

EXAMPLE 5–34

Verify $\sin^{-1} 0.5 = 30°$.

Solution Enter 0.5 on a calculator, depress the INV key and SIN^{-1} to obtain 30. A phase point with height-to-radius ratio 0.5 is 30° from the horizontal axis.

Notice that the inverse superscript is placed immediately after the operator. The expression $\sin^{-1}$ stands for an inverse operation, but $(\sin \phi)^{-1}$ means $1/\sin \phi$.

There are some exceptions to the "-1" convention. For example, the exponentiation operation 10^x means base 10 raised to the power x, but the inverse operation is called log or logarithm. The -1 convention is not used for the inverse of exponentiation, since the -1 would be confused with the power of a base.

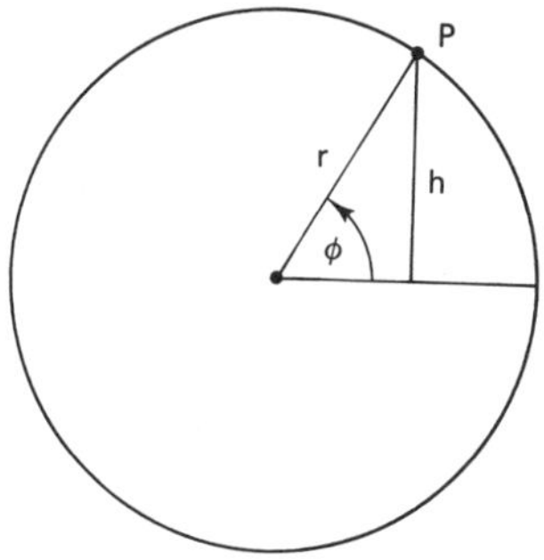

Figure 5–30 Definition of the sine function, $\sin \phi = h/r$.

EXAMPLE 5–35

Find x if $10^x = 10{,}000$.

Solution Enter 10,000 on a calculator and depress LOG to obtain 4. We have $\log 10^x = \log 10{,}000 = \log 10^4 = 4$.

In Example 5–35 the inverse operation returns the operand to its starting value: $\log \circ 10^x = x$; operation LOG is the inverse of exponentiation.

EXAMPLE 5–36

Use a calculator to verify that $10^{\log 3.2} = 3.2$.

Solution Enter 3.2 and depress LOG to obtain 0.505. Now depress keys INV and "10^x" to obtain 3.2.

A *matrix* is an operator that stretches and rotates an operand. Matrix algebra was used in the eighteenth and nineteenth centuries by physicists to apply Hooke's law to elastic materials in two or three dimensions. Recall that Hooke's law relates the tension in a spring to its stretch, $f = kx$ or $1/k \circ f = x$. If f is the applied force, then operator $1/k$ produces the stretch x. Here $1/k$ is the inverse of the spring constant k. For the electromagnetic ether the reciprocals of the spring constants are called *permittivity* and *permeability*.

There can be only stretch, no rotation, in the one-dimensional case. For a two-dimensional example, consider a window screen that has been cut on the bias (Fig. 5–31). Since the spring constant on the bias is weaker than the spring constant on the selvedge, the screen will stretch sideways. If the applied force is vertical, the stretch is rotated through angle phi. We need a two-dimensional operator representation of the spring constants which, operating on the force, will produce the angle and magnitude of the stretch. A matrix is such an operator.

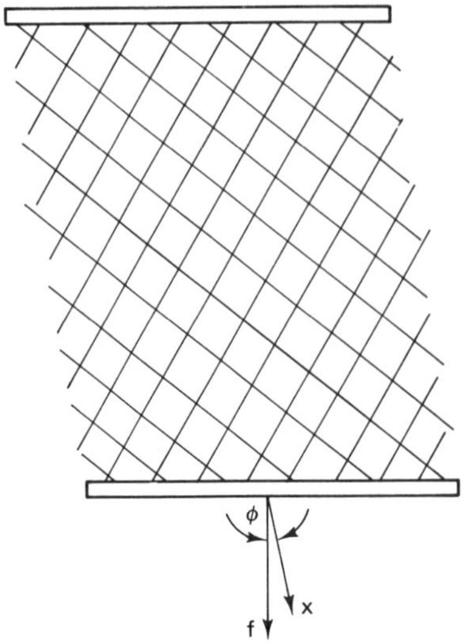

Figure 5–31 A window screen cut on the bias will not hang straight. f, Applied force x, stretch; ϕ, angle of rotation.

Our interest in matrix operators for electronics is now evident. The action of a transistor can be described by an operation in matrix algebra. A transistor, for example, amplifies, and changes the phase of an input: the output is a "stretched" and "rotated" input.

An operator is defined by what it does, not what it looks like. Unfortunately, a matrix is sometimes defined as a square or rectangular array of elements. Three examples of representations of a third-order matrix are shown in Fig. 5–32. Evidently, the definition of a matrix has nothing to do with its representation on a piece of paper. The only reason for our preference for the square array is that it provides easily remembered algorithms for carrying out the operations of matrix addition, subtraction, multiplication, and division. In the rectangular or square representation we can use subscripts to show the position of an element; for example, the element a_{23} is found in the second row, third column.

Matrix addition or subtraction is done by adding or subtracting element by element.

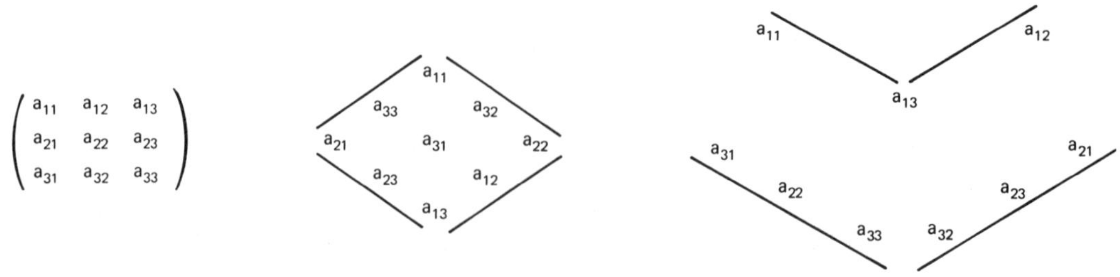

Figure 5–32 A matrix is defined by what it does, not what it looks like.

EXAMPLE 5–37

Add

$$\begin{pmatrix} 2 & 3 \\ 4 & 1 \end{pmatrix} \text{ and } \begin{pmatrix} 1 & -3 \\ 4 & 1 \end{pmatrix}.$$

Solution

$$\begin{pmatrix} 2 & 3 \\ 4 & 1 \end{pmatrix} + \begin{pmatrix} 1 & -3 \\ 4 & 1 \end{pmatrix} = \begin{pmatrix} 3 & 0 \\ 8 & 2 \end{pmatrix}$$

In matrix multiplication we multiply columns of the operand by rows of the operator, element by element. The first element of the product is the sum of the element-by-element products of the first column of the operand and the first row of the operator.

EXAMPLE 5–38

Multiply

$$\begin{pmatrix} 4 & -3 \\ -6 & 7 \end{pmatrix} \text{ by } \begin{pmatrix} 2 & 3 \\ 4 & 1 \end{pmatrix}.$$

Solution

$$\begin{pmatrix} 2 & 3 \\ 4 & 1 \end{pmatrix} \circ \begin{pmatrix} 4 & -3 \\ -6 & 7 \end{pmatrix} = \begin{pmatrix} p_{11} & p_{12} \\ p_{21} & p_{22} \end{pmatrix}$$

$$= \begin{pmatrix} 2 \times 4 + 3 \times (-6) & 2 \times (-3) + 3 \times 7 \\ 4 \times 4 + 1 \times (-6) & 4 \times (-3) + 1 \times 7 \end{pmatrix} = \begin{pmatrix} -10 & 15 \\ 10 & -5 \end{pmatrix}$$

In Example 5–38 the element p_{11} was obtained by multiplying the first row of the operator and the first column of the operand, element by element, and adding the products:

$$2 \;\; 3 \circ \begin{pmatrix} 4 \\ -6 \end{pmatrix} = 8 - 18 = -10$$

The next element, p_{12}, was found by multiplying the first row of the operator and the second column of the operand:

$$2 \;\; 3 \circ \begin{pmatrix} -3 \\ 7 \end{pmatrix} = -6 + 21 = 15$$

The element $p_{21} = 10$ was found by multiplying the second row of the operator and the first column of the operand; finally, p_{22} was found by multiplying the second row of the operator and the second column of the operand.

Matrix multiplication is not commutative, $A \circ B \neq B \circ A$.

EXAMPLE 5–39

Change $\mathbf{A} \circ \mathbf{B}$ in Example 5–38 to $\mathbf{B} \circ \mathbf{A}$ to show that the result is different.

Solution

$$\begin{pmatrix} 4 & -3 \\ -6 & 7 \end{pmatrix} \circ \begin{pmatrix} 2 & 3 \\ 4 & 1 \end{pmatrix} = \begin{pmatrix} -4 & 9 \\ 16 & -11 \end{pmatrix} \neq \begin{pmatrix} -10 & 15 \\ 10 & -5 \end{pmatrix}$$

Because matrix multiplication is not commutative we shall avoid error by speaking of multiplication from the left or from the right.

The *unit matrix* $\mathbf{I} = \begin{pmatrix} 1 & 0 \\ 0 & 1 \end{pmatrix}$ leaves the operand unchanged.

EXAMPLE 15–40

Multiply

$$\begin{pmatrix} 1 & 0 \\ 0 & 1 \end{pmatrix} \text{ and } \begin{pmatrix} 1 & 2 \\ -3 & 4 \end{pmatrix}.$$

Solution

$$\begin{pmatrix} 1 & 0 \\ 0 & 1 \end{pmatrix} \circ \begin{pmatrix} 1 & 2 \\ -3 & 4 \end{pmatrix} = \begin{pmatrix} 1 & 2 \\ -3 & 4 \end{pmatrix}$$

Since an operation followed by its inverse operation leaves the operand unchanged, we have $\mathbf{A}^{-1} \circ \mathbf{A} = \mathbf{I}$.

The multiplication of a scaler and a matrix is done element by element.

$$b \circ \begin{pmatrix} a_{11} & a_{12} \\ a_{21} & a_{22} \end{pmatrix} = \begin{pmatrix} ba_{11} & ba_{12} \\ ba_{21} & ba_{22} \end{pmatrix}$$

EXAMPLE 5–41

Multiply

$$\begin{pmatrix} 1 & 2 \\ 3 & 4 \end{pmatrix} \text{ by } 2.5.$$

Solution

$$2.5 \circ \begin{pmatrix} 1 & 2 \\ 3 & 4 \end{pmatrix} = \begin{pmatrix} 2.5 & 5 \\ 7.5 & 10 \end{pmatrix}$$

Multiplication of a matrix and a number set can be done by imagining the number set to have an extra column of zeros and carrying out the operation as for two matrices.

EXAMPLE 5–42

Multiply the number set

$$\begin{pmatrix}2\\1\end{pmatrix} \text{ by } \begin{pmatrix}1 & 2\\3 & 4\end{pmatrix}.$$

Solution

$$\begin{pmatrix}1 & 2\\3 & 4\end{pmatrix} \circ \begin{pmatrix}2\\1\end{pmatrix} = \begin{pmatrix}1 & 2\\3 & 4\end{pmatrix} \circ \begin{pmatrix}2 & 0\\1 & 0\end{pmatrix} = \begin{pmatrix}4\\10\end{pmatrix}$$

We divide matrices by using an inverse operation. If $\mathbf{A} \circ \mathbf{B} = \mathbf{C}$; we can find $\mathbf{B}$ by multiplying from the left by $\mathbf{A}^{-1}$:

$$\mathbf{A}^{-1} \circ \mathbf{A} \circ \mathbf{B} = \mathbf{A}^{-1}\mathbf{C},$$
$$\mathbf{B} = \mathbf{A}^{-1}\mathbf{C}$$

Before doing a numerical example we shall have to know how to find the inverse of a matrix. This is a tedious task, especially if the *order* is greater than 2, that is, if the array is greater than 2 × 2.

Fortunately, most minicomputers and some microcomputers have a software capability for matrix algebra, which reduces the task to simplicity. For a second-order matrix, however, it is easy to find the inverse of a matrix. The inverse of

$$\mathbf{A} = \begin{pmatrix}a_{11} & a_{12}\\a_{21} & a_{22}\end{pmatrix} \quad \text{is} \quad \mathbf{A}^{-1} = \frac{1}{|\mathbf{A}|}\begin{pmatrix}a_{22} & -a_{12}\\-a_{21} & a_{11}\end{pmatrix}$$

where $|\mathbf{A}|$ is the determinant of $\mathbf{A}$.

The determinant of a second-order matrix is the difference of the products of the diagonal elements:

$$|\mathbf{A}| = \begin{vmatrix}a_{11} & a_{12}\\a_{21} & a_{22}\end{vmatrix} = a_{11}a_{22} - a_{21}a_{12}$$

EXAMPLE 5–43

Use matrix algebra to solve the mesh problem shown in Fig. 5–33.

Solution We use Kirchhoff's voltage law to obtain the equations

$$10v - 2i_1 - 4i_1 + 4i_2 = 0$$
$$-5v - 4i_2 - 6i_2 + 4i_1 = 0$$

We collect terms and arrange the equations in an orderly manner:

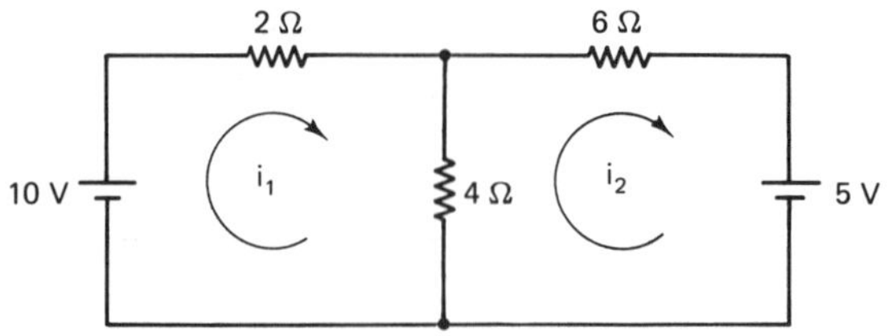

Figure 5–33 Mesh problem: given the source voltages, find the mesh currents i_1, i_2; Example 5–43.

$$\begin{aligned} -10 &= -6i_1 + 4i_2 \\ 5 &= 4i_1 - 10i_2 \end{aligned}$$

which can be written as the matrix equation

$$\begin{pmatrix} -6 & 4 \\ 4 & -10 \end{pmatrix} \circ \begin{pmatrix} i_1 \\ i_2 \end{pmatrix} = \begin{pmatrix} -10 \\ 5 \end{pmatrix}$$

Note that the order of the matrix is equal to the number of the independent variables.

We solve for the currents by multiplying from the left by the inverse matrix:

$$\begin{pmatrix} -6 & 4 \\ 4 & -10 \end{pmatrix}^{-1} \circ \begin{pmatrix} -6 & 4 \\ 4 & -10 \end{pmatrix} \circ \begin{pmatrix} i_1 \\ i_2 \end{pmatrix} = \begin{pmatrix} -6 & 4 \\ 4 & -10 \end{pmatrix}^{-1} \circ \begin{pmatrix} -10 \\ 5 \end{pmatrix}$$

The problem is reduced to finding the inverse of the matrix:

$$\begin{pmatrix} -6 & 4 \\ 4 & -10 \end{pmatrix}^{-1} = \frac{1}{D} \begin{pmatrix} -10 & -4 \\ -4 & -6 \end{pmatrix}$$

where the determinant D is $60 - 16 = 44$. The answer is

$$\begin{aligned} \begin{pmatrix} i_1 \\ i_2 \end{pmatrix} &= \frac{1}{44} \begin{pmatrix} -10 & -4 \\ -4 & -6 \end{pmatrix} \begin{pmatrix} -10 \\ 5 \end{pmatrix} \\ &= \frac{1}{44} \begin{pmatrix} 80 \\ 10 \end{pmatrix} = \begin{pmatrix} 1.82 \\ 0.23 \end{pmatrix} \end{aligned}$$

$i_1 = 1.82$ A $\quad i_2 = 0.23$ A

To show the ease of matrix calculations with the aid of a computer, we shall solve a network problem involving four mesh currents. The circuit is shown in Fig. 5–34. By applying Kirchhoff's voltage law to each mesh, we find

$$\begin{aligned} -5v - i_1 - 2i_1 - 0.5i_1 + 0.5i_2 &= 0 \\ 5v - 0.5i_2 - 1.5i_2 - 3i_2 - 2v + 0.5i_1 + 3i_4 &= 0 \\ 2v - 9v - i_3 - 2.5i_3 + 2.5i_4 &= 0 \\ 9v - 3i_4 - 2i_4 + 3i_2 - 2.5i_4 + 2.5i_3 &= 0 \end{aligned}$$

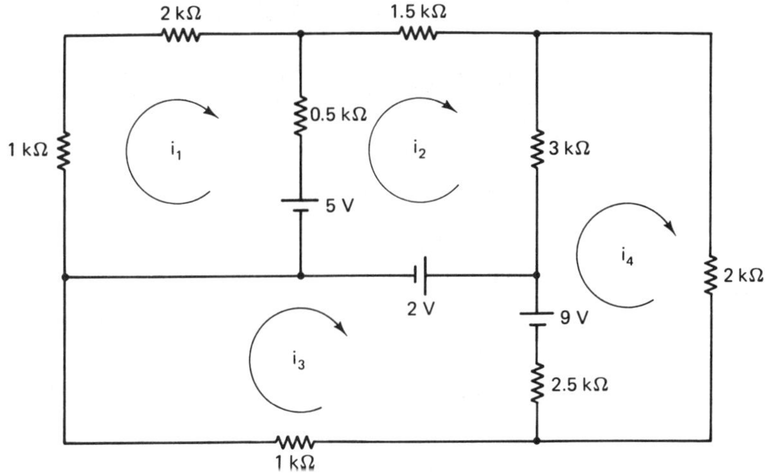

Figure 5–34 Problem with four mesh currents.

We arrange the equations in an orderly manner,

$$\begin{aligned} -3.5i_1 + 0.5i_2 \qquad\qquad\qquad &= 5 \\ 0.5i_1 - 5.0i_2 \qquad\quad + 3i_4 &= -3 \\ -3.5i_3 + 2.5i_4 &= 7 \\ 3i_2 + 2.5i_3 - 7.5i_4 &= -9 \end{aligned}$$

and write the result as a matrix equation,

$$\begin{pmatrix} -3.5 & 0.5 & 0 & 0 \\ 0.5 & -5.0 & 0 & 3 \\ 0 & 0 & -3.5 & 2.5 \\ 0 & 3 & 2.5 & -7.5 \end{pmatrix} \circ \begin{pmatrix} i_1 \\ i_2 \\ i_3 \\ i_4 \end{pmatrix} = \begin{pmatrix} 5 \\ -3 \\ 7 \\ -9 \end{pmatrix}$$

To solve for the currents we must find the inverse of the fourth-order matrix and multiply from the left to obtain $\mathbf{i} = \mathbf{z}^{-1}\mathbf{v}$.

A simple program in BASIC for the solution is presented below, and an explanation follows the program.

```
]LIST

200  REM  4X4 MESH PROBLEM
230  DIM A(4,4),B(4,4),V(4,1),I(4,1)
240  FOR J = 1 TO 4: READ X:A(1,J) = X: NEXT
250  DATA  -3.5, 0.5, 0, 0
260  FOR J = 1 TO 4: READ X:A(2,J) = X: NEXT
270  DATA  .5, -5, 0, 3
280  FOR J = 1 TO 4: READ X:A(3,J) = X: NEXT
290  DATA  0, 0, -3.5, 2.5
```

```
300  FOR J = 1 TO 4: READ X:A(4,J) = X: NEXT
310  DATA  0, 3, 2.5, -7.5
320  FOR J = 1 TO 4: READ X:V(J,1) = X: NEXT
330  DATA  5, -3, 7, -9
340  & B = INV(A)
350  & I = B.V
355  PRINT ''THE CURRENTS IN MILLIAMPS:''
360  & I = PNT(16)
365  & A = PNT(8)
370  END

]
```

```
RUN
THE CURRENTS IN MILLIAMPS:

    -1.24174654
     1.30777423
    -1.00958467
     1.38658147

    -3.5       .5       0       0
      .5      -5        0       3
      0        0     -3.5     2.5
      0        3      2.5    -7.5

]PR#3
```

200 Title identifies the program

230 The DIMENSION command "DIM" reserves space in the computer memory for entry of the elements of matrices and number sets.

240–310 Enter matrix data row by row

320–330 Enter data for voltage number set

340 The computer will perform the tedious task of finding the inverse matrix

350 The computer now contains the answer to the problem

355–365 We would like it to tell us the answer, and printing out the original matrix helps us to check our work.

In our final examples of the use of matrices, we remind ourselves that a matrix operator produces a stretch and a rotation.

EXAMPLE 5–44

Perform the matrix operation

$$\begin{pmatrix} x_1 \\ x_2 \end{pmatrix} = \begin{pmatrix} 1 & 4 \\ 3 & 2 \end{pmatrix} \circ \begin{pmatrix} 1 \\ 2 \end{pmatrix}$$

and determine the stretch and angle of rotation.

Solution

$$\begin{pmatrix} x_1 \\ x_2 \end{pmatrix} = \begin{pmatrix} 1 + 8 \\ 3 + 4 \end{pmatrix} = \begin{pmatrix} 9 \\ 7 \end{pmatrix}$$

The magnitudes and phases of the original and final number sets can be found by transforming from rectangular to polar coordinates:

$$\begin{pmatrix} 1 \\ 2 \end{pmatrix} = 2.24\angle 63.4^\circ \qquad \begin{pmatrix} 9 \\ 7 \end{pmatrix} = 11.4\angle 37.9^\circ$$

The rotation is 37.9 − 63.4 = −25.5°, the stretch is 11.4 − 2.24 = 9.16, and the stretch factor is 11.4/2.24 = 5.09 (Fig. 5–35).

EXAMPLE 5–45

Rotate the phasor $1 + j = \sqrt{2}\angle 45^\circ$ through an angle of 30° and stretch it by a factor of 3.

Solution The phasor $1 + j$ can be represented as a column number set, $\begin{pmatrix} 1 \\ 1 \end{pmatrix}$, and the required matrix operation is

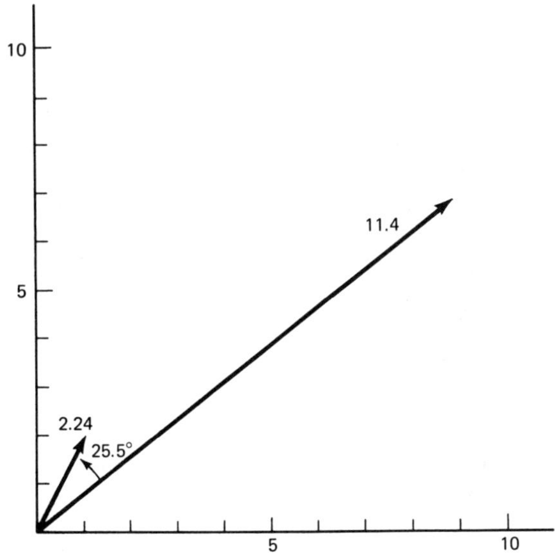

Figure 5–35 Rotation and stretch of $2.24\angle 63.4^\circ$ to $11.4\angle 37.9^\circ$.

$$\begin{pmatrix} 2.6 & -1.5 \\ 1.5 & 2.6 \end{pmatrix} \circ \begin{pmatrix} 1 \\ 1 \end{pmatrix} = \begin{pmatrix} 1.096 \\ 4.098 \end{pmatrix} = 1.096 + j4.098 \quad \text{or} \quad 4.24\underline{/75^\circ}$$

The rotation is from 45° to 75°, or 30°, and the stretch factor is $4.24/2^{1/2} = 3.0$ (Fig. 5–36).

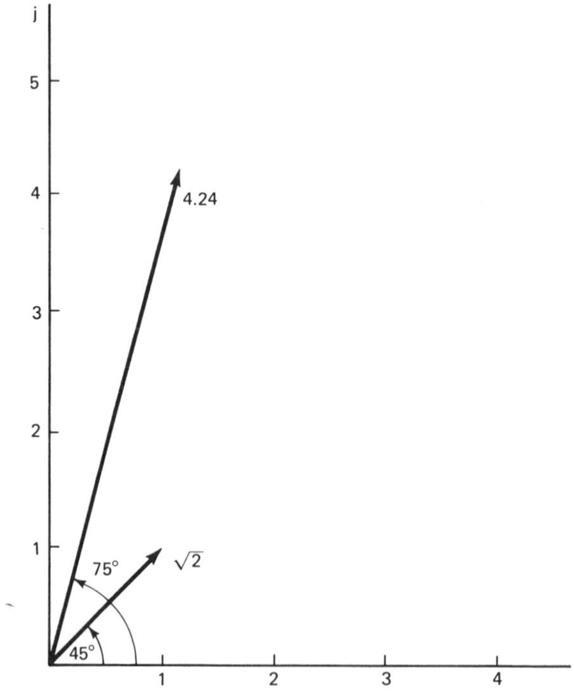

Figure 5–36 Rotation of a phasor $(2)^{1/2}\underline{/45^\circ}$ through an angle of 30° to $4.24\underline{/75^\circ}$, while stretching it by a factor of $3 = 4.24/(2)^{1/2}$.

The operator matrix in Example 5–45 was found in the following manner. The unimodular matrix

$$\mathbf{A} = \begin{pmatrix} \cos\phi & -\sin\phi \\ \sin\phi & \cos\phi \end{pmatrix}$$

rotates an operand through angle ϕ but leaves the magnitude of the operand unchanged. It is called "unimodular" because its determinant is +1: $|\mathbf{A}| = \cos^2\phi + \sin^2\phi = 1$. It rotates without stretching. To stretch by a factor of 3, we multiplied the matrix by 3:

$$3 \circ \begin{pmatrix} \cos 30^\circ & -\sin 30^\circ \\ \sin 30^\circ & \cos 30^\circ \end{pmatrix} = \begin{pmatrix} 2.6 & -1.5 \\ 1.5 & 2.6 \end{pmatrix}$$

Two matrices are equal if they are equal element by element:

$$\begin{pmatrix} a_{11} & a_{12} \\ a_{21} & a_{22} \end{pmatrix} = \begin{pmatrix} b_{11} & b_{12} \\ b_{21} & b_{22} \end{pmatrix}$$

if

$$\begin{aligned} a_{11} &= b_{11} \\ a_{12} &= b_{12} \\ a_{21} &= b_{21} \\ a_{22} &= b_{22} \end{aligned}$$

EXAMPLE 5–46

The S-parameters for a Hewlett-Packard HXTR-5104 transistor at an operating frequency of 3 GHz are

$$\begin{aligned} S_{11} &= 0.45\underline{/153^\circ} \\ S_{21} &= 1.78\underline{/16^\circ} \\ S_{12} &= 0.11\underline{/16^\circ} \\ S_{22} &= 0.59\underline{/-86^\circ} \end{aligned}$$

If the forward and reverse incident amplitudes are $a_1 = 20$ mV and $a_2 = 10$ mV, what is the voltage incident on a 50-Ω load?

Solution The matrix equation is

$$\begin{pmatrix} b_1 \\ b_2 \end{pmatrix} = \begin{pmatrix} 0.45\underline{/153^\circ} & 1.78\underline{/16^\circ} \\ 0.11\underline{/16^\circ} & 0.59\underline{/-86^\circ} \end{pmatrix} \begin{pmatrix} 20 \\ 10 \end{pmatrix} \times 10^{-3}$$

$$= \begin{pmatrix} 9.0\underline{/153^\circ} + 17.8\underline{/16^\circ} \\ 2.2\underline{/16^\circ} + 5.9\underline{/-86^\circ} \end{pmatrix} \times 10^{-3}$$

$$= \begin{pmatrix} -8.02 + j4.09 + 17.11 + j4.91 \\ 2.11 + j0.606 + 0.412 - j5.09 \end{pmatrix} \times 10^{-3}$$

$$= \begin{pmatrix} 9.09 + j9.00 \\ 2.52 - j4.48 \end{pmatrix} \times 10^{-3}$$

$$= \begin{pmatrix} 12.8\underline{/44.7^\circ} \\ 5.14\underline{/-60.6^\circ} \end{pmatrix} \times 10^{-3}$$

and the voltage incident on the load is

$$b_1 = 12.8\underline{/44.7^\circ} \text{ mV}.$$

APPLICATION 5–1: Lumped-Parameter Transmission Line

A lumped-parameter transmission line is a low-pass filter. We shall use it to remove the high-frequency Fourier harmonics of a train of rectangular pulses. In this way we shall be able to visualize the formation of a waveform by the repeated summation of sinusoidal harmonics. We shall also measure the phase delay of waves propagating on the line, the cutoff frequency, and the attenuation beyond the cutoff frequency.

The delay line described here has a low cutoff frequency in the audio range. This was chosen because of a lack of high-frequency instrumentation. If UHF or microwave equipment is available, it would be interesting to construct a distributed-parameter delay line using multiple stubs as lattice parameters.

Equipment and Materials

dual-trace oscilloscope with camera
2-Hz to 2-MHz function generator
assorted capacitors, resistors, and inductance coils
soldering iron

Procedure

Build a delay line of 5 to 10 lattice cells; for example, use five 10-mH coils and five 0.022-μF capacitors. Choose a convenient cell length between 10 and 20 cm. An example of a five-cell line with a lattice constant of 10 cm is shown in Fig. 5–37. Terminate the line in a resistance equal to the characteristic impedance of the delay line, and connect the input to the signal generator. A series resistance of about 40 kΩ or a directional coupler should be inserted between the generator and delay line when making phase measurements.

Use a dual-channel oscilloscope to determine the cutoff frequency by comparing voltage measurements at points *A* and *B* (Fig. 5–37). Plot the measurements on semilog graph paper, as shown in Fig. 5–8. The cutoff frequency is the frequency at which the voltage drops to $2^{-1/2} = 0.707$ of its average value. Amplitude measure-

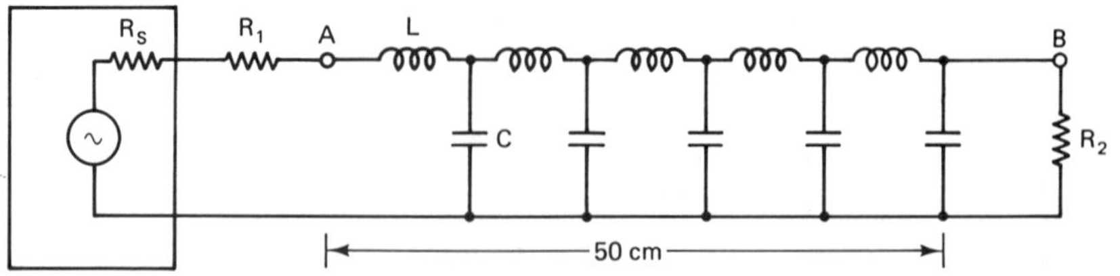

Figure 5–37 Example of a five-cell lumped-parameter transmission line, or delay line.

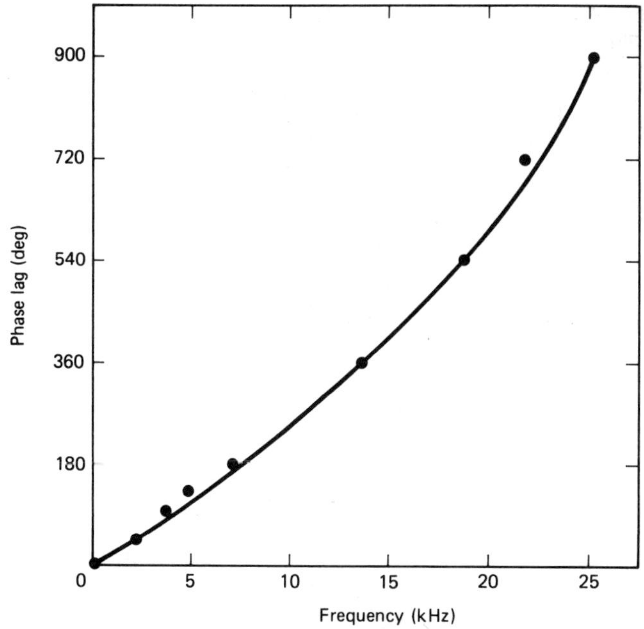

Figure 5–38 Measured phase delay as a function of frequency for the delay line described in the text.

ments are made by comparing the voltages on the two channels, phase measurements by comparing the time displacements. In your report compare the measured cutoff frequency to that calculated from Eq. (5–3).

Extrapolate the measurements beyond cutoff, as shown in Fig. 5–8, to determine the attenuation per octave change in frequency, and compare your result with that described in Section 5–5.

Measure the phase delay between points *A* and *B* by comparing the waveforms on channels A and B of the oscilloscope. Make a graph of phase delay versus frequency, as shown in Fig. 5–38.

Calculate the velocity of propagation from Eq. (5–1). Calculate the phase delay at any convenient frequency, well below cutoff, from 360° nd/v, where nd is the overall length of the line. Compare this result with what you would expect from the graph of phase delay versus frequency.

Start at the cutoff frequency and lower the operating frequency until the fundamental first completely appears on the oscilloscope. Record the frequency at which this occurs. Continue to lower the operating frequency until the third harmonic first completely appears. Record the frequency and take an oscilloscope photograph of the first and third harmonics. Compare this with what you would expect theoretically. Continue the procedure for several more harmonics and tabulate the results. Compare the measured frequencies with those calculated from

$$f_n = \frac{f_{n-1}}{n+2} \tag{5–64}$$

An example is shown in Table 5–2.

TABLE 5–2 COMPARISON OF MEASURED AND CALCULATED FREQUENCIES

Harmonic	Measured frequency (kHz)	Frequency calculated from Eq. (5–64) or (5–3) (kHz)
Cutoff	22.5	21.8
1	8.10	$21.8\left(\frac{1}{1+2}\right) = 7.27$
3	4.70	$7.27\left(\frac{3}{3+2}\right) = 4.36$
5	3.15	3.11
7	2.48	2.42
9	2.04	1.98
11	1.67	1.68
13	1.398	1.453
15	1.240	1.282
27	0.705	0.752
39	0.529	0.532

Discuss dispersion, as evidenced in a graph of phase delay versus frequency. If there is asymmetry in the pulse forms obtained in the oscilloscope photographs, attempt to account for it by dispersion.

SUMMARY

A lumped-parameter transmission line is markedly different from a distributed line; however, at frequencies well below resonance or cutoff its characteristics are the same. At frequencies below cutoff the lumped-parameter line can be used as a delay line or a compact substitute for a distributed-parameter line; also, it serves as a mathematical model for transmission-line theory. The cutoff frequency is

$$f_c = [\pi(LC)^{1/2}]^{-1}$$

and the corresponding equation for the cutoff wavelength

$$\lambda_c = 2d$$

is known as the Bragg condition.

For a lumped-parameter line the propagation velocity is given by the Baden-Powell equation,

$$v = v_0 \frac{\sin(\pi d/\lambda)}{\pi d/\lambda}$$

where $v_0 = d/(LC)^{1/2}$ is the velocity well below cutoff, $\lambda >> d$. The characteristic impedance is

$$Z_c = Z_0 \frac{\sin(\pi d/\lambda)}{\pi d/\lambda}$$

with $Z_0 = (L/C)^{1/2}$, $\lambda >> d$.

Since propagation velocity is a function of frequency, the various harmonics of a pulse or waveform propagate at different speeds, which results in dispersion. To understand dispersion quantitatively we introduce the beginnings of Fourier analysis, a mathematical technique in which a function or waveform is expressed as a series of sine or cosine terms. After a pulse is expressed as a sum of circular functions or "harmonics," the propagation time and attenuation of each harmonic can be computed. The modified harmonics are then summed or "reassembled" so that we can see what the pulse will look like after it has propagated through a dispersive, lossy medium or transmission line.

A short, lumped-parameter line is a form of low-pass filter. Above cutoff, each octave increase in frequency reduces the signal amplitude to $(\frac{1}{2})^n$ of its initial value, where n is the number of inductors and capacitors in the line.

With the addition of series and shunt resistances, we find that the lumped-parameter line is a mathematical model for lossy lines. The characteristic impedance of a lossy line is

$$Z_c = \left(\frac{r + jX_L}{g + jb_C}\right)^{1/2}$$

At the optical frequencies lossiness is expressed by Lambert's law,

$$\frac{I}{I_0} = \frac{P}{P_0} = \exp(-2\alpha x)$$

where 2α is the absorption coefficient and α is the attenuation constant or attenuation factor. The distinction between 2α and α is blurred in different fields of study, and one way to avoid mistakes is to distinguish between electrical and optical power:

$$1 \text{ dB (electrical)} = 2 \text{ dB (optical)}$$

By combining Lambert's law with our model of a lossy transmission line we obtain theoretical results that are widely applicable, from audio to optical frequencies, from coaxial to fiber optic cable, from lossy dielectrics to conducting media. The development of the theory requires the introduction of several new concepts and definitions:

Wave number, $n = \dfrac{1}{\lambda}$

Angular wave number, $k = \dfrac{1}{\lambda}$

Phase constant, $\beta = \frac{1}{\lambda}$

Propagation constant,

$$\gamma = (u + jv)^{1/2}$$
$$= \beta_0\left[\left(1 + \frac{g^2}{\omega^2C^2}\right)\left(1 + \frac{r^2}{\omega^2L^2}\right)\right]^{1/4}\left[-\sin\frac{\theta}{2} + j\cos\frac{\theta}{2}\right] = -\alpha + j\beta$$

The equation for the propagation constant can be used throughout a large frequency domain, leading to results such as the following for

Attenuation in coaxial cable at low frequencies.

Skin depth in conducting media,

$$\delta_s = \left(\frac{2}{\omega\mu g}\right)^{1/2}$$

Extinction coefficient,

$$K_0 = \frac{\lambda_0}{2}\left(\frac{\omega\mu g}{2}\right)^{1/2}$$

Absorption index,

$$K = K_o\epsilon_r^{-1/2}$$

Complex reflection coefficient,

$$r = \frac{\eta(1 - jK) - 1}{\eta(1 - jK) + 1}$$

Reflecting power,

$$\Gamma = \frac{(\eta - 1)^2 + K_0^2}{(\eta + 1)^2 + K_0^2}$$

Loss tangent of a dielectric medium,

$$\tan\delta = \frac{g}{\omega C}$$

Dielectric power factor,

$$\sin\delta$$

Dielectric quality,

$$Q_d = \cot \delta$$

We can use lumped-parameter components to construct a transmission line; the turnabout is that we can use short lengths of distributed-parameter lines as lumped-parameter components. With the aid of a Smith chart we can readily find the required length of an open- or short-circuited stub to provide a reactance at a given point on a transmission line.

When the input and output ports of a device are connected to transmission lines, the device is said to be embedded. For a passive device or medium the transmission coefficient is

$$t_{12} = \frac{2z_2}{z_2 + 1}$$

If a passive device or medium is of finite size or thickness, multiple reflections can occur between input and output ports. The coefficients for the resultant reflection and transmission are

$$R = r_{12} + \frac{t_{12}t_{21}r_{23}}{1 - r_{23}r_{21}}$$

$$T = \frac{t_{12}t_{23}}{1 - r_{23}r_{21}}$$

and the quality of the device is

$$Q = \text{Re}\,\frac{2\pi z_1 z_3}{z_2}\,\frac{1 + |r_{23}|^2}{z_1|t_{23}|^2 + z_3|t_{21}|^2|r_{23}|^2}.$$

The resultant forward and reverse reflection and transmission coefficients for an embedded device, active or passive, are called scattering parameters:

- S_{11} forward reflection coefficient
- S_{21} forward transfer ratio
- S_{12} reverse transfer ratio
- S_{22} reverse reflection coefficient

The S-parameters are elements of a matrix,

$$\begin{pmatrix} S_{11} & S_{12} \\ S_{21} & S_{22} \end{pmatrix}$$

Matrix multiplication can be used to find the effect of cascading two or more embedded devices. The S-parameters of microwave active devices often are provided in manufacturers' catalogs. If the S-parameters are known, we can calculate the transducer gain,

$$G = \frac{(1 - |r_s|^2)|S_{21}|^2(1 - |r_L|^2)}{|(1 - r_s S_{11})(1 - r_L S_{22}) - S_{21}S_{12}r_L r_s|^2}$$

A matrix can be defined as an operator that rotates and stretches; for example, a transistor output is an amplification (stretch) and a phase change (rotation) of the input. A review of elementary matrix operations is given in Section 5–15.

PROBLEMS

5–1. What is the cutoff frequency of an artificial delay line constructed from 200-μH inductors and 1000-pF capacitors spaced 5 cm apart?

5–2. What is the characteristic impedance, at frequencies well below cutoff, of the lumped-parameter transmission line described in Problem 5–1?

5–3. For the artificial delay line described in Problem 5–1, what is the wavelength on the line and in free space at the cutoff frequency?

5–4. For the line described in Problem 5–1, what is the refractive index at the cutoff frequency?

5–5. The shortest wavelength of x-rays that gives a maximum reflection from rock salt is about 11.2×10^{-8} cm. What is the lattice constant of rock salt?

5–6. If the delay line described in Problem 5–1 consists of five cells, find the approximate power attenuation at a frequency of $1.5f_c$, where f_c is the cutoff frequency. A graphical solution is acceptable.

5–7. Rectangular pulses at a frequency of 2 MHz are transmitted through a transmission line to a receiver. Approximately what bandwidth is required of the line and receiver if all harmonics up to the eleventh are to be detected at the output of the receiver?

5–8. A coaxial line has an attenuation constant of 0.015 Np/m. Find the power loss in decibels for a length of 20 m.

5–9. Find the square root of $3 - j5$. Give the answer in polar and rectangular form.

5–10. The conductivity of seawater is about 4 S/m. Calculate the skin depth in seawater at an operating frequency of 2 kHz. Do you think it would be practical to broadcast to a submarine that is 20 ft below the surface?

5–11. Estimate the attenuation factor an RG-8/U CATV cable will have at an operating frequency of 50 MHz. Assume the outer braid to be sandwiched in aluminum foil.

5–12. A short-circuited coaxial line, $Z_c = 50\ \Omega$, is to be used as a quarter-wave stub. Find the physical length of the stub if the operating frequency is 250 MHz and the dielectric constant of the insulation is 2.25.

5–13. A half-wavelength, 50-Ω coaxial line is used as a feedback loop from the output to the input of a transistor. The output impedance has been reduced to 25 Ω and the input impedance to the transistor is 10 Ω. Find the Q of the feedback network.

5–14. Find the S-parameters for $j15\ \Omega$ of inductance shunted across a 50-Ω transmission line.

5–15. The S-parameters for a transistor are

$$S_{11} = 0.98\angle 134^\circ$$
$$S_{21} = 1.97\angle -76^\circ$$
$$S_{12} = 0.061\angle 113^\circ$$
$$S_{22} = 1.10\angle -59^\circ$$

and the source and load reflection coefficients are $0.2\angle 30^\circ$ and $0.33\angle 165^\circ$. Find the transducer gain.

6

Waveguide and the Speed of Light

6–1 INTRODUCTION

Any type of transmission line or medium, even free space, can be considered to be a waveguide, but we shall generally use "waveguide" to refer to hollow, cylindrical, or rectangular pipe. Cylindrical guide has the disadvantage of allowing the plane of polarization to wander. Short sections of cylindrical guide are used where a mechanical rotating joint is required, as at the base of a scanning radar antenna. Waveguide, like coaxial line, has the advantage of being shielded from external interference; in addition, the power-handling capability of waveguide is greater than that of coax.

In Chapter 5 we used lumped parameters and circuit concepts to develop transmission-line theory. This approach would be awkward for waveguide; instead, we shall base waveguide theory on optics and wave concepts. For example, to find the characteristic impedance of rectangular guide we shall compare the speed of light in free space to its speed in guide, and then define characteristic impedance in terms of refractive index, as we did in Section 2–3.

6–2 WAVE PROPAGATION IN RECTANGULAR GUIDE

Propagation in rectangular guide occurs as a series of reflections as the waves bounce down the pipe from source to load. To analyze this mechanism we first shall study the breaking of water waves on a beach.

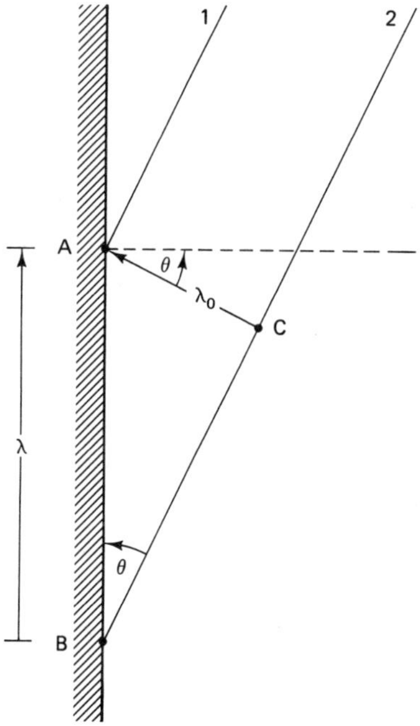

Figure 6–1 Plan view of waves breaking obliquely on a beach.

In Fig. 6–1 let the shaded line represent a beach and the oblique lines the crests of waves approaching the beach at an angle of incidence of θ. Points A and B show where the waves are breaking. As the second wave approaches the beach, point C moves to point A through a distance of one wavelength, λ_0. At the same time, point B also moves to point A, a distance of one wavelength, λ, as measured along the beach. If the wave period is T, the velocity along the beach, $v = \lambda/T$, is greater than the wave speed, $c = \lambda_0/T$, since $\lambda > \lambda_0$. For this reason a surfer, riding a wave crest obliquely and parallel to the beach, can attain a speed greater than the wave speed.

If the shaded line in Fig. 6–1 represents a sea wall rather than a beach, we can imagine obliquely reflected waves combining with the incident waves to form an interference pattern of "standing waves" that move parallel to the wall (Fig. 6–2). If we provide a second sea wall to form a channel, we can trap one or more columns of crests in the channel, and we have the water wave analogue of electromagnetic wave propagation in rectangular guide.

A principal difference between water waves and electromagnetic waves is the boundary condition at the walls of the channel. Water waves slosh upward to form a crest at a wall, but electric waves form nulls at the conducting walls of a guide, since voltage is zero along the surface of a conductor. For this reason the oblique lines in Fig. 6–3 represent electric field nulls, not crests, and are spaced a half-wavelength apart.

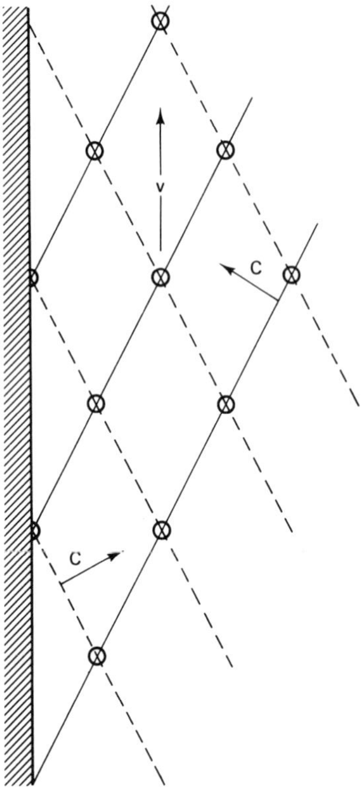

Figure 6–2 Interference pattern formed by water waves obliquely reflected from a sea wall.

From the triangles formed in Fig. 6–3 we see that

$$\sin\theta = \frac{\lambda_0}{\lambda} \tag{6–1}$$

and

$$\cos\theta = \frac{\lambda_0}{2w} \tag{6–2}$$

$$= \frac{\lambda_0}{\lambda_c}$$

where

$$\lambda_c = 2w \tag{6–3}$$

is the *cutoff wavelength* and w is the width of the waveguide. When the angle of incidence is zero, $\lambda_0 = \lambda_c$, a transverse standing wave is formed and there is no progress down the guide.

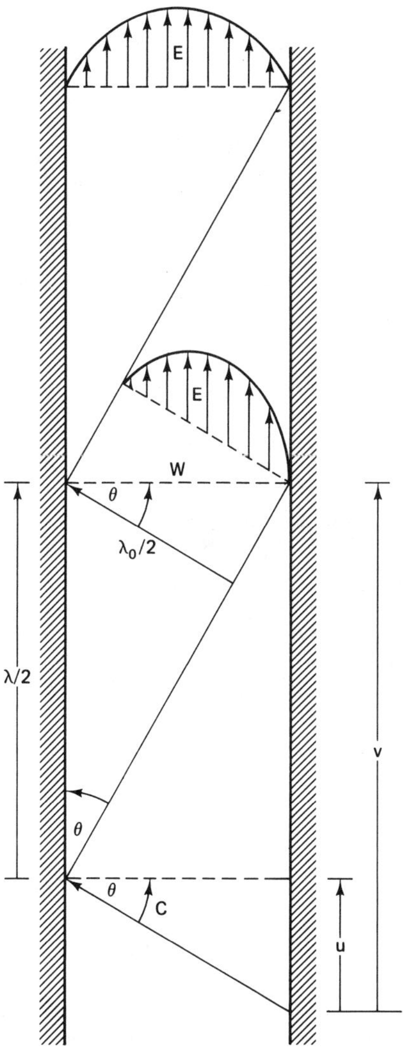

Figure 6–3 *H*-plane section in rectangular waveguide showing the geometry of the internal reflections. The *E*-plane—the plane containing the electric vector—is perpendicular to the plane of the figure.

We can find an expression for the guide wavelength in terms of the free-space wavelength by combining Eqs. (6–1) and (6–2).

$$\lambda = \frac{\lambda_0}{\sin \theta} = \frac{\lambda_0}{(1 - \cos^2 \theta)^{1/2}}$$

or

$$\lambda = \frac{\lambda_0}{[1 - (\lambda_0/\lambda_c)^2]^{1/2}} \tag{6–4}$$

and by dividing each side of Eq. (6–4) by the wave period we can relate velocity in the guide to the velocity in free space,

$$v = \frac{c}{\sin\theta} = \frac{c}{[1 - (\lambda_0/\lambda_c)^2]^{1/2}} \tag{6–5}$$

The refractive index of the waveguide is $\eta = c/v = \sin\theta$, and the characteristic impedance is $Z_c = Z_0/\eta$, or

$$Z_c = \frac{Z_0}{\sin\theta} = \frac{377}{[1 - (\lambda_0/\lambda_c)^2]^{1/2}} \tag{6–6}$$

EXAMPLE 6–1

Calculate the characteristic impedance and phase velocity in X-band waveguide when the operating frequency is 10 GHz. The width of X-band guide is 2.286 cm (Table 6–1).

Solution At an operating frequency of 10 GHz the free-space wavelength is $\lambda_0 = c/f = (3 \times 10^{10})/(10 \times 10^9) = 3$ cm, and the cutoff wavelength is $\lambda_c = 2w = 2 \times 2.286 = 4.572$ cm. From Eq. (6–6) we have

$$Z_c = \frac{377}{[1 - (3/4.572)^2]^{1/2}} = 500\ \Omega$$

and from Eq. (6–5) we find

$$v = \frac{3 \times 10^8}{[1 - (3/4.572)^2]^{1/2}} = 3.98 \times 10^8 \text{ m/s}$$

TABLE 6–1 RECTANGULAR WAVEGUIDE CHARACTERISTICS

Band	Frequency range (GHz)	Inside dimensions (cm)	Cutoff frequency (GHz)	Attenuation (dB/100 m)
L	1.12–1.70	16.51 × 8.255	0.908	1.353
S	2.60–3.95	7.21 × 3.40	2.078	2.546
G	3.95–5.85	4.755 × 2.215	3.152	5.849
C	4.90–7.05	4.039 × 2.019	3.71	10.15
X	8.20–12.40	2.286 × 1.016	6.56	27.45
P	12.4–18.00	1.580 × 0.790	9.49	40.92
K	18.0–26.5	1.067 × 0.432	14.1	87.51

6–3 WAVEGUIDE MODES AND WAVE PROPAGATION

If we refer again to Figs. 6–2 and 6–3 we see that with a wider channel we can include two columns of channeled waves, and the transverse electric field distribution would look like that shown in Fig. 6–4b. These modes resemble those we see on rope lines: the higher the frequency, the higher the order of the mode that can be confined in a limited space. The modes shown in Fig. 6–4 are called the TE_{10} and TE_{20}, where TE stands for *transverse electric*, and the first subscript designates the number of antinodes in the transverse electric field standing wave. The second subscript refers to the number of antinodes in the magnetic field distribution. The reflection of waves inside the guide induces currents in the walls of the guide, and the currents in turn produce magnetic fields with longitudinal as well as transverse components. Since only the electric field has no longitudinal components, these modes are called "transverse electric."

There are other possibilities. If the electric field has longitudinal components, due to the bunching of charge on the inside surfaces of the guide, the mode is TM or *transverse magnetic*. In Figs. 6–5 and 6–6 we show sketches of some of the simpler modes in rectangular guide. The TE_{10} mode is the most commonly used and often is called the *fundamental*. There is no TEM mode in waveguide. We are tempted to say that the TEM mode requires two conductors but are reminded that it is the TEM mode that propagates in empty space.

A qualitative study of mode distributions in waveguide provides an opportunity to gain an intuitive insight into the propagation of waves on transmission lines and in free space. Consider, for example, the field distributions shown in Fig. 6–5. The region with a maximum of electric field lines, marked A, can be regarded as a capacitor, or capacitance per unit length, in which electric charge has momentarily been displaced to the top and bottom walls of the guide. A quarter-cycle later—a quarter-wavelength farther down the guide, at point B—we see the capacitance has discharged into an inductor, or inductance per unit length. The magnetic field lines circulate around a region of maximum displacement current. The direction of the displacement current is perpendicular to the plane of the lower figure, and the direction of the magnetic field lines is given by the usual right-hand rule. The return circuit for the displacement current is via conduction currents in the walls of the guide.

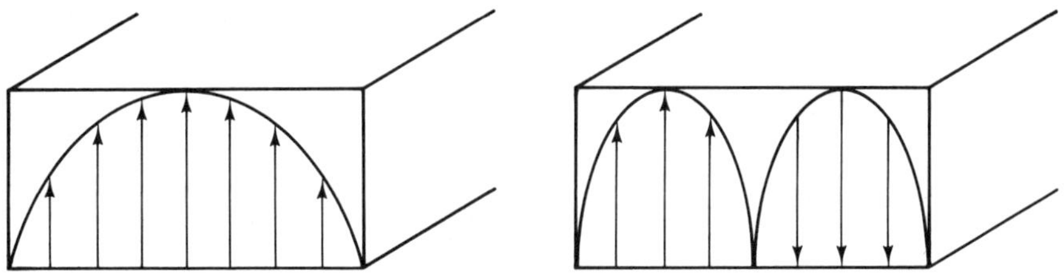

(a) (b)

Figure 6–4 Transverse electric field distributions in rectangular guide: (a) TE_{10} mode; (b) TE_{20} mode.

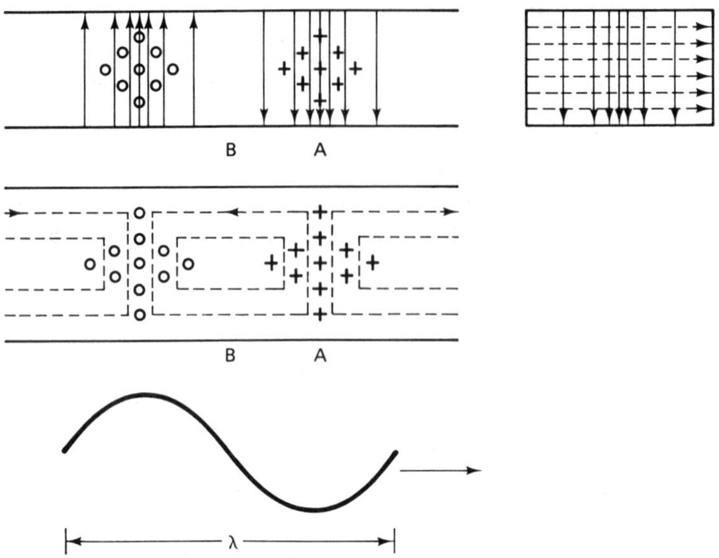

Figure 6–5 Side, end, and bottom views of the fundamental mode in rectangular guide. Solid lines, electric field; dashed lines, magnetic field.

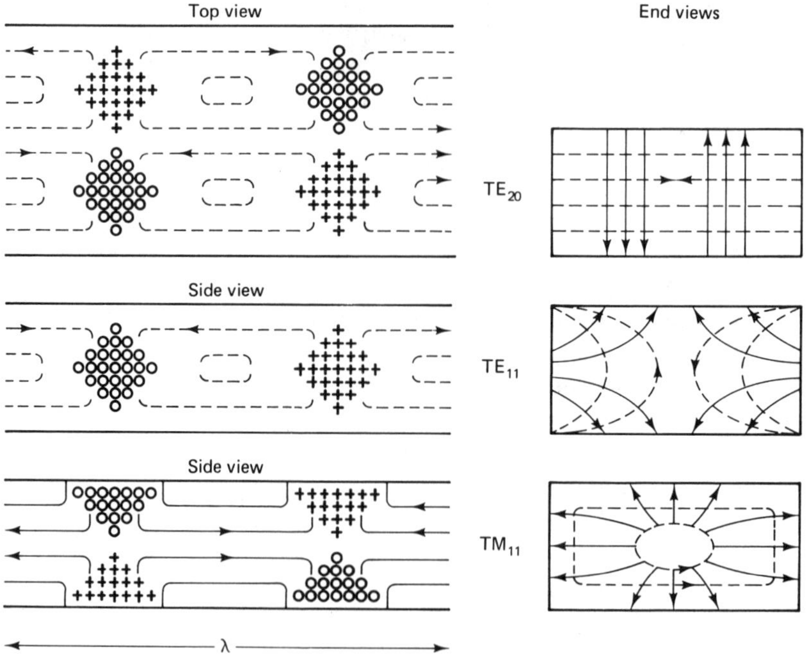

Figure 6–6 Electric and magnetic field configurations for higher-order modes in rectangular guide. Solid lines, electric field; dashed lines, magnetic field. Field lines out of the plane of the figure are shown as o's and those into the plane of the figure as +'s.

By "displacement current" we mean the ac current that appears to flow through empty space. Part of James Clerk Maxwell's contribution to our understanding of electromagnetic theory is his assumption that a displacement current exists in empty space and is proportional to the time rate of change of the electric displacement,

$$i_D = \frac{\Delta D}{\Delta t} \tag{6–7}$$

The electric displacement is defined by the equation

$$E = \frac{1}{\epsilon} D \tag{6–8}$$

where E is the electric field strength or electric force per unit charge, measured in volts per meter or newtons per coulomb, ϵ is the permittivity of the medium, and the displacement, D, is measured in coulombs per meter^2. The term "displacement" probably comes from the early history of electromagnetic theory, in which an attempt was made to imitate the theory of mechanics. It was known centuries ago that the stretch of a spring is proportional to the force applied to it:

$$f = kx \tag{6–9}$$

where f is the force, x is the stretch or *displacement*, and k, the spring constant, is a measure of the stiffness of the spring. In comparing Eqs. (6–8) and (6–9) we see that the reciprocal of the permittivity is the "spring constant" of the medium.

In the case of free space we may wonder what it is that is being stretched, where ϵ_0 is a measure of its stretchiness. In the nineteenth century it was believed that all space was permeated with an electromagnetic *ether* which was acted on by electromagnetic forces but not by mechanical forces. Early in the twentieth century the concept was abandoned by many philosophers when it was found to be impossible to measure the speed of the earth with respect to the ether.

EXAMPLE 6–2

A parallel-plate air capacitor with a plate area of 0.01 m^2 and a plate separation of 0.001 m is charged to 50 V in 0.1 ms. Find the electric displacement and the displacement current.

Solution The capacitance is

$$C = \frac{\epsilon_0 A}{d} = \frac{(8.85 \times 10^{-12}) \times 0.01}{0.001}$$

$$= 88.5 \text{ pF}$$

The charge is

$$Q = CV = (88.5 \times 10^{-12}) \times 50$$
$$= 4.43 \times 10^{-9} \text{ C}$$

and the displacement is

$$D = \frac{Q}{A} = \frac{4.43 \times 10^{-9}}{0.01}$$

$$= 4.43 \times 10^{-7} \mathrm{C/m^2}$$

From Eq. (6–7) we find the average displacement current,

$$i_D = \frac{\Delta D}{\Delta t} = \frac{4.43 \times 10^{-9}}{0.1 \times 10^{-3}}$$

$$= 0.0443 \ \mathrm{mA/m^2}$$

We shall find the displacement current concept useful in understanding the radiation of waves by a dipole antenna (Chapter 9).

The useful bandwidth of waveguide is limited on the low side by the cutoff frequency of the fundamental mode and on the high side by the frequency at which higher-order modes can appear. Near cutoff the angle of incidence is small and the number of reflections required for appreciable progress through the guide is excessive. To avoid excessive reflection losses and dispersion, the lower limit of the operating frequency is taken to be about 20% higher than the cutoff frequency. Normally, only the fundamental mode is used, and energy that appears in higher-order modes is wasted. Although higher-order modes are not used, we should be aware that irregularities, discontinuities, and other imperfections in the guide can generate higher-order modes and cause a loss of useful energy. The useful bandwidths and other characteristics of a few of the standard waveguide sizes are listed in Table 6–1.

6–4 MEASUREMENT TECHNIQUES WITH WAVEGUIDE

Waveguide, like coax, is shielded from the external world, and to find out what is happening inside we have to cut holes in the walls of the guide. *Couplers* are made by cutting holes in the common wall separating two waveguides. In this manner we can measure or analyze a small sample of energy in an auxiliary guide to monitor what is happening in the main guide. Another technique is the cutting of a narrow slot in the top of a waveguide to allow the insertion of a voltage probe. This allows a measurement of the electric field inside the guide without appreciably disturbing the original conditions.

A section of waveguide with a slot on the top side is called a *slotted line*. Slotted lines are available with a traveling carriage that allows changing the position of the probe along the length of the guide; in addition, provision is made to control the depth of the probe insertion into the guide. For accurate measurements the signal in the guide is modulated, the modulation frequency typically is 1 kHz, and the output from the probe is connected to a diode detector mounted in a tunable cavity. The output of the detector is connected to a high-gain tuned amplifier, tuned to 1 kHz, and the output is displayed on a meter calibrated in VSWR. The tuned amplifier and its meter often are simply called a *VSWR meter*.

By moving the probe along the length of the slot we can determine the maximum and minimum voltages in the guide and their locations. In a sense, we can ''see'' the standing waves inside the waveguide, much as we see the standing waves on a rope line. VSWR can be determined by calculating the ratio of maximum and minimum voltages; however, the manufacturers of VSWR meters usually provide a convenient gain control to allow the operator to adjust the maximum voltage to unity on the meter. Subsequent movement of the probe to a minimum allows the operator to read VSWR directly from the meter without further calculation. The traveling probe on a slotted line allows us to measure VSWR, wavelength in the guide, and phase change when the termination of the guide is replaced by a short circuit. These three measurements together allow us to determine the complex impedance of a load terminating the end of the guide.

We consider first the use of the slotted line to measure the wavelength in the guide. From this and the measurement of the operating frequency, we can determine the speed of light in free space.

6–5 THE SPEED OF LIGHT

The speed of light is an important physical constant. We use it to determine the refractive index of various media and the impedance of transmission lines. We use it to determine the rate at which information can be transferred from one place to another; we use it to determine the size of the universe. By measuring standing wavelength in a slotted line and by measuring the operating frequency, we can determine the speed of light to an accuracy that was undreamed of in former centuries.

An early attempt to measure the speed of light is attributed to Galileo. His idea was to station two men with lanterns on distant hilltops. The first man would cover his lantern, and the second would cover his own lantern when he had perceived that the first had been extinguished. Galileo, viewing the lanterns from a distance, would time the difference between extinctions. His conclusion was that light is instantaneous or very fast indeed.

The first successful measurement of the speed of light was accomplished through astronomical observations. After observing the periods of the eclipses of the moons of Jupiter, the astronomer Olaf Römer concluded that the longer periods resulted from the greater distance traveled by light waves when Earth was most distant from Jupiter. From Fig. 6–7 we see that the increase in distance is approximately the diameter of Earth's orbit. From his observations, and from the seventeenth-century knowledge of the astronomical constants, Römer determined a speed of light appreciably smaller than the value accepted today, but nonetheless a valuable contribution to knowledge at that time.

In the late nineteenth and early twentieth centuries the speed of light was

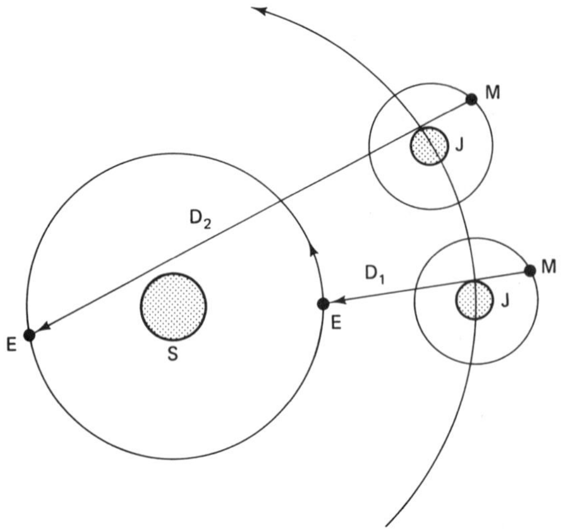

Figure 6–7 Astronomical determination of the speed of light: S, sun; E, Earth; J, Jupiter; M, a satellite of Jupiter. By observing the eclipse of Jupiter's satellite at different times of the year, a time delay is observed corresponding to the extra distance, $D_2 - D_1$, traveled by the light. The speed of light can be estimated from a knowledge of the diameter of Earth's orbit and the observed time delay.

determined with greater accuracy by observers who used various terrestrial methods. One of these was done by Heinrich Hertz, using microwaves. Hertz generated the first microwaves in the last decade of the nineteenth century by discharging a capacitor through a spark gap. The reflections of the waves from metal surfaces form standing waves. By measuring frequency and standing wavelength, Hertz was able to calculate the speed of propagation. Hertz's results were of special significance in that they confirmed the identity of electromagnetic waves and light. We can repeat Hertz's results by measuring standing wavelength in a slotted line.

EXAMPLE 6–3

A slotted line and probe are used to measure a distance of 4.61 cm between consecutive nulls in the standing-wave pattern of a short-circuited waveguide. The guide width is 7.21 cm and the operating frequency is a measured 3.90 GHz. Calculate the speed of light in free space.

Solution Since the distance between nulls is half a wavelength, the wavelength in the guide is $\lambda = 2 \times 4.61 = 9.22$ cm. The cutoff wavelength is $\lambda_c = 2w = 2 \times 7.21 = 14.42$ cm.

With this information we can find the free-space wavelength by solving Eq. (6–4) for λ_0: $\lambda_0^2 = \lambda^2[1 - (\lambda_0/\lambda_c)^2]$, $\lambda_0^2[1 + (\lambda/\lambda_c)^2] = \lambda^2$, or

$$\lambda_0 = \frac{\lambda}{[1 + (\lambda/\lambda_c)^2]^{1/2}} \qquad (6\text{–}10)$$

$$= \frac{9.22}{[1 + (9.22/14.42)^2]^{1/2}}$$

$$= 7.77 \text{ cm}$$

With the free-space wavelength and the frequency we now determine the speed of light in free space as

$$c = \lambda_0 f = 7.77 \times (3.90 \times 10^9)$$
$$= 3.03 \times 10^{10} \text{ cm/s}$$

a result about 1% greater than the presently accepted value.

6–6 IMPEDANCE MEASUREMENTS

We can also use a slotted line to measure the impedance of a load at the end of a waveguide. The VSWR in the guide can be measured with the traveling probe and VSWR meter, and the result can be plotted on a Smith chart as a VSWR circle. This gives the magnitude of the normalized load impedance, and we now need only to determine the phase angle. We can measure the phase angle by positioning the probe at a null in the standing-wave pattern. We then replace the load by a short circuit and measure the distance to the nearest null in the new standing-wave pattern. In this way we refer the phase of the wave reflected from the load impedance to the phase of a wave reflected from a short circuit, which is 180° and shown on the extreme left side of a Smith chart.

The circuitry required for the impedance measurements is shown in Fig. 6–8, and the step-by-step procedure is explained in the following example.

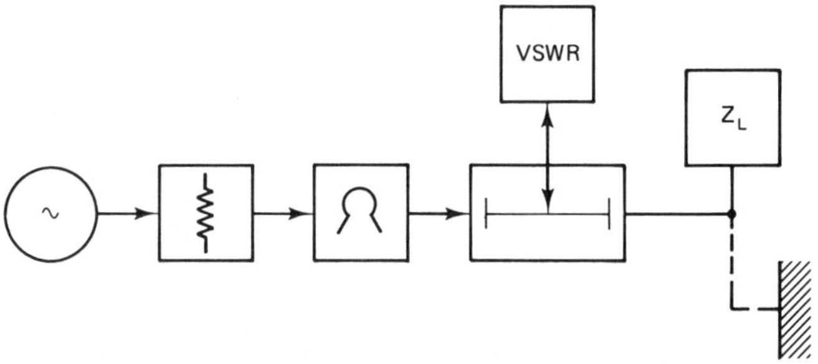

Figure 6–8 Circuit for the measurement of complex load impedance. From left to right: signal generator, pad, cavity frequency meter, slotted line and VSWR meter, load or short.

EXAMPLE 6–4

The following measurements are made in S-band guide (w = 7.21 cm) at an operating frequency of 3.90 GHz:

The VSWR is 3.20.

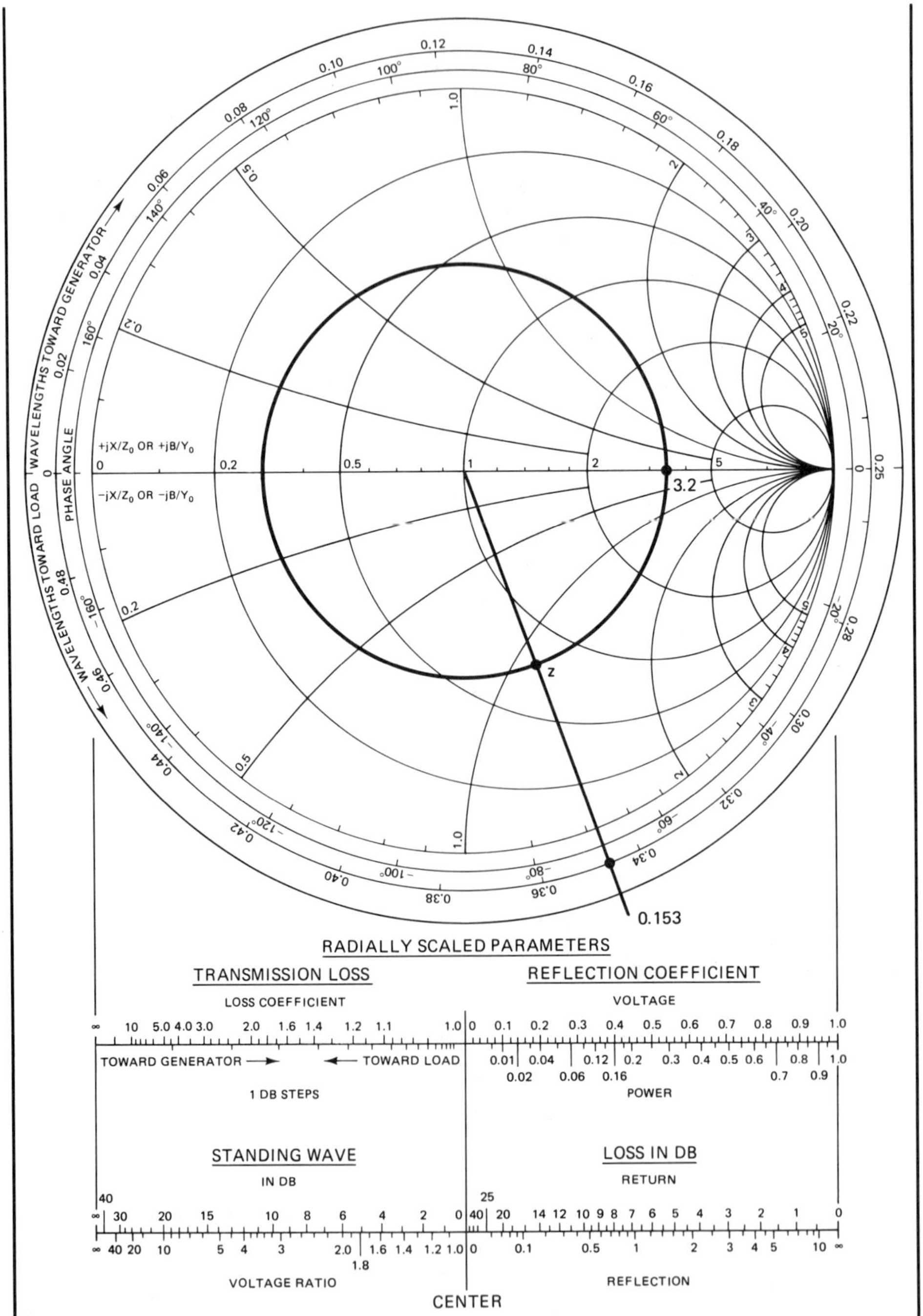

Figure 6–9 Use of the Smith chart for determining the load impedance from slotted-line measurements.

The distance between consecutive nulls in the standing-wave pattern is 4.61 cm.

When the load is replaced by a short circuit the null is shifted 1.41 cm *toward the load.*

Find the complex impedance of the load.

Solution We start by setting a pencil compass at 3.2 on the horizontal axis of a Smith chart and draw a VSWR circle (Fig. 6–9).

The wavelength in the guide is twice the distance between nulls in the standing-wave pattern, $2 \times 4.61 = 9.22$ cm. The shift in null position, when the load is replaced by a short, is 1.41 cm, or $1.41/9.22 = 0.153$ wavelength, toward the load.

We start at the left side of the Smith chart and use the circular scale labeled "wavelengths toward the load" to measure a distance of 0.153, as shown in the figure. We draw a radius from the center of the chart to the point 0.153; the intersection of the radius and the VSWR circle determines the normalized load impedance, $z = 0.8 - j1.05$.

To find the unnormalized load impedance, we first have to calculate the characteristic impedance of the waveguide at the operating frequency. The operating wavelength in free space is $\lambda_0 = c/f = 3 \times 10^{10}/3.9 \times 10^9 = 7.69$ cm. From Eq. (6–6) we have

$$Z_c = \frac{377}{[1 - (7.69/14.42)^2]^{1/2}}$$
$$= 446\ \Omega$$

The absolute load impedance is

$$Z_L = zZ_c = (0.8 - j1.05)(446)$$
$$= 357 - j468\ \Omega$$

Alternative procedure The characteristic impedance of the guide can be determined directly from the standing wavelength measurement, rather than from an independent frequency measurement.

From Eq. (6–10) we have

$$\eta = \frac{\lambda_0}{\lambda} = \frac{1}{[1 + (\lambda/\lambda_c)^2]^{1/2}}$$

and since $Z_c = Z_0/\eta$, the characteristic impedance of the guide is

$$Z_c = Z_0\left[1 + \left(\frac{\lambda}{\lambda_c}\right)^2\right]^{1/2} \qquad (6\text{–}11)$$

Since Eq. (6–11) expresses the characteristic impedance in terms of the guide wavelength, we can use the wavelength measurements to find Z_c directly:

$$Z_c = 377\left[1 + \left(\frac{9.22}{2 \times 7.21}\right)^2\right]^{1/2}$$
$$= 447\ \Omega$$

In Example 6–4 the null shift is toward the load, and the load impedance is capacitive. If the shift to the nearest null were toward the generator, the load impedance would have been inductive. A common mistake is to forget to record the direction of the null shift.

6–7 SWEEP-FREQUENCY TECHNIQUES

If a portion of the wave energy in a transmission line is transferred to an auxiliary line, the arrangement is called a coupler. An example of a multihole coupler in waveguide is sketched in Fig. 6–10. The coupler transfers waves from the main line in either direction, but it can be made into a *directional coupler* by terminating one end of the auxiliary line with a load matched to the characteristic impedance of the auxiliary line. In the example shown in Fig. 6–10, backward-moving wave energy is coupled into the auxiliary termination and is dissipated without reflection; the wave energy delivered to the exit port of the coupler is proportional to the forward-moving wave energy in the main line. Multihole directional couplers are manufactured to provide a given ratio of sample to main-line amplitude or power over a wide range of frequencies.

As an example of the use of directional couplers for broadband measurements,

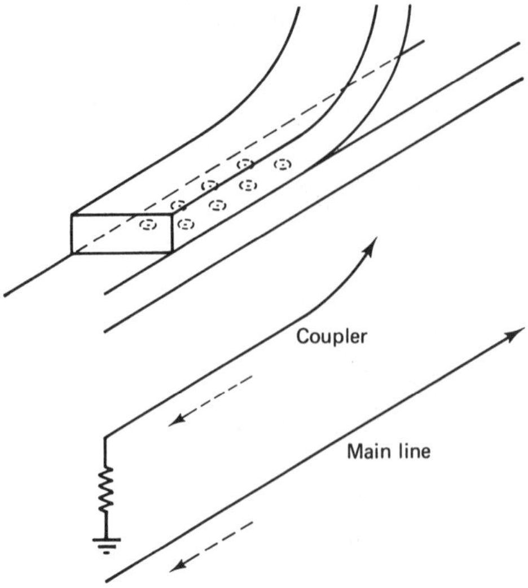

Figure 6–10 Sketch and schematic representation of a multihole directional coupler.

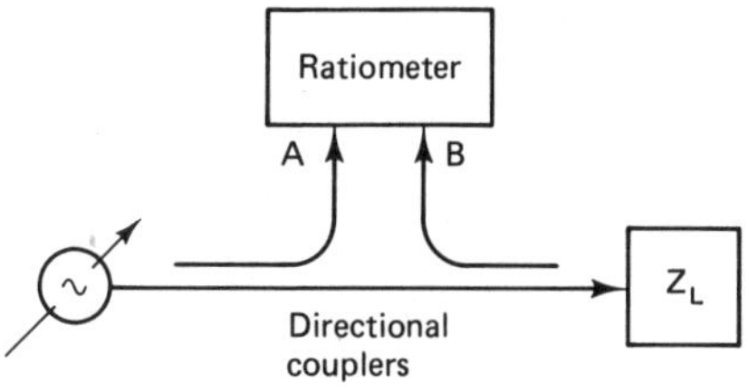

Figure 6–11 Use of broadband, back-to-back directional couplers to measure the magnitude of the load reflection coefficient across a swept-frequency band.

we shall consider combining two directional couplers in order to measure the magnitude of the reflection coefficient of a load. In Fig. 6–11 we show two identical directional couplers placed back to back in a main line between a sweep-frequency generator and a load under test. The first coupler delivers a sample at port *A* that is proportional to the amplitude of the wave incident on the load and the second coupler a sample proportional to the amplitude of the wave reflected from the load. The ratio, by definition, is the magnitude of the reflection coefficient of the load.

The advantage of this method of measuring reflection coefficient is that it can be done quickly over a large bandwidth through the use of a sweep-frequency generator. Recorders are available, *ratiometers*, that output the ratio on a CRT as the generator sweeps through a given frequency band. One measurement, and data are recorded that would take many hours invested in ''point-by-point'' measurements.

There are many other types of equipment that permit rapid, broadband measurements. Such equipment is a necessity in industry, where the saving of labor hours is an important economic factor. Unfortunately, such equipment is not generally available at small schools or to amateur experimenters; fortunately, the principles required for an understanding of microwave techniques can be learned with simpler instrumentation.

6–8 PHASE AND GROUP VELOCITY AND DISPERSION

If we follow the progress of a wave cresting obliquely on a beach, as suggested in Fig. 6–1, we are marking the progress of a particular phase of the wave, in this case, the crest. For this reason the velocity along the beach is called the *phase velocity*. Unlike the wave velocity, which is determined by the properties of the medium in which the wave propagates, the phase velocity can have arbitrarily large values.

If we follow the progress of a wave as it reflects from the walls of a guide, as indicated in Fig. 6–3, we see that the speed at which wave energy progresses down the guide must be less than the wave speed, since it follows a zigzag path. This is the *group velocity*, and from the triangle at the bottom of Fig. 6–3 we see that its relation to wave speed is

$$u = c \sin \theta \qquad (6\text{–}12)$$

If we multiply Eqs. (6–5) and (6–12) we obtain a simple relation between phase and group speeds in waveguide,

$$uv = c^2 \tag{6–13}$$

EXAMPLE 6–5

Find the group speed when the guide wavelength is 6.0 cm in X-band waveguide.

Solution From Table 6–1 we find the width of X-band guide is 2.286 cm, and from Eq. (6–3) we have

$$\lambda_c = 2 \times 2.286 = 4.572 \text{ cm}$$

If we divide each side of Eq. (6–10) by the wave period, we obtain a convenient relation between phase and wave speed:

$$v = c\left[1 + \left(\frac{\lambda}{\lambda_c}\right)^2\right]^{1/2} \tag{6–14}$$

$$= 3 \times 10^8\left[1 + \left(\frac{6.00}{4.572}\right)^2\right]^{1/2}$$

$$= 4.95 \times 10^8 \text{ m/s}$$

Now that we know the phase speed we can obtain the group speed from Eq. (6–13):

$$u = \frac{c^2}{v} = \frac{(3 \times 10^8)^2}{4.95 \times 10^8}$$

$$= 1.82 \times 10^8 \text{ m/s}$$

We see, from Eqs. (6–5) and (6–12), that phase and group speed depend on the angle of reflection. At a low frequency, the large wavelength "turns sideways" to fit into the guide, the angle of incidence is small, and the wave must undergo many reflections to progress down the guide. Evidently, the group speed is slower for lower frequencies, the phase speed higher, and this is the cause of *waveguide dispersion*. Dispersion occurs when phase speed changes with change in frequency. A pulse or *wave packet* can be Fourier analyzed as an assemblage of different wavelengths, limited in distance and time; for example, consider the Fourier components of a rectangular pulse discussed in Chapter 5. If such pulse propagates in waveguide, the higher-frequency components will arrive later than the lower, and the pulse will be spread out or *dispersed*.

Waveguide dispersion is distinguished from dispersion in bulk materials, the latter being called *material dispersion*. If the refractive index of a bulk material changes with frequency, the Fourier components of a wave packet will travel at different speeds and the packet will undergo material dispersion. In general, if the high-frequency components travel more slowly than the low, the dispersion is called *normal*. If the low-frequency components travel more slowly, the dispersion is *anomalous*. Evidently waveguide dispersion is normal.

Dispersion in bulk materials is caused by absorption: for example, the absorption caused by an atomic or molecular resonance. Usually, dispersion is greater the closer the operating frequency is to the absorption or attenuation region. We saw in Chapter 5, for example, that the lumped-parameter delay line showed increasing dispersion near the cutoff frequency.

If waveguide and material dispersion are both present, the combined effect is called *chromatic dispersion*. Chromatic dispersion occurs in fiber optic cables. Multiple reflections from the sides of the fiber produce waveguide dispersion, and material dispersion is caused by a resonance of the $Si0_2$ molecule at infrared frequencies. We shall discuss dispersion further in Chapter 10.

Until now we have not explained why the speed of energy transport is called "group velocity," nor have we related group velocity to dispersion in a quantitative manner. An intuitive insight into group velocity is provided in the observation of ocean waves. On the Pacific Ocean, waves typically propagate in groups of seven, as sketched in Fig. 6–12. Since the dispersion is anomalous, the individual wavelets travel more slowly than the group. A surfer riding wave *A* cannot derive a free ride across the ocean, because eventually wave *A* will become wave *B*, and the surfer finally will be dropped into still water at *C*. It is because of the groups of seven that sailors, when approaching a shoreline, will count the cadence of the waves and attempt to beach their craft at the optimum time.

For a quantitative relationship between dispersion and group velocity we have

$$u = v - \lambda \frac{\Delta v}{\Delta \lambda} \tag{6–15}$$

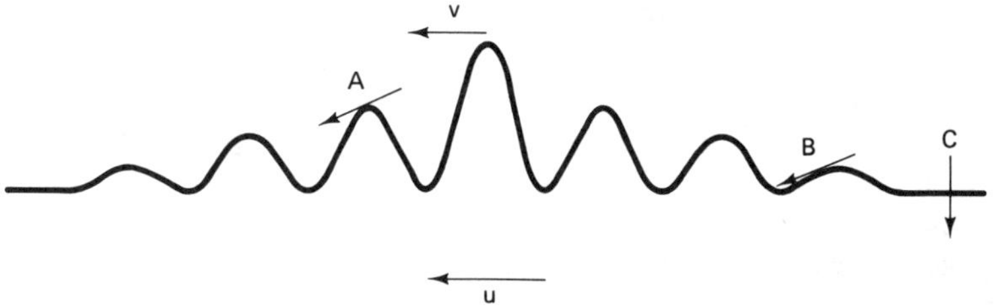

Figure 6–12 Why a surfer cannot ride across the ocean: the phase speed, *v*, is less than the group speed, *u*.

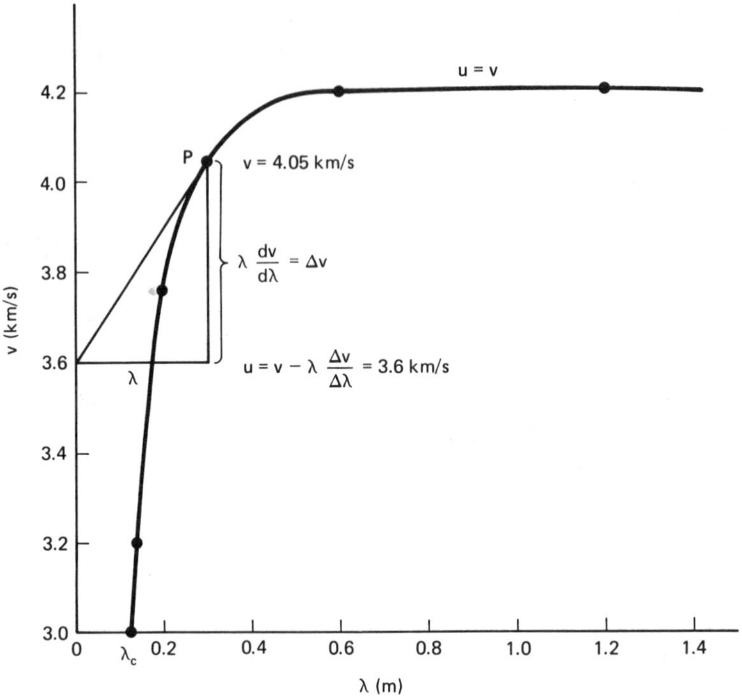

Figure 6–13 Graph of phase speed versus wavelength for a lumped-parameter transmission line, showing the geometrical construction by which group speed is derived from phase speed.

where Δv is the change in phase speed corresponding to a change in wavelength $\Delta\lambda$. As an example of the application of Eq. (6–15), we shall consider the artificial delay line studied in Chapter 5.

In Fig. 6–13 we show a graph of phase velocity, for various wavelengths, for a delay line constructed of 10-mH coils and 0.022-μF capacitors. The cutoff wavelength is $\lambda_c = 2d = 0.12$ m. As the frequency increases (shorter wavelengths) the velocity decreases, and the dispersion is normal.

The rate of change of phase speed with change in wavelength, $\Delta v/\Delta\lambda$, is the slope of the curve, and it is shown at point P as a tangent at that point. Since the slope is positive, we see from Eq. (6–15) that the group speed is less than the phase speed. This is characteristic of normal dispersion.

The right side of Eq. (6–15) can be constructed as the triangle shown in Fig. 6–13. From the graphic construction we find that if the phase speed is 4.05 km/s, the group speed is 3.6 km/s.

EXAMPLE 6–6

Determine, graphically, the group speed in S-band waveguide when the phase speed is 4.2×10^8 m/s.

Solution We start with Eq. (6–14) and tabulate phase velocities for various wavelengths; for example,

$$\frac{v}{c} = \left[1 + \left(\frac{\lambda}{\lambda_c}\right)^2\right]^{1/2}$$
$$= [1 + (0.2)^2]^{1/2}$$
$$= 1.02$$

where, since $\lambda/\lambda_c = 0.2$, $\lambda = 0.2\lambda_c = 0.2 \times 2 \times 7.21$ cm $= 0.0284$ m. We now find $v = 1.02 \times (3 \times 10^8) = 3.06 \times 10^8$ m/s. These results are entered in the first row of Table 6–2. A graph of the entries in Table 6–2 is shown in Fig. 6–14.

TABLE 6–2 VELOCITY FOR VARIOUS WAVELENGTHS

λ/λ_c	v/c	λ(m)	$v(10^8$ m/s)
0.2	1.02	0.0284	3.06
0.4	1.08	0.0568	3.24
0.6	1.17	0.0853	3.51
0.8	1.28	0.114	3.84
1.0	1.41	0.142	4.23
1.2	1.56	0.171	4.68
1.4	1.72	0.199	5.16
1.6	1.88	0.227	5.64
1.8	2.06	0.256	6.18
2.0	2.23	0.284	6.69

We draw a tangent to the curve at a point corresponding to a phase speed of 4.2×10^8 m/s, and at the intersection of the tangent line and the $\lambda = 0$ axis we determine a group speed of 2.12×10^8 m/s.

Check: We can verify our result with Eq. (6–13):

$$(uv)^{1/2} = c$$
$$[(2.12 \times 10^8) \times (4.2 \times 10^8)]^{1/2} = 3 \times 10^8$$
$$2.98 \times 10^8 = 3 \times 10^8$$

The relationship between absorption and dispersion will be discussed further in Chapters 9 and 10.

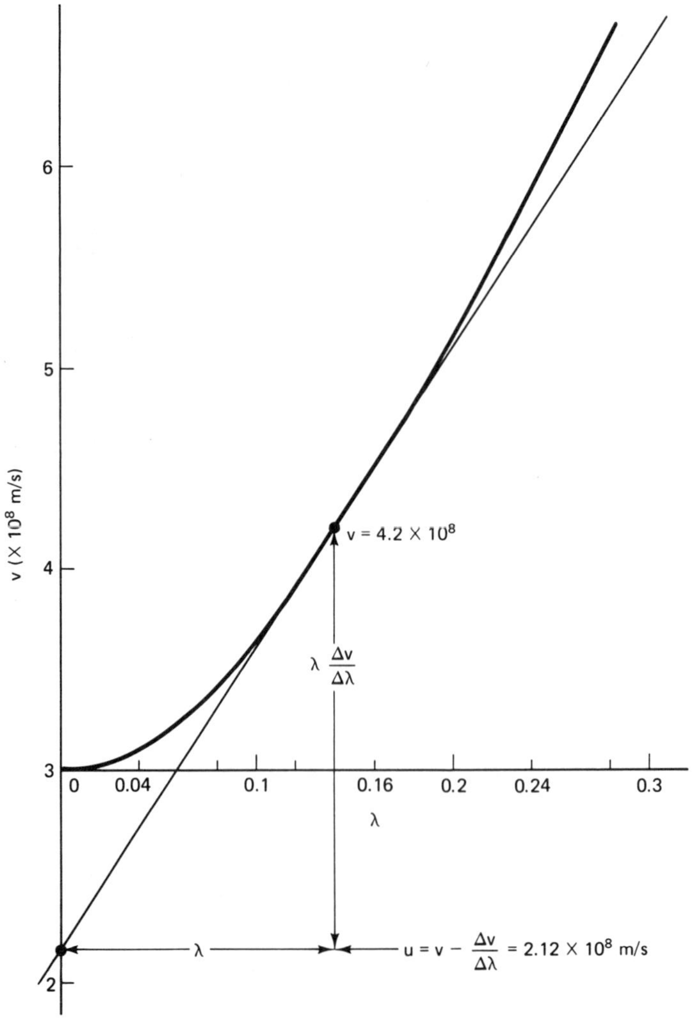

Figure 6–14 Graph of phase speed versus wavelength in rectangular waveguide, showing the geometrical construction by which group speed is derived from phase speed.

6–9 WAVEGUIDE HORN ANTENNA

The waveguide horn antenna is easy to understand because the complicated process of radiation is assumed to have already occurred somewhere back in the waveguide. The electrical design of a horn is simply the choice of a flare angle such that an emerging ray is in the forward direction (Fig. 6–15).

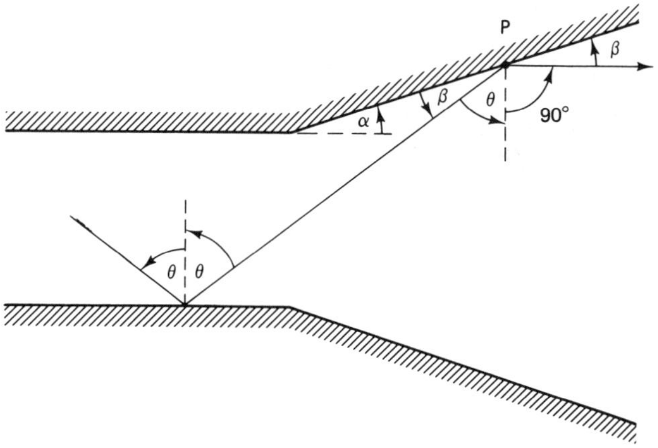

Figure 6–15 Wave reflections in an optimum sectoral horn. θ, Angles of incidence and reflection in the guide; β, grazing angle of incidence and reflection in the horn; α, flare angle.

From the geometry shown in the figure we see that

$$\alpha = \beta \tag{6–16}$$

and at the point P the sum of the angles around a straight line is 180°,

$$2\beta + \theta + 90° = 180°$$

or

$$\beta = ½(90° - \theta) \tag{6–17}$$

By combining Eqs. (6–17), (6–16), and (6–2) we obtain the design equation for a waveguide sectoral horn,

$$\alpha = ½\left(90° - \cos^{-1}\frac{\lambda_0}{\lambda_c}\right) \tag{6–18}$$

EXAMPLE 6–7

Design a waveguide sectoral horn for operation at 3.5 GHz.

Solution Since the operating frequency is within the range of S-band guide (Table 6–1) we have $\lambda_c = 2w = 14.42$ cm. The free-space operating wavelength is $\lambda_0 = c/f = 3 \times 10^{10}/3.5 \times 10^9 = 8.57$ cm, and the ratio of the free-space and cutoff wavelengths is $\lambda_0/\lambda_c = 8.57/14.42 = 0.594$. Now that we have this ratio we can use Eq. (6–18) to find the flare angle:

$$\alpha = ½(90° - \cos^{-1} 0.594)$$
$$= 18.2°$$

Figure 6–15 is drawn to scale for this design.

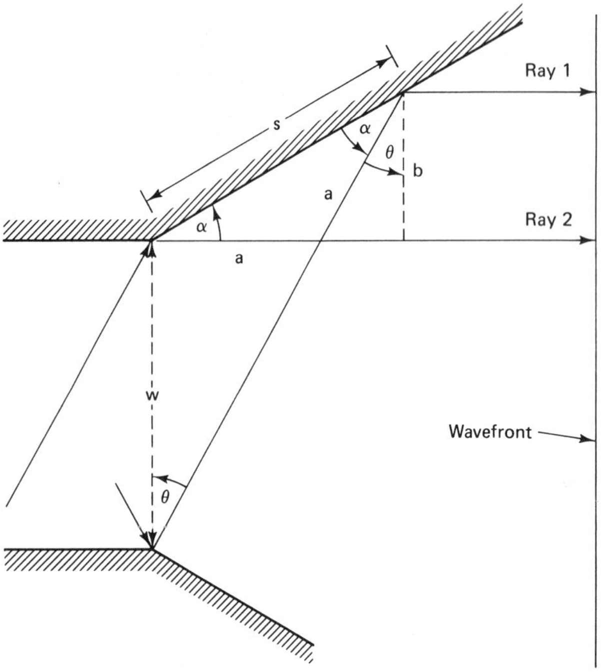

Figure 6–16 Section through a sectoral horn, showing the first and last reflected rays, which determine the minimum length of the sides of the horn.

The design in Example 6–7 is not complete in that we have not decided on the length of the sides of the horn. In Fig. 6–16 we show a sketch of the extreme rays—the first and last rays to be reflected from the surface of the horn. The length of the sides certainly should be long enough to reflect these rays. From the first right triangle drawn across the face of the guide we see that

$$a = w \tan \theta \tag{6–19}$$

and from the upper, smaller right triangle we have $b = a \cos \theta = w \tan \theta \cos \theta$, or

$$b = w \sin \theta \tag{6–20}$$

From the upper, larger right triangle we have $s = b/\sin \alpha$; or, with Eq. (6–20),

$$s = w \frac{\sin \theta}{\sin \alpha} \tag{6–21}$$

EXAMPLE 6–8

Determine the minimum length of the sides of the sectoral horn design in Example 6–7.

Solution The angle in the waveguide is, from Eq. (6–2),

$$\theta = \cos^{-1}\frac{\lambda}{\lambda_c} = \cos^{-1}(0.594)$$

$$= 53.6°$$

From Eq. (6–21) we now find

$$s = 7.21\left(\frac{\sin 53.6°}{\sin 18.2°}\right)$$

$$= 18.6 \text{ cm}$$

The horn designed in Example 6–8 is optimum in the sense that it produces a narrow beam for its size and for the amount of material used in its construction. By "beam" we mean the distribution of radiated energy in space, that is, the *radiation pattern*. A sketch of a radiation pattern is shown in Fig. 6–17. A cut or

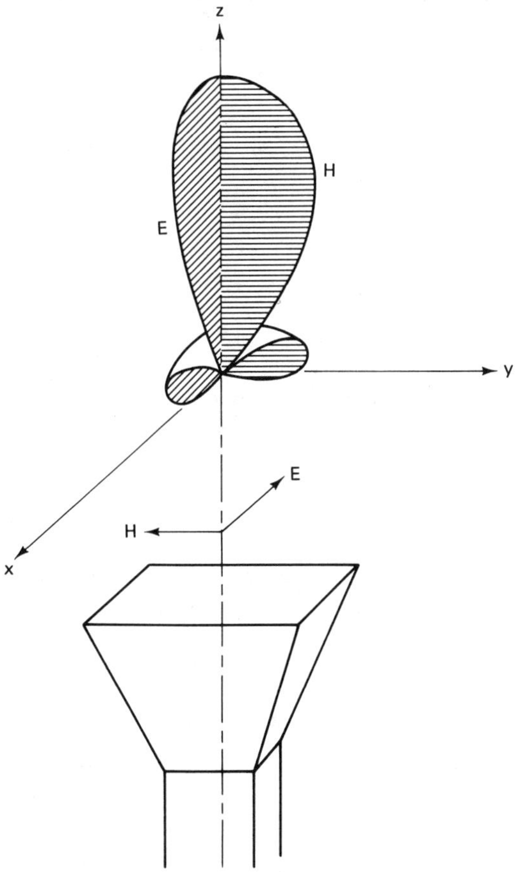

Figure 6–17 *E*- and *H*-plane sections through the radiation pattern of a pyramidal horn antenna; the corresponding orientation of the horn is shown in the lower part of the figure.

section through the three-dimensional radiation pattern that contains the electric field vector is called the *E-plane radiation pattern*, and a section containing the magnetic field vector is called an *H-plane radiation pattern*. Radiation patterns often are measured by placing the test antenna on a turntable, and the received power is recorded automatically on a chart recorder synchronized with the rotation of the turntable. The pattern may be recorded on polar or rectangular coordinates, as shown in Fig. 6–18.

Our choice of the exact length of the sides of a sectoral horn design depends on the sidelobe level we desire. Further increase in the length of the sides beyond the minimum length does not markedly narrow the main beam, but does reduce the sidelobe level by decreasing the edge illumination. A compromise would be an increase of 25% beyond the minimum length.

If we wish to control the beam in both the E- and H-planes, we can modify our design by flaring all four sides, as shown at the bottom of Fig. 6–17. The dimension of the aperture in the H-plane partly determines the flare angle in the H-plane. If a completely symmetrical pattern is desired, the aperture should be rectangular, with the H-plane dimension about 10% larger than the E-plane.

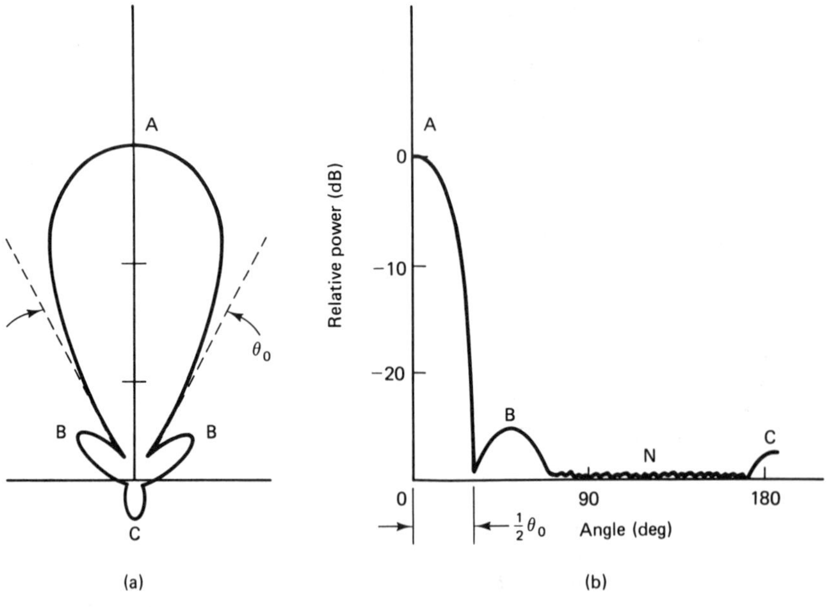

Figure 6–18 Radiation pattern in (a) polar and (b) rectangular coordinates. A, Main lobe; B, sidelobes; C, back lobe; N, noise; θ_0, beam width between nulls.

EXAMPLE 6–9

Modify the sectoral horn design to that of a pyramidal horn with a symmetrical radiation pattern.

Solution We shall take the length of the side in the E-plane to be 25% greater than the minimum length calculated in Example 6–8,

$$s_H = 1.25 \times 18.6 = 23.2 \text{ cm}$$

From the H-plane section of the horn (Fig. 6–19), we see that the width of the aperture in the H-plane is

$$\begin{aligned} L_H &= w + 2s \sin \alpha \\ &= 7.21 + 2 \times 23.2 \sin 18.2^\circ \\ &= 21.7 \text{ cm} \end{aligned}$$

We shall take the E-plane dimension of the aperture to be 10% shorter,

$$\begin{aligned} L_E &= 0.9L_H = 0.9 \times 21.7 \\ &= 19.5 \text{ cm} \end{aligned}$$

We complete the design by referring to the E-plane section through the horn (Fig. 6–20). The narrow dimension of the guide is $h = 3.40$ cm, from Table 6–1. The height of the upper right triangle is

$$\begin{aligned} d &= \tfrac{1}{2}(L_E - h) = \tfrac{1}{2}(19.5 - 3.40) \\ &= 8.05 \text{ cm} \end{aligned}$$

The base of the same triangle is, from Fig. 6–19,

$$\begin{aligned} s_H \cos \alpha &= 23.2 \cos 18.2^\circ \\ &= 22.0 \text{ cm} \end{aligned}$$

The E-plane flare angle is

$$\gamma = \tan^{-1} \frac{d}{s_H \cos \alpha} = \tan^{-1} \frac{8.05}{22.0}$$

$$= 20.1^\circ$$

and the length of the sides in the E-plane is

$$s_E = \frac{d}{\sin \gamma} = \frac{8.05}{\sin 20.1^\circ}$$

$$= 23.2 \text{ cm}$$

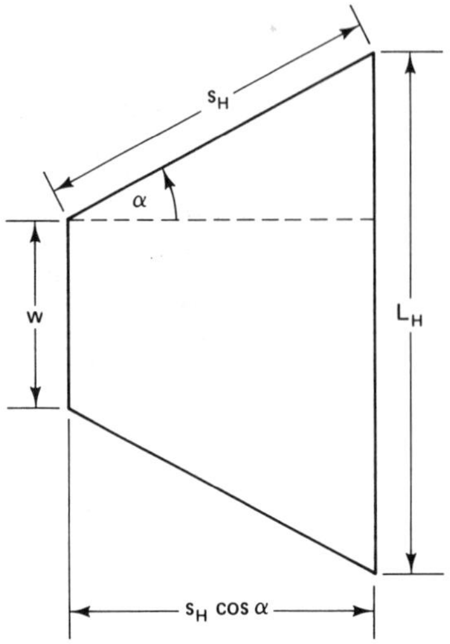

Figure 6–19 *H*-plane section through a pyramidal horn (not to scale).

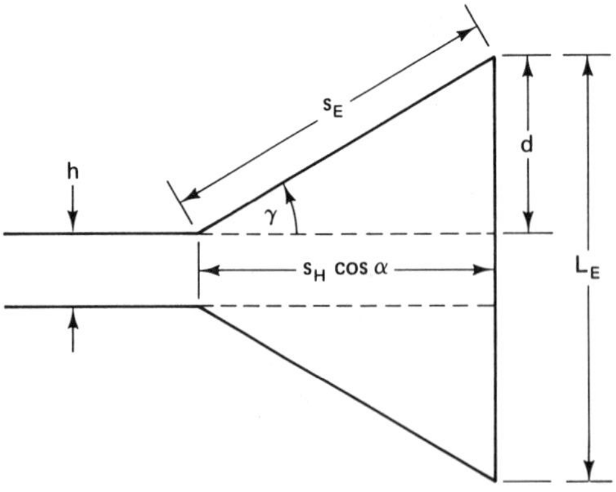

Figure 6–20 *E*-plane section through a pyramidal horn (not to scale).

Pyramidal horns often are used in combination with parabolic reflectors in radar antennas; also, they are sometimes manufactured to provide an alternative gain standard to the dipole. The relationships between radiation pattern, gain, and aperture size are discussed in Chapter 9.

6–10 WAVEGUIDE HARDWARE

The purpose of various types of waveguide sections is usually obvious from their appearance. Waveguide bends in the *E*-plane or *H*-plane, called "*E*-bends" and "*H*-bends," serve the same purpose as the plumber's "elbow." These are shown in Fig. 6–21a. A waveguide "twist" (Fig. 6–21b) rotates the plane of polarization—

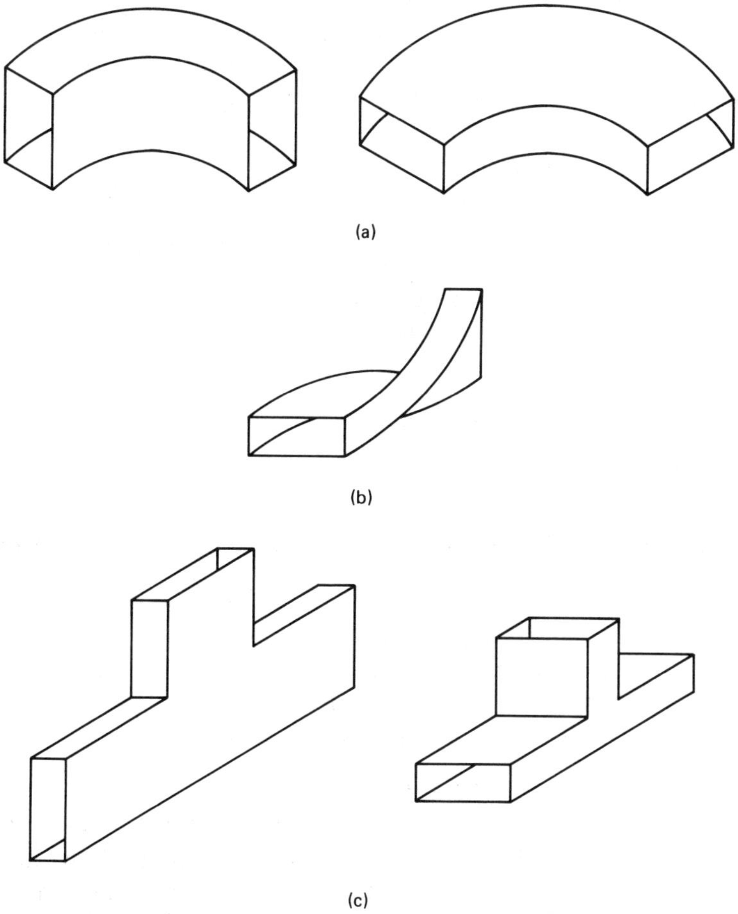

Figure 6–21 Waveguide sections, shown without flanges or wall thicknesses: (a) *E*-bend and *H*-bend; (b) twist; (c) series tee and shunt tee.

the E-vector—through an angle of 90°. Flexible guide can be used for the same purpose. E-plane and H-plane tees (Fig. 6–21a) serve a purpose similar to a plumber's tee; however, one is a series connection and the other a shunt connection. For this reason they are called "series" and "shunt" tees. Other devices will be discussed in the measurements section as they are needed.

APPLICATION 6–1: VSWR Measurement

In this series of measurements we shall determine VSWR in a waveguide terminated in a load, the load comprising a variable attenuator followed by a short circuit. This arrangement allows us to use the attenuator to control the VSWR and so obtain experience in the measurement of a wide range of values; in addition, we can check the accuracy of our measurements by comparing our results with the VSWR calculated from the attenuator readings.

Equipment and Materials

X- or S-band slotted line, probe carriage, and tunable diode detector
modulated microwave signal generator
pad, if not included in the generator
tuned-amplifier/VSWR meter
waveguide precision variable attenuator
shorting plate or switch

Procedure

Assemble the equipment as shown in Fig. 6–22. The turn-on procedure for the equipment is as follows:

1. Turn on the signal generator power switch.
2. Turn the modulation switches to "internal" and 1 kHz, and rectangular pulse, rather than sinusoidal.
3. Turn on the VSWR tuned amplifier.
4. Set the variable attenuator to 0 dB.
5. Set the generator frequency control to the desired operating frequency.
6. Adjust the generator modulation frequency to match the tuned amplifier frequency. This can be done by adjusting the signal generator modulation frequency until there is a maximum response on the VSWR meter.
7. Adjust the tunable cavity of the diode detector for a maximum response on the VSWR meter.

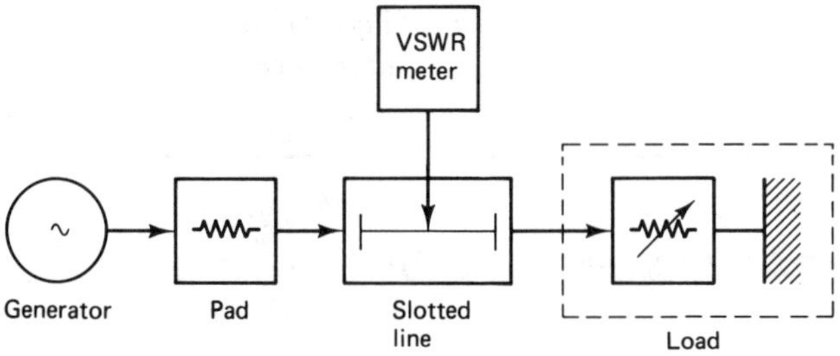

Figure 6–22 Variable attenuator and short used as a resistive load for generating reflected waves of various amplitudes.

If there is no indication other than noise, move the slotted-line carriage to a different position and repeat the foregoing procedure.

When the equipment is operating satisfactorily for accurate measurements, there should be a ratio of maximum signal to minimum signal (noise) of at least 20 to 30 dB in voltage, as the carriage is moved. The slotted-line probe depth should be decreased if the signal-to-noise ratio is greater than 30 dB.

To measure VSWR, move the traveling carriage to a maximum voltage reading and adjust the VSWR amplifier gain until the meter reading is 1.000 or 0 dB. Move the carriage to a null and read VSWR directly. If a VSWR meter is not available, use a voltmeter or oscilloscope to measure voltage at maximum and minimum and calculate the ratio V_{max}/V_{min}.

Make a series of VSWR measurements for various settings of the variable attenuator, from 0 to 15 dB. Arrange the results in tabular form and compare them with VSWR as calculated from the attenuator readings.

EXAMPLE 6–10

The measured VSWR is 1.35 when the attenuator is set at 8 dB. Compare measured and calculated VSWRs.

Solution Since the signal reflected from the short has passed through the attenuator twice, the total power attenuation is $2 \times 8 = -16$ dB, or -8 dB in voltage. We have

$$r = \frac{V_r}{V_0} = \log^{-1}(-0.8)$$
$$= 0.158$$

From Eq. (2–5) we have

$$\text{VSWR} = \frac{1 + r}{1 - r} = \frac{1 + 0.158}{1 - 0.158}$$
$$= 1.376$$

and the comparison is

Measured VSWR	*Attenuator*	*Calculated VSWR*	*Percentage difference*
1.35	−8.0 dB	1.376	1.9

APPLICATION 6–2: Measurement of the Speed of Light

Equipment and Materials

Use the same equipment as in Application 6–1; in addition, insert a precision cavity frequency meter between the generator and the slotted line.

Procedure

The turn-on procedure is described in Application 6–1. Make a series of 10 guide wavelength measurements by using the slotted-line probe to measure the distance between the nulls in a standing-wave pattern. Make each measurement at a slightly different frequency; select frequencies well within the bandwidth of the waveguide.

Follow the procedure in Example 6–3 to determine the speed of light in free space from each set of measurements. Tabulate the results and calculate an average value of the speed of light from all the measurements. The average value should agree with the accepted value to within a fraction of a percent.

APPLICATION 6–3: Impedance Measurements

Equipment and Materials

The required equipment is the same as in Application 6–1; in addition, a load impedance is required. A load having a complex impedance can be made conveniently by inserting an iris between the end of the slotted line and a waveguide termination. An *iris* is an aperture in a thin conductor. If the width of the aperture is less than the guidewidth, the load will appear to be inductive; if the height or E-plane width is less than that of the guide, the load will appear to be capacitive (Fig. 6–23).

Procedure

Measure VSWR with the load in place, as explained in Application 6–1. While the probe is still positioned at a null, replace the load with a short and determine the null shift, Δx, by moving the probe to the nearest null when the short is in place. The null shift between load and short is illustrated in Fig. 6–24.

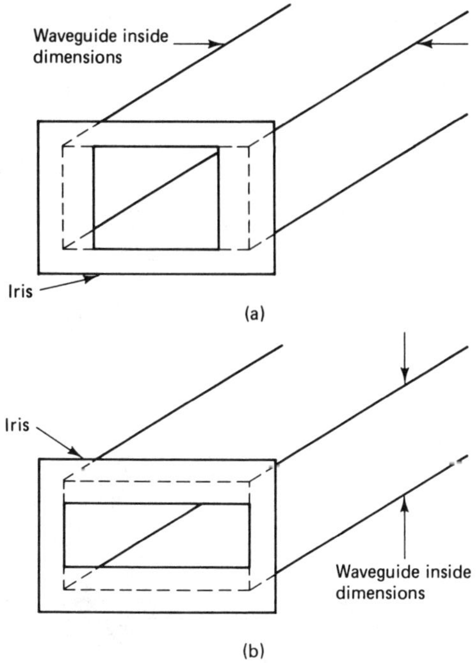

Figure 6–23 (a) Inductive iris; (b) capacitive iris.

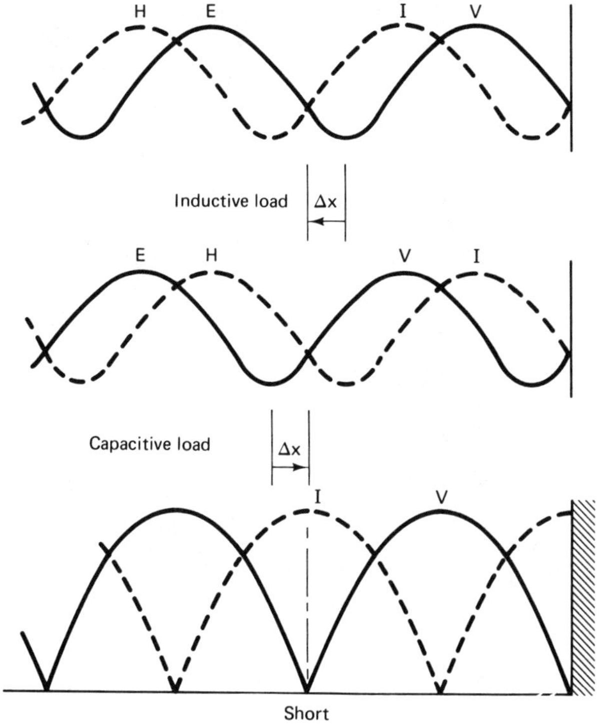

Figure 6–24 Voltage (solid lines) and current (dashed lines) distributions on transmission lines terminated with inductive or capacitive loads or a short. The shift of a voltage null, when a load is replaced by a short, is shown as Δx.

Move the probe to the next null to determine the guide wavelength. Remember, the guide wavelength is twice the distance between nulls.

Follow the procedure given in Example 6–4 to determine the load impedance from the measurements.

Measure the impedances of a capacitive load at five different frequencies and an inductive load at five different frequencies. Report your results in tabular form. It might be advisable to measure the width of the slotted line.

APPLICATION 6–4: Waveguide Horn Antenna

The fabrication of a pyramidal horn from sheet metal requires skills not usually expected of an electronics technician or engineer; however, it is instructive to design and construct a horn of thin cardboard, adhesive tape, and aluminum foil (Fig. 6–25).

Figure 6–25 A pyramidal horn antenna can be constructed quickly from foam and foil and used for class instruction or as a prototype for a final design. The polystyrene foam, because of its low refractivity, affects performance only slightly.

Equipment and Materials

microwave signal generator
slotted line
VSWR amplifier/meter
thin cardboard or polystyrene foam, adhesive tape, glue, aluminum foil
scissors and small hand tools

Procedure

Select an operating frequency appropriate to the waveguide you are going to use (refer to Table 6–1). Design a horn according to the procedure given in Section 6–9. Lay out the design on thin cardboard or a block of polystyrene foam; provide tabs and flaps on the edges for ease in assembly with adhesive tape or glue.

After the horn is assembled, use rubber cement and aluminum foil to cover the surfaces. The design should include a short section of cardboard or polystyrene-foam waveguide, and the foil should cover the flange of the waveguide to provide a good electrical connection when the horn is clamped to the flange of a slotted line. If rigidity is a problem, a block of polystyrene foam, cut to the interior dimensions of the waveguide, can be inserted into the cardboard waveguide and slotted line.

Clamp the horn to the flange of the slotted line and measure its impedance. The space in front of the horn should be free of obstructions that might reflect waves back into the aperture.

Caution. Although a laboratory signal generator generates only a few milliwatts of power, it is nevertheless important to practice normal safety procedures, so that they may become habitual. Never look into the aperture of an active waveguide or antenna. When making radiation pattern measurements, be sure to be far enough away from the antenna to ensure that the power density in your location is less than 0.1 mW/cm^2.

EXAMPLE 6–11

An S-band signal generator is operated at +12 dBm and is used to excite a waveguide pyramidal horn. Assume that the gain of the horn is less than 13 dB and estimate the power density in the aperture and at a distance of 2.5 m.

Solution A power level of +12 dBm is INV log 1.2 = 15.8 mW, and a gain of 13 dB is INV log 1.3 = 20. The area of the aperture is 14 × 11.5 = 161 cm^2. The power density in the aperture is

$$p_A = \frac{P}{A} = \frac{15.8}{161} = 0.098 \text{ mW/cm}^2$$

To estimate the power density at a distance of 2.5 m, we first assume the power to be radiated equally in all directions and calculate the area it covers, the surface of a sphere of radius 2.5 m:

$$A = 4\pi r^2 = 4\pi(250)^2$$
$$= 7.85 \times 10^5 \text{ cm}^2$$

If the energy were radiated equally in all directions, the power density would be

$$p_0 = \frac{P}{A} = \frac{15.85}{7.85 \times 10^5}$$
$$= 2.02 \times 10^{-5} \text{ mW/cm}^2$$

Since the horn has a gain of 13 dB = 20, the power density at a distance of 2.5 m in front of the horn is

$$p = 20p_0 = 20 \times (2.02 \times 10^{-5})$$
$$= 4.04 \times 10^{-4} \text{ mW/cm}^2$$

We see from Example 6–11 that the power density in the aperture of the horn is at the conservative limit for safety and much less at a distance of 2.5 m. If, however, the power were increased 60-fold to 1 W, the power density in the aperture would exceed the safety limit but would remain safe at a distance of 2.5 m. Antenna gain and power densities are discussed in Chapter 9.

Make an approximate measurement of the *E*- and *H*-plane radiation patterns of the horn at a distance of 3 or 4 m. Use a probe or small dipole antenna connected to a cavity-tuned diode detector and VSWR meter to measure the relative power at various angles to the forward direction. If reflections from surrounding objects are a problem, note the maximum and minimum values at slightly different ranges and record the average.

Make rectangular or polar graphs of the results, and from the graphs estimate the half-power beamwidths. Save the results for further consideration after Chapter 9.

SUMMARY

A waveguide is, literally, anything that guides waves. It can be anything from a rough surface to a fiber optic cable, but for now we restrict the meaning to hollow conducting cylinders, circular or rectangular in cross section.

We can visualize the progress of a wave through rectangular guide as a zigzag path produced by multiple reflections at oblique incidence on the sidewalls. The shortest wavelength is measured in a direction normal to a wavefront; wavelengths are longer if measured at oblique angles, angles of incidence less than 90°. Since wavefronts are reflected at oblique incidence on the guide walls, the wavelength in the direction of the waveguide axis is greater than the normal wavelength,

$$\lambda = \frac{\lambda_0}{[1 - (\lambda_0/\lambda_c)^2]^{1/2}}$$

where $\lambda_c = 2w$ is the cutoff wavelength. The guide wavelength λ can be measured with a slotted line. If the free-space wavelength is to be determined, the equation above must be solved for λ_0,

$$\lambda_0 = \frac{\lambda}{[1 + (\lambda/\lambda_c)^2]^{1/2}}$$

The phase velocity in the guide is obtained by dividing the wavelength equation by the period,

$$v = \frac{c}{[1 - (\lambda_0/\lambda_c)^2]^{1/2}}$$

For the TE_{10} mode we define waveguide impedance as the impedance of free space normalized by the refractive index of the guide, $Z_c = Z_0/\eta = Z_0 v/c$,

$$Z_c = \frac{377}{[1 - (\lambda_0/\lambda_c)^2]^{1/2}}$$

Waveguide modes are classified as transverse electric (TE) or transverse magnetic (TM), with subscripts indicating the number of electric and magnetic antinodes contained in a cross section of the guide. Usually, only the TE_{10} mode is used. Waveguide modes can be visualized in three-dimensional space by introducing Maxwell's concept of a displacement current, a current that flows through dielectrics and empty space,

$$i_D = \frac{\Delta D}{\Delta t}$$

The useful bandwidth of waveguide is determined on the low-frequency side by the cutoff wavelength and on the high-frequency side by the possible generation of modes higher than the TE_{10}.

With the use of a slotted line we can measure guide wavelength, VSWR, and phase change when a load is replaced by a short. This is sufficient information to determine the impedance of the load, and the determination is made easily by plotting the information on a Smith chart. Broadband measurements of reflection coefficient and impedance can be made with the use of directional couplers and sweep-frequency generators.

We can combine measurements of frequency and guide wavelength to make a precision determination of the speed of light.

Since waves proceed through a guide by following a zigzag path, energy flows down the guide at a speed less than the speed of light, $u = c \sin \theta$. The speed of energy flow is the group speed u and is related to the phase speed by

$$uv = c^2$$

This equation is characteristic of waveguide dispersion.

From the equations presented so far we see that waveguide speeds are a function of wavelength or frequency; thus the various harmonic components of a pulse or waveform propagate at different speeds and cause the pulse to be spread out or dispersed. Since waveguide phase speed is greater than group speed, the dispersion is called normal. Dispersion in bulk dielectrics is usually caused by absorption. Outside an absorption region, higher-frequency components propagate more slowly, and the dispersion is normal. Inside an absorption region the higher frequencies propagate more rapidly than the lower, and the dispersion is anomalous. If both waveguide and bulk dispersion are present, as in a waveguide filled with a dielectric, the dispersion is called chromatic.

A general equation for group velocity of wide applicability is

$$u = v - \lambda \frac{\Delta v}{\Delta \lambda}$$

which can be used to find group speed graphically from a dispersion curve.

An optimum waveguide horn antenna can be designed by determining a flare angle such that the last reflections from the sides of the horn direct rays in the forward direction. We measure the radiation pattern of an antenna by recording the power received at distant points at various angles to the forward direction. Since a radiation pattern is three-dimensional, it is common practice to prepare at least two graphs, one in the plane of the *E*-vector, called the *E*-plane radiation pattern, and one in the *H*-plane.

The absolute gain of an antenna is defined as the ratio of the power received at a distant point to the power that would be received if the antenna were removed and the power were radiated uniformly in all directions, that is, as if the antenna were replaced by an isotropic radiator. The gain concept is discussed further in Chapter 9.

PROBLEMS

6–1. The following data were obtained with an X-band slotted line terminated in a load:
Operating frequency, 9.5 GHz
VSWR, 1.293
Null position with load, 19.75 cm
Null position with load replaced by a short, 19.45 cm toward the load
Position of next null, 21.47 cm

Use a Smith chart to find the normalized load impedance and the absolute load impedance in ohms.

6–2. Slotted-line measurements at 10 GHz in X-band rectangular guide show a VSWR of 5.20, a shift toward the load of 0.3976 cm when the load is replaced by a short, and a distance between nulls of 1.988 cm. Find the normalized load impedance and the complex reflection coefficient.

6–3. The following data are obtained for an X-band slotted line connected to a section of X-band, open-ended waveguide:

Operating frequency, 10.00 GHz
VSWR, 1.60
Distance between nulls, 2.00 cm
Null shift toward the generator when the line is short-circuited, 0.50 cm

Find the normalized and absolute impedances of the open-ended guide. If the open-ended guide were used as a feed for a dish antenna, what would be the mismatch loss?

6–4. Use the distance between nulls in Problem 6–3 to calculate the speed of light in free space, and calculate the percentage difference between the result and the generally accepted value.

6–5. Calculate the angle of reflection of the waves in X-band guide for the conditions described in Problem 6–3.

6–6. Calculate the phase and group speeds of waves in X-band guide under the conditions described in Problem 6–3.

6–7. A signal generator delivers a power of 5 W to an open-ended, S-band waveguide. Find the power density at the end of the guide.

If the open guide has a gain of 1.6 dB, at approximately what distance is it safe to stand in front of the guide? Assume a safety limit of 0.1 mW/cm^2.

6–8. A parallel-plate capacitor is made with a plate area of 0.05 m^2. The relative permittivity of the dielectric is 2.2 and its thickness is 2 mm. Estimate the average displacement current if the capacitor is charged to 10 V in 2 ms.

6–9. Estimate the useful bandwidth of waveguide if the guide width is 5.00 cm.

6–10. Calculate the cutoff frequency of the TE_{10} mode in C-band waveguide.

6–11. Use the data you obtained in the delay-line assignment of Chapter 5 to determine graphically the group velocity corresponding to a selected phase velocity. Select a phase velocity at a frequency above the cutoff frequency.

7

Amplifiers

7–1 INTRODUCTION

Amplification frequently is achieved by varying a circuit parameter, by changing the parameter from a passive to an active component. The parameter may be a resistance or a reactance. The transistor is a familiar example; indeed, the word comes from *trans*fer res*istor*. In the *CE* configuration, an increase in transistor base current controls a larger change in collector current, as though the resistance in the collector circuit had been decreased.

The name *parametric amplifier*, or *paramp*, is usually restricted to amplifiers in which the active component is a reactance. The reactance can be inductive or capacitive, but now is usually capacitive because of the convenience and availability of varactors. A *varactor* or *varicap* is a diode that has been manufactured to enhance the variation of the width of the depletion layer with change in bias voltage. The capacitance of a diode is approximately $C = \epsilon A/d$, where d is the width of the depletion layer. Since reverse bias increases the width of the depletion layer, an increase in reverse bias voltage decreases the capacitance (Fig. 7–1).

If a capacitor is charged to a voltage $V = Q/C$, a subsequent decrease in capacitance C will further increase the voltage across the capacitor; the work done in decreasing the capacitance now appears as an increase in voltage. This is the basic idea in parametric amplification.

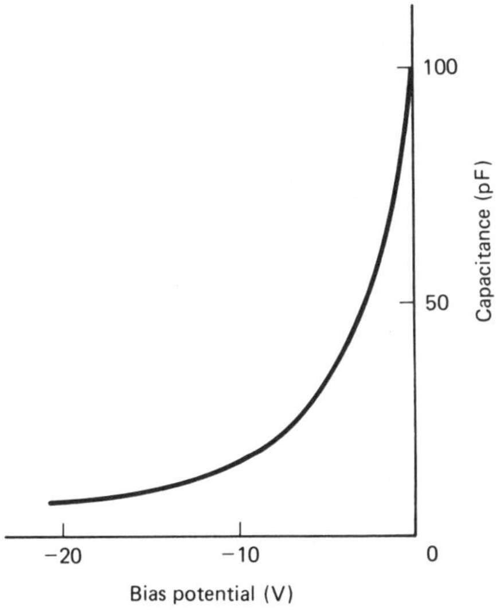

Figure 7–1 Varactor capacitance as a function of bias voltage.

DEMONSTRATION 7–1

Parametric amplification can be demonstrated with a simple pendulum. Tie a weight or bob to a string about 1 m long and suspend it through a fixed guide (Fig. 7–2). Allow the bob to swing with an amplitude of about ±10 cm and observe the period. Pull the string vertically up and down about ±15 cm at about twice the frequency of the bob. There is a tendency to overestimate the required pumping frequency, if this occurs gradually decrease the pumping frequency. After several attempts the correct phase and frequency will be found and the amplitude of the swinging bob will increase rapidly to an extreme state.

In Demonstration 7–1 the active parameter is the length of the pendulum. Since the period of a pendulum is proportional to the square root of its length, its frequency is increased as the string is pulled up through the guide. The work done in lifting the bob is not entirely returned when the bob is subsequently lowered. Part of the work goes into increasing the frequency of the bob. When the bob is lowered, its increased velocity produces an increase in amplitude.

The pendulum demonstration helps us to understand how a child "pumps" a swing. At the proper time (phase) the child grasps the ropes and pulls them toward the backward direction. The swing momentarily has a shorter length to the fulcrum at the child's hands, and the period is shortened, the frequency increased (Fig. 7–3). As the additional speed is added to the swing the child relaxes the pull on the

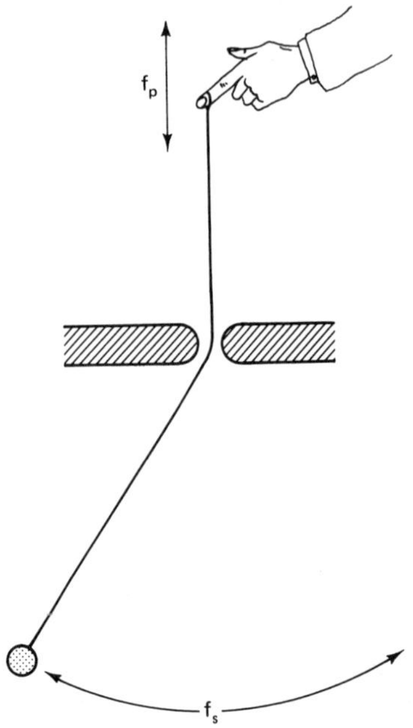

Figure 7–2 Parametric amplification demonstrated mechanically: pump frequency, f_p; signal frequency, f_s.

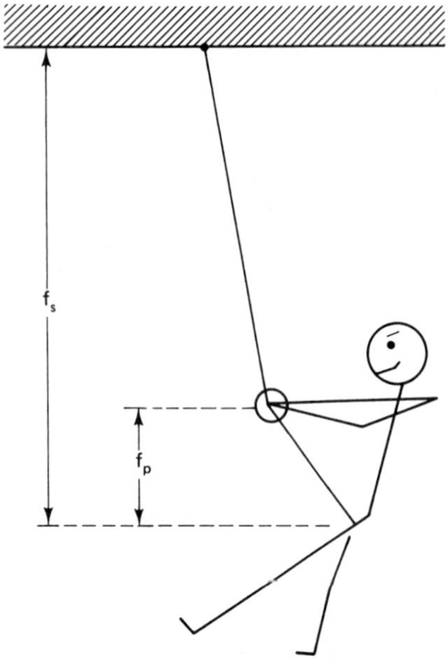

Figure 7–3 Parametric amplification: the swing frequency is determined by the total length of the supporting lines, the pump frequency by the line length from seat to hands.

ropes and the swing or signal amplitude is increased. The PRF is usually the same as the swing frequency because children only pump at one end of a cycle.

For a capacitive paramp we need to devise circuitry that will do automatically what a child on a swing does intuitively: we have to decrease the capacitance when the signal voltage across the capacitor is near a maximum or minimum. The output signal will not be decreased in amplitude if the bias voltage is reduced to its original value when the capacitor charge is zero.

7–2 PARAMETRIC AMPLIFIERS

Parametric amplification using an active reactance has a noise advantage over active resistance devices. The diode varactor is a good choice for an active reactance at microwave to millimeter frequencies because its capacitance variation is small, in the range of 10 to 100 pF (Fig. 7–1). We use a pump frequency bias voltage to decrease the capacitance at the proper time (phase), and so increase the signal amplitude (Fig. 7–4). A circuit that can be made to satisfy our requirements is shown in Fig. 7–5.

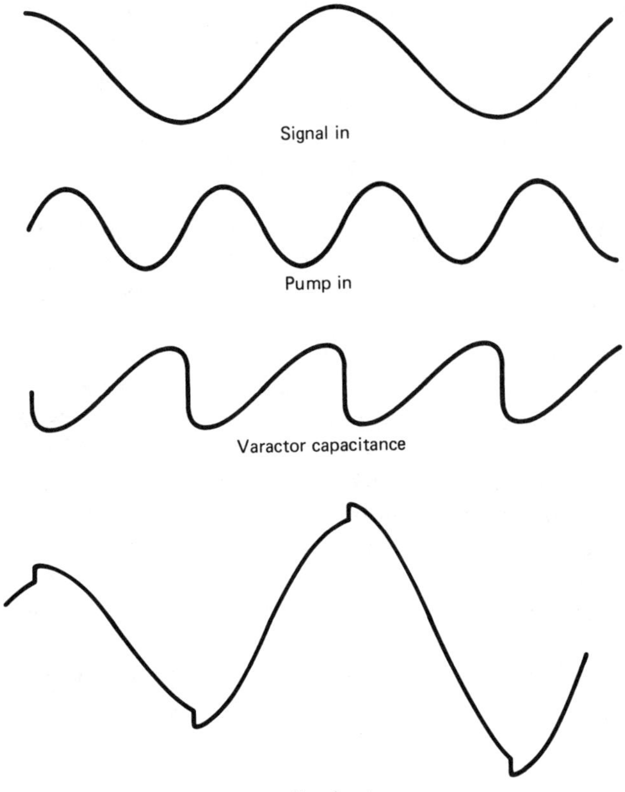

Figure 7–4 Phase relations between signal and pump voltages are shown as waveforms for a paramp in which the pump frequency is double the signal frequency.

The lumped-parameter representation shown in Fig. 7–5 comprises two circuits resonant at frequencies f_s and f_i that are coupled by a varactor pumped at $f_p = f_s + f_i$. Since the varactor is nonlinear, there are frequency components to drive the tank circuits at f_s and f_i and deliver energy to them.

If the pump amplitude is reduced so that oscillations are not quite sustained, the circuit can be used as an amplifier. A signal impressed on the first tank at frequency f_s will be increased in amplitude as pump energy is delivered to the circuit. In this case the second tank is called an *idler circuit*. The idler circuit

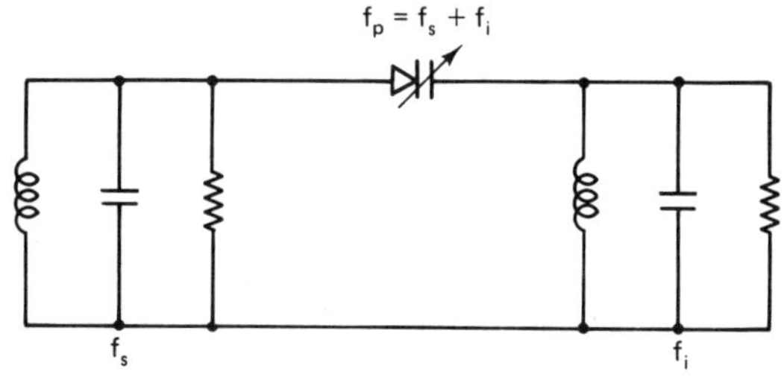

Figure 7–5 Parametric amplifier circuit. Pump input frequency f_p, signal input f_s, idler input f_i.

accommodates the f_i frequency resulting from heterodyning by the nonlinear varactor. If frequency f_i were suppressed, there would be no gain in signal amplitude. The reduction in pump amplitude to avoid sustained oscillations means that we now have *forced oscillations*, which are in phase and remain in phase.

Since the output signal is from the same port as the input, directional couplers or a circulator are required to separate them. We show an example of the use of a circulator in Fig. 7–6.

Output may be taken from either the signal or idler circuits. If output is taken from the idler circuit and the idler frequency is greater than the signal frequency, we have an *up-converter* with gain. If the idler is lower, we have a *down-converter*. The special case illustrated in Fig. 7–4 shows a pump frequency double the signal frequency, the idler and signal frequencies are equal, $f_p = f_s + f_s$, and the amplifier is called *degenerate*. In this case the idler circuit is redundant and may be removed.

In the construction of microwave paramps, the lumped-parameter circuits shown in Fig. 7–5 may be realized as cavities or lengths of transmission lines. A design difficulty occurs in that input and output are from the same port. If reflection occurs from a following stage, the low-noise advantage of the paramp may be lost as a noisy signal is returned to the input. This problem is avoided by the use of a *circulator*, a form of directional coupler. A circulator has three or more ports. Energy entering any port emerges only at the adjacent port in the clockwise direction (Fig. 7–6).

Circulator design frequently is based on optical activity. Substances that manifest *optical activity* have the ability to rotate the plane of polarization of an electromagnetic

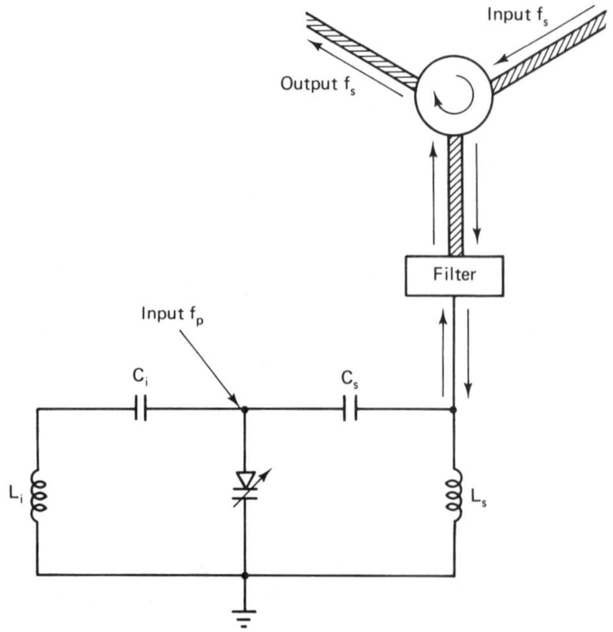

Figure 7–6 Use of a circulator to isolate input and output at the port of a paramp.

wave. A sugar solution, for example, will rotate the plane of polarization in proportion to the concentration of sugar. By passing polarized light through a sugar solution and measuring the rotation, the concentration can be determined. This technique is sometimes used to automatically monitor food processing at a cannery. Some substances become optically active if a magnetic field is imposed, and the rotation, now called the *Faraday rotation*, is proportional to the applied magnetic field strength. The Faraday effect in ferrite frequently is used in microwave devices.

The basic design principle in many circulators is the incorporation of an optically active medium between the ports. An EM wave entering one port is rotated sufficiently to exit at the next, but is rotated to cross-polarization at other ports and cannot exit.

The paramp is a low-noise device with a narrow bandwidth, less than 10%. Greater bandwidth, with less gain, can be obtained by using a terminated, distributed- or lumped-inductance transmission line in place of the tuned circuits or cavities (Fig. 7–7). This is a form of *traveling-wave amplifier*. Since the line is a low-pass filter, the pump, signal, and idler frequencies must be less than the cutoff frequency of the line and the sum frequency must be greater. A wave at the pump frequency propagating down the line is attenuated as its energy is transferred to waves at the signal and idler frequencies. Line losses limit the length of the line and realizable amplification. Bandwidths as large as 50% may be obtained.

A *maser* amplifier is one in which the active parameter is atomic or molecular energy levels. Maser is an acronym for *m*icrowave *a*mplification by *s*timulated *e*mission *r*adiation. When we say that a gas, liquid, or solid is "pumped" at microwave or optical frequencies, we mean that energy has been delivered to the substance to raise its atoms or molecules to energy levels higher than normal. When the atoms or molecules subsequently decay to a lower energy level, radiation may be emitted at a signal frequency different from the pumping frequency. If the number of atoms or molecules in excited states (that will decay by emitting radiation at the signal frequency) is greater than the number of ground state atoms or molecules (that will absorb energy from the signal), more energy is released by stimulated emission than is absorbed, and the signal wave will be amplified.

A schematic representation of a maser is shown in Fig. 7–8a. The cavity is resonant at both the pumping and signal frequencies, f_p and f_s. In the energy-level diagram (Fig. 7–8b) we show the pumping frequency, which is related to the energy

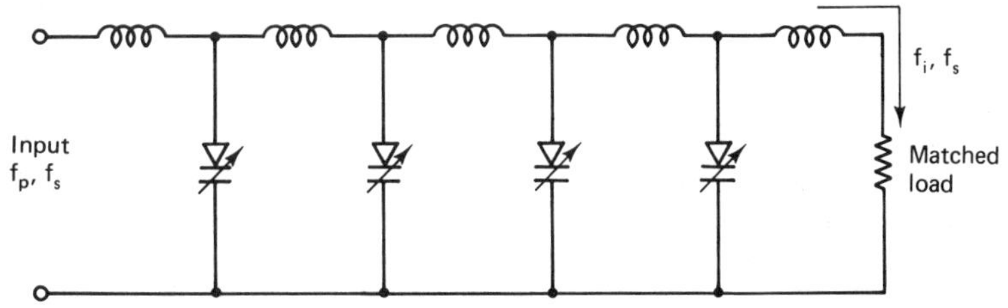

Figure 7–7 Traveling-wave parametric amplifier.

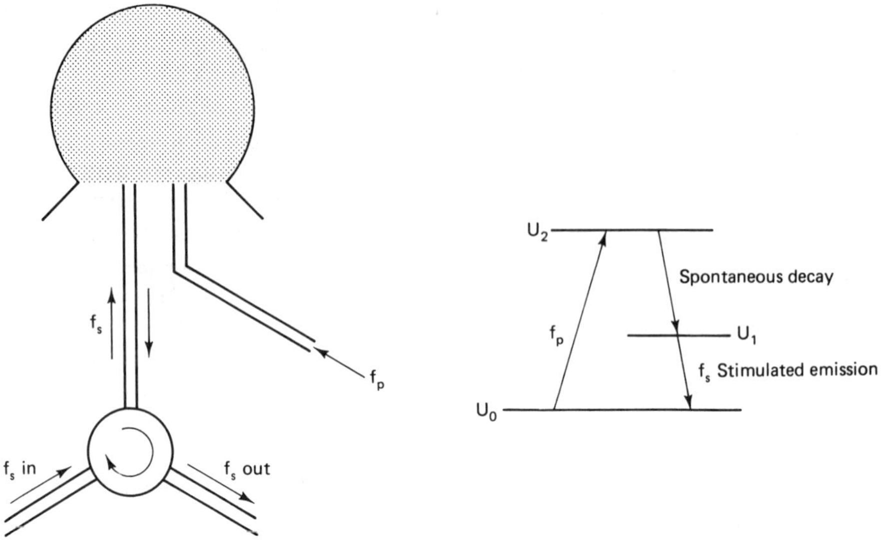

(a) (b)

Figure 7–8 Maser amplifier: (a) representation as a cavity containing an active medium; (b) energy-level diagram, spontaneous decay from excitation level U_2 produces a population inversion at U_1, stimulated emission from U_1 back to U_0 generates coherent radiation.

levels by the equation $U_2 - U_0 = hf_p$, where h is Plank's constant; similarly, the signal frequency is $U_1 - U_0 = hf_s$.

In the classical theory of radiation, transitions between energy levels are represented as radiation from dipole oscillators. We shall see in Chapters 8 and 9 that linear oscillators, dipole oscillators in particular, can be represented by resonant *LCR* circuits; thus there is a close correspondence between a varactor paramp and a maser or laser.

In some active media the relative positions of the energy levels can be changed by imposing a magnetic field, and by changing the field strength we can tune the signal frequency. The active medium in one of the earlier maser amplifiers is a ruby rod (Al_2O_3 doped with chromium). The rod is polished and half-silvered at each end and is contained in a longer cavity. By changing the imposed magnetic field, various frequencies can be tuned in a range of 1 to 6 GHz.

Lasers and masers are distinguished principally by their operating frequencies; the laser operates at optical frequencies. In a laser the pump frequency need not be tuned; the pump energy need only be intense radiation with a spectrum that overlaps the pump frequency. Lasers are discussed in Chapter 10.

7–3 TRANSISTOR AMPLIFIERS

In a transistor amplifier the active parameter is a resistance. In recent years the development of low-cost UHF-microwave transistors has brought inexpensive microwave experimentation and communication to the classroom, individual experimenter, and radio amateur. A transistor amplifier can be built and tested in a few laboratory periods, and for this reason we shall present considerable design detail and treat the subject quantitatively. Amplifier projects are assigned in Application 7–1. We shall confine the designs to the UHF to lower microwave frequency range so that the projects can be completed within a few hours.

The reader may here decide to concentrate only on the design results and examples, reserving the derivations and theoretical background to a second reading.

We begin with a review of the study of multiple reflections and scattering parameters that was presented in Chapter 5. As we review the material we shall change our point of view slightly and show the connection between multiple reflections, loop gain, and feedback.

7–4 SCATTERING PARAMETERS

A circuit or device may have one or more ports, that is, one or more pairs of terminals for input or output. We shall consider only two-port examples: one input port, one output. The device under test, or DUT, may be active, a transistor for example, or passive, such as a matching network. For a worker in physical optics, the DUT could be a slab of dielectric in which light is incident at the input port and transmitted from the output port (Fig. 7–9). In electronics we often do not consider all the multiple reflections in Fig. 7–9, but only their sums or resultants (Fig. 7–10). The resultant output amplitudes are related to the inputs by a pair of simultaneous linear equations,

$$\begin{aligned} b_1 &= s_{11}a_1 + s_{12}a_2 \\ b_2 &= s_{21}a_1 + s_{22}a_2 \end{aligned} \qquad (7\text{–}1)$$

Here a_1 is the input amplitude; b_1 is the reflected amplitude, including reverse transfer from a_2; b_2 is the forward transfer amplitude, including reflected amplitude from a_2; a_2 is the reverse input amplitude. The coefficients in the equations above are called *S-parameters*. We can define and measure them in the following way. We connect the output port to a matched 50-Ω load so that $a_2 = 0$; then from Eqs. (7–1) we have

$$s_{11} = \left.\frac{b_1}{a_1}\right|_{a_2=0} \qquad \text{and} \qquad s_{21} = \left.\frac{b_2}{a_1}\right|_{a_2=0} \qquad (7\text{–}2)$$

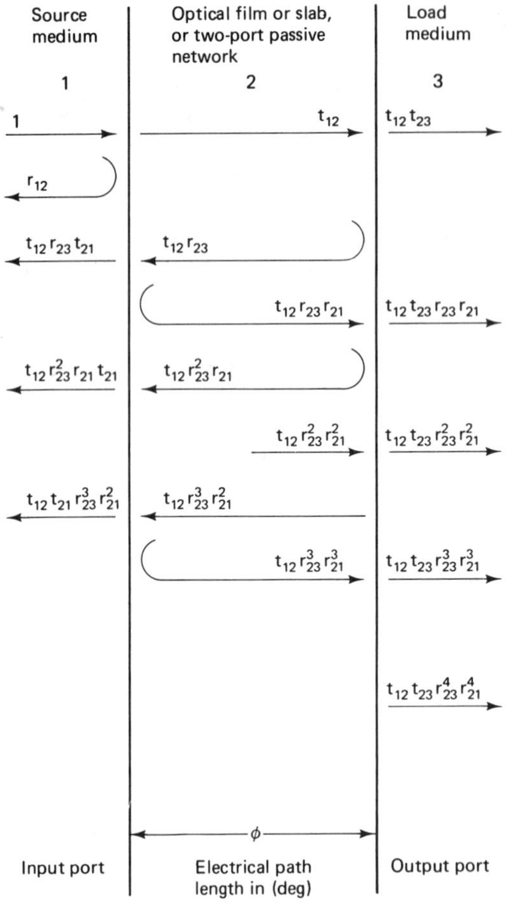

Figure 7–9 Multiple reflections in a dielectric film or slab.

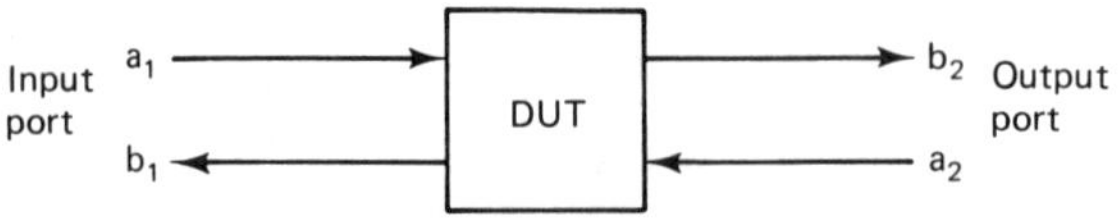

Figure 7–10 Input and output amplitudes in a two-port network.

Similarly, by connecting the source to the output and matching a load to the input terminals, we find that

$$s_{12} = \frac{b_1}{a_2}\bigg|_{a_1=0} \quad \text{and} \quad s_{22} = \frac{b_2}{a_2}\bigg|_{a_1=0} \tag{7–3}$$

EXAMPLE 7–1

Find the scattering parameters at an air–Duroid interface. The dielectric constant of the Duroid is 2.25.

Solution As a practical matter, we can assume that the second face of the Duroid is matched to a terminated transmission line, or fitted with a quarter-wave matching layer, so that reflections from a second face are eliminated, $a_2 = 0$. We now have, from the first of Eqs. (7–2) and (2–8),

$$s_{11} = \left.\frac{b_1}{1}\right|_{a_2=0} = \frac{1-\sqrt{\epsilon_r}}{1+\sqrt{\epsilon_r}} = \frac{1-\sqrt{2.25}}{1+\sqrt{2.25}} = \underline{0.200\angle 180^\circ}$$

The second of Eqs. (7–2) becomes

$$s_{21} = \left.\frac{b_2}{1}\right|_{a_2=0}$$

and we have to determine b_2. Since the network is passive we can assume unit incident amplitude and find the magnitude of b_2 from conservation of energy, $b_1b_1^* + b_2b_2^* = 1$, or $b_2 = \sqrt{1-|b_1|^2} = \sqrt{1-0.200^2} = 0.980$. The phase of b_2 is 0°, from Eq. (5–45). If Eq. (5–45) is used to calculate b_2, the result must be normalized by dividing by $z_2^{1/2} = \eta^{1/2}$. We now have $s_{21} = \underline{0.980\angle 0^\circ}$.

To find the remaining parameters we reverse the situation by placing a unit amplitude source on the other side of the interface we obtain

$$s_{22} = \left.\frac{b_2}{1}\right|_{a_1=0} = \frac{\sqrt{\epsilon_r}-1}{\sqrt{\epsilon_r}+1} = \underline{0.200\angle 0^\circ}$$

and

$$s_{12} = \left.\frac{b_1}{1}\right|_{a_1=0} = \underline{0.980\angle 0^\circ}$$

7–5 MULTIPLE REFLECTIONS

We now turn to the more complicated problem of finding the sum of all the multiple reflections in a slab of dielectric, as shown in Fig. 7–9. From a network point of view, this is a feedback problem. We shall find the net energy transmitted into the load, first by multiple reflections, then by feedback.

The many transmitted and reflected rays shown in Fig. 7–9 are labeled in the following manner:

1: unit-amplitude incident ray

$r_{12} = |r_{12}| \angle \psi_{12}$: amplitude and phase of the first ray reflected from the first interface

$t_{12} = |t_{12}| \angle \theta_{12} + \phi$: amplitude and phase change on transmission through interface 1–2, plus propagation phase, of the first transmitted ray

$t_{12}t_{23} = |t_{12}| \angle \theta_{12} + \phi \, |t_{23}| \angle \theta_{23}$: amplitude and phase of the first ray transmitted into the load

$t_{12}r_{23} = |t_{12}| \angle \theta_{12} + \phi \, |r_{23}| \angle \psi_{23} + \phi$: amplitude and phase of the first ray reflected from the 2–3 interface

$t_{12}r_{23}r_{21} = t_{12}r_{23}|r_{21}| \angle \psi_{21} + \phi$: amplitude of the first ray reflected from both the 2–3 and 2–1 interfaces

$t_{12}t_{23}r_{23}r_{21}$: amplitude and phase of the second ray transmitted into the load

$t_{12}t_{23}r_{23}^2r_{21}^2$: amplitude and phase of the third ray or wave transmitted into the load

$t_{12}t_{23}r_{23}^3r_{21}^3$: amplitude and phase of the fourth ray

Our task, which is easier than it looks, is to find the sum of all the rays transmitted into the load: $t_L = t_{12}t_{23}(1 + r_{23}r_{21} + r_{23}^2r_{21}^2 + r_{23}^3r_{21}^3 + \cdots)$. By the binomial theorem, $1/(1 - x) = 1 + x + x^2 + x^3 + \cdots$, and we have

$$t_L = \frac{t_{12}t_{23}}{1 - r_{23}r_{21}} \tag{7–4}$$

The total transmitted intensity into the load is

$$t_L t^*_L = |t_L|^2 = \frac{|t_{12}|^2|t_{23}|^2}{|1 - r_{23}r_{21}|^2} \tag{7–5}$$

At the interfaces we have, by conservation of energy, $t_{12}^2 = 1 - |r_{12}|^2$ and $t_{23}^2 = 1 - |r_{32}|^2 = 1 - |r_{23}|^2$; accordingly, we can write Eq. (7–5) in the form

$$|t_L|^2 = \frac{(1 - |r_{12}|^2)(1 - |r_{23}|^2)}{|1 - r_{23}r_{21}|^2} \tag{7–6}$$

We now can compute the gain realized from having a matching layer between a source and a load. Suppose the characteristic impedances of media 2 and 3 to be the same, as though there were no matching layer; then the energy transmitted into the load would be $1 - |r_{13}|^2$. The gain is the ratio of the load intensities with and without the matching layer:

$$G = \frac{(1 - |r_{12}|^2)(1 - |r_{23}|^2)}{|1 - r_{23}r_{21}|^2 \, (1 - |r_{13}|^2)} \tag{7–7}$$

Maximum gain is achieved by making the source reflection coefficient equal to the complex conjugate of the load reflection coefficient, $r_{21} = r_{23}^*$; then Eq. (7–7) is simplifed to

$$G_{\max} = \frac{1}{1 - |r_{13}|^2} \tag{7–8}$$

In Eqs. (7–7) and (7–8) we have the *loop gain* realized by using a passive matching network.

EXAMPLE 7–2

Calculate the gain realized by using a 100-Ω quarter-wave matching layer between a 300-Ω source and a 50-Ω load.

Solution The interface reflection coefficients are

$$r_{12} = \frac{100 - 300}{100 + 300}\underline{/0^\circ} = -0.5$$

$$r_{13} = \frac{50 - 300}{50 + 300} = -0.714$$

$$r_{21} = \frac{300 - 100}{300 + 100}\underline{/90^\circ} = 0.5$$

$$r_{23} = \frac{50 - 100}{50 + 100}\underline{/90^\circ} = 0.333\underline{/-90^\circ}$$

From Eq. (7–7) we find

$$G = \frac{(1 - 0.5^2)(1 - 0.333^2)}{|1 - (0.333\underline{/-90^\circ})(0.5\underline{/90^\circ}|^2(1 - 0.714^2)} = 2.92 \text{ dB}$$

7–6 S-PARAMETERS AND MATCHING NETWORKS

In Section 7–4 we defined *S*-parameters as reflection coefficients for an active device when the source and load impedances are 50 Ω. Impedances of 50 Ω are commonly used in industry, except in cable networks, where 75 Ω is used. *S*-parameter notation follows the matrix notation used for *z*-, *y*-, and *h*-parameters. Where we have used r_{12} as the input reflection coefficient from medium 1 to 2, the electronics worker would use s_{11}. The parameters s_{21} and s_{12} are the forward and reverse amplitude transfer coefficients.

We shall apply the matching-layer gain formula to the problem of using input and output networks to match a 50-Ω source and 50-Ω load to a transistor amplifier. We assume that the scattering parameters for the transistor are available from the manufacturer and formulate the problem as shown in Fig. 7–11.

The maximum gain for the input network is, from Eq. (7–8), $G_1 = 1/(1 - |s_{11}|^2)$, and for the output network, $G_2 = 1/(1 - |s_{22}|^2)$. We assume that the transistor

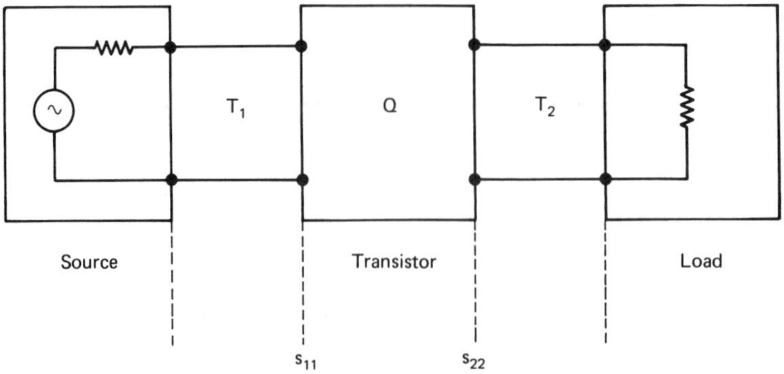

Figure 7–11 One-stage transistor amplifier with impedance-matching networks T_1, T_2.

is a unilateral device; that is, s_{12} is small enough to neglect. With this assumption the transistor gain is just s_{21}. The maximum gain of the entire system is $G = G_1 G_Q G_2$, or

$$G = \frac{1}{1 - |s_{11}|^2} \times |s_{21}|^2 \times \frac{1}{1 - |s_{22}|^2} \tag{7–9}$$

which is the *maximum* (unilateral) *transducer gain*. Equation (7–9) also follows from Eq. (5–63) with $r_S = S_{11}^*$ and $r_L = S_{22}^*$.

EXAMPLE 7–3

Calculate the maximum transducer gain for an HXTR 3101 transistor matched to a 50-Ω source and a 50-Ω load. Assume an operating frequency of 500 MHz, $V_{CE} = 10$ V, and $I_C = 10$ mA.

Solution From the manufacturer's catalog (see Fig. 7–12) we find

$$s_{11} = 0.565\angle{-139^\circ}$$
$$s_{12} = 0.036\angle{41^\circ}$$
$$s_{21} = 10.116\angle{101^\circ}$$
$$s_{22} = 0.499\angle{-40^\circ}$$

From Eq. (7–9) we have

$$G = \frac{1}{1 - 0.565^2} \times 10.116^2 \times \frac{1}{1 - 0.499^2} = 200 \quad \text{or} \quad 23 \text{ dB}$$

We should remember that s_{11} and s_{22} are nothing more than reflection coefficients. The parameter s_{21} is the gain in wave amplitude in an active device. In wave theory s_{21} is the increase in amplitude when a wave propagates through an

HEWLETT PACKARD

LOW COST GENERAL PURPOSE TRANSISTOR

HXTR-3101

Features

HIGH GAIN
19.5 dB Typical at 1 GHz

LOW NOISE FIGURE
1.8 dB Typical F_{MIN} at 1 GHz

LOW COST HERMETIC PACKAGE

Description

The HXTR-3101 is a low cost NPN bipolar transistor designed for high gain and wide dynamic range up to 4000 MHz. To achieve excellent uniformity and reliability, the manufacturing process utilizes self-alignment and ion implantation techniques. The chip is provided with a dielectric scratch protection over its active area.

The HXTR-3101 is supplied in the HPAC-100X, a rugged metal/ceramic package, and is capable of meeting the environmental requirements of MIL-S-19500 and the electrical test conditions of MIL-STD-750.

Absolute Maximum Ratings *

(T_{CASE} = 25° C)

Symbol	Parameter	Value
V_{CBO}	Collector to Base Voltage	30V
V_{CEO}	Collector to Emitter Voltage	18V
V_{EBO}	Emitter to Base Voltage	1.5V
I_C	DC Collector Current	50mA
P_T	Total Device Dissipation	600 mW
T_J	Junction Temperature	300° C
T_{STG}	Storage Temperature	-65°C to 150°C

*Operation in excess of any one of these conditions may result in permanent damage to this device.

Notes:

1. A Θ_{JC} maximum of 180° C/W should be used for derating and junction temperature calculations ($T_J = P_D \times \Theta_{JC} + T_{CASE}$).
2. A MTTF of 1 x 10^7 hours will be met or exceeded when the junction temperature is maintained under T_J = 200° C (based on an activation energy of 1.1 eV). For operation above this condition, refer to page 108. "Reliability Performance of Bipolar Transistors".

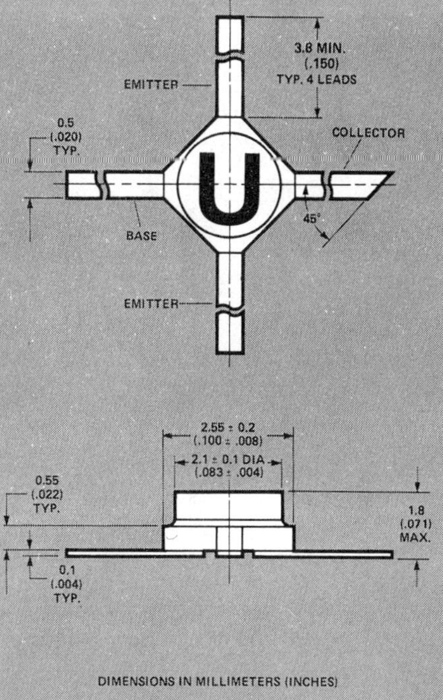

Outline HPAC-100X

(continued)

Figure 7–12 Courtesy of Hewlett-Packard.

Electrical Specifications at $T_{CASE} = 25°C$

Symbol	Parameters and Test Conditions	MIL-STD-750 Test Method	Units	Min.	Typ.	Max.
BV_{CBO}	Collector-Base Breakdown Voltage at $I_C = 100\ \mu A$	3001.1*	V	30		
I_{CBO}	Collector-Base Cutoff Current at $V_{CB} = 15$ V	3036.1**	nA			500
h_{FE}	Forward Current Transfer Ratio at $V_{CE} = 10$ V, $I_C = 10$ mA	3076.1*		50		180
f_T	Gain Bandwidth Product at $V_{CE} = 10$ V, $I_C = 15$ mA		GHz		6	
$\|S_{21E}\|^2$	Transducer Gain at 1000 MHz at $V_{CE} = 10$ V, $I_C = 15$ mA		dB		15.0	
F_{MIN}	Minimum Noise Figure at 1000 MHz $V_{CE} = 10$ V, $I_C = 10$ mA	3246.1	dB		1.8	
MAG	Maximum Available Gain at 1000 MHz $V_{CE} = 10$ V, $I_C = 15$ mA		dB		19.5	

*300 μs wide pulse measurement ≤ 2% duty cycle.
**Measured under low ambient light conditions.

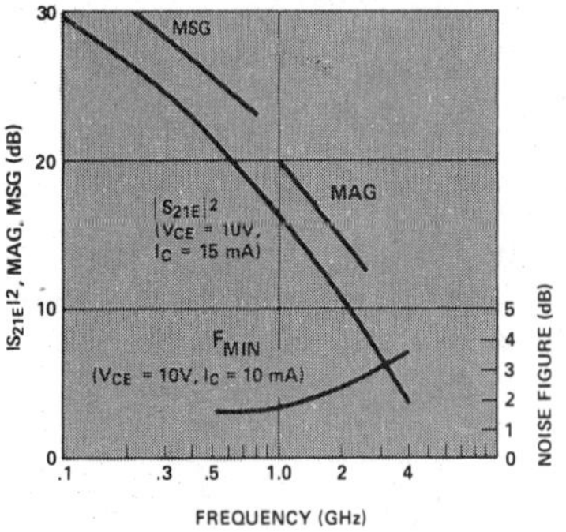

Figure 1. Typical $|S_{21E}|^2$, MAG, Maximum Stable Gain (MSG), and Noise Figure vs. Frequency

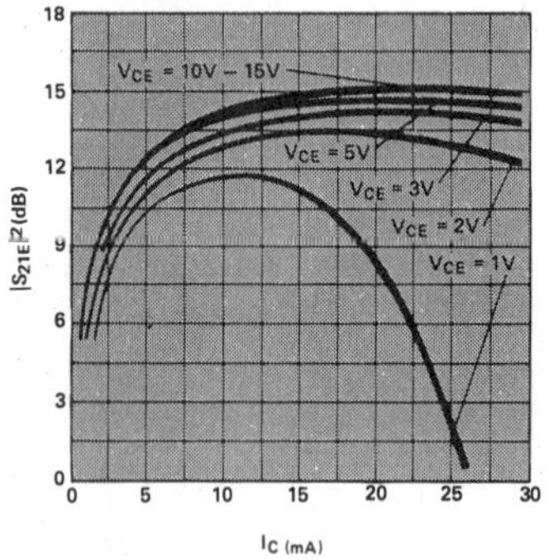

Figure 2. Typical $|S_{21E}|^2$ vs. Current at 1000 MHz

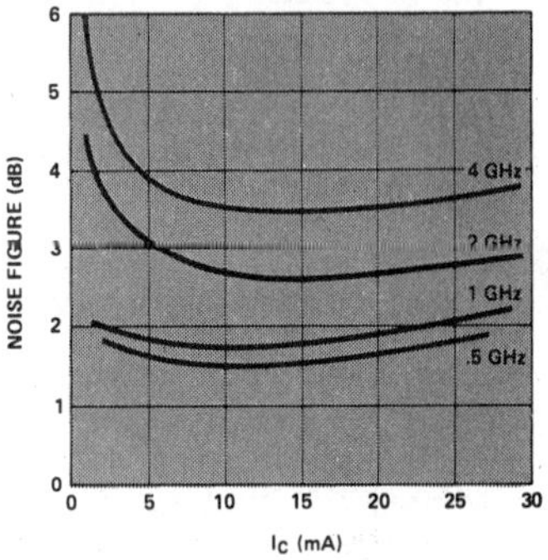

Figure 3. Typical Noise Figure vs. Collector Current (V_{CE} = 10V)

Typical S-Parameters ($V_{CE} = 10$ V, $I_C = 10$ mA)

Freq. (MHz)	S_{11} Mag.	S_{11} Ang.	S_{21} (dB)	S_{21} Mag.	S_{21} Ang.	S_{12} (dB)	S_{12} Mag.	S_{12} Ang.	S_{22} Mag.	S_{22} Ang.
100	0.705	-50	27.7	24.266	149	-36.7	0.015	60	0.903	-20
300	0.606	-110	23.5	14.962	116	-30.4	0.030	43	0.624	-36
500	0.565	-139	20.1	10.116	101	-28.9	0.036	41	0.499	-40
800	0.559	-162	16.5	6.683	89	-27.4	0.043	43	0.430	-41
1000	0.571	-169	14.5	5.330	78	-25.7	0.052	44	0.408	-43
1500	0.574	174	11.2	3.627	63	-23.6	0.066	48	0.394	-48
2000	0.591	161	8.9	2.774	49	-21.9	0.080	48	0.392	-57
3000	0.619	143	5.7	1.936	25	-18.8	0.115	45	0.427	-81
4000	0.639	125	3.4	1.488	1	-16.2	0.155	39	0.470	-107

($V_{CE} = 10$ V, $I_C = 15$ mA)

Freq. (MHz)	S_{11} Mag.	S_{11} Ang.	S_{21} (dB)	S_{21} Mag.	S_{21} Ang.	S_{12} (dB)	S_{12} Mag.	S_{12} Ang.	S_{22} Mag.	S_{22} Ang.
100	0.641	-60	29.5	29.854	144	-37.6	0.013	57	0.863	-23
300	0.565	-122	24.5	16.788	112	-31.8	0.026	44	0.556	-38
500	0.551	-149	20.7	10.839	98	-30.1	0.031	44	0.444	-40
800	0.553	-168	17.1	7.161	87	-28.1	0.039	50	0.387	-41
1000	0.560	-175	15.1	5.709	77	-26.4	0.048	49	0.363	-42
1500	0.564	171	11.8	3.869	62	-23.6	0.066	54	0.356	-47
2000	0.583	159	9.4	2.955	49	-21.6	0.083	52	0.354	-56
3000	0.611	142	6.3	2.058	26	-18.4	0.120	47	0.389	-81
4000	0.633	124	4.0	1.587	2	-15.9	0.160	39	0.431	-106

Figure 7–12 *(Continued).*

active medium; for example, when a sound wave propagates through a medium in which chemical or nuclear reactions are occurring. If the reaction rates are proportional to the density of the medium, the wave will cause an increase in energy release and the wave will grow in amplitude. In the last half of the twentieth century, self-amplifying waves frustrated the efforts of engineers attempting to build electric generators from active media contained by magnetic fields.

We said that we would also solve the multiple reflection problem as a feedback problem. We shall start by deriving an equation for gain with feedback. The reader already familiar with the feedback formula may wish to skip the next section.

7–7 GAIN WITH FEEDBACK

We assume that a device has a voltage or current gain, A_0. A network or loop is added which returns to the input a fraction of the output. The problem is to calculate the gain now that there is feedback: find the *closed-loop gain*, A.

Let f be the fraction of the output that is fed back, as shown in Fig. 7–13, where a_1 is the input amplitude, b_2 the output amplitude, and $a_1 + fb_2$ the input with feedback. From the definition of gain as the ratio of output to input, we have $A_0 = b_2/(a_1 + fb_2)$. We solve this equation for b_2/a_1 to obtain

$$\frac{b_2}{a_1} = \frac{A_0}{1 - fA_0}$$

By definition, the ratio b_2/a_1 is the gain with feedback, and we have

$$A = \frac{A_0}{1 - fA_0} \tag{7–10}$$

EXAMPLE 7–4

A transistor stage of an amplifier has a voltage gain of 100. When an emitter bypass capacitor is removed, 10% of the output voltage appears across the emitter resistor as a negative feedback voltage. Under this condition, what is the gain?

Solution From Eq. (7–10) we have

$$A = \frac{100}{1 + 0.1 \times 100} = 9.1$$

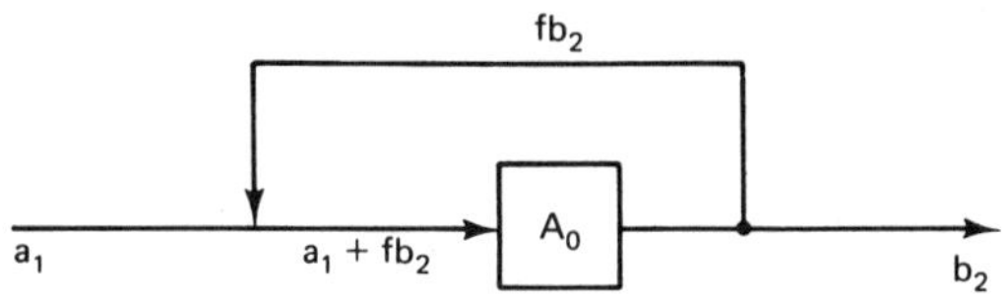

Figure 7–13 Active device with a feedback loop.

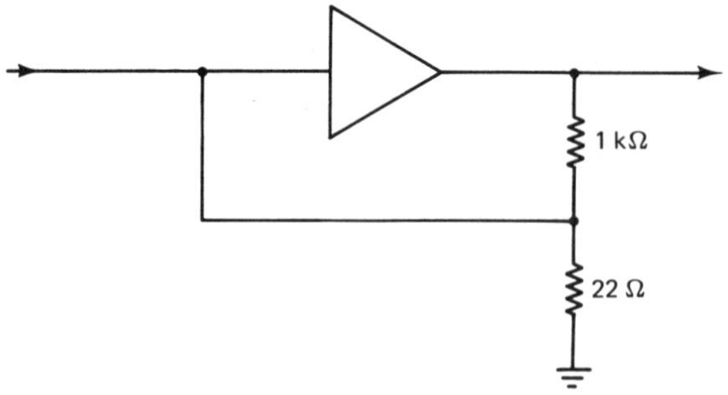

Figure 7–14 Amplifier with positive feedback, Example 7–5.

In Example 7–4 we see that the gain of the transistor is controlled mostly by the negative feedback rather than by the open-loop gain of the device. The next example shows that a positive feedback factor as small as a few percent is effective in producing a large closed-loop gain.

EXAMPLE 7–5

The noninverting amplifier in Fig. 7–14 has an open loop gain of 35. Find the closed-loop gain.

Solution The feedback factor is approximately 22/1000, or 0.022. From Eq. (7–10) we have

$$A = \frac{35}{1 - 0.022 \times 35} = 152$$

7–8 MULTIPLE REFLECTIONS AS FEEDBACK

We now visualize the matching layer with multiple reflections as a passive network with feedback (Fig. 7–15). With unit amplitude incident on the first interface, the wave amplitudes transmitted through the first and second interfaces are b_{12} and $b_{12}b_{23}$. The doubly reflected amplitude, $b_{12}r_{23}r_{21}$, provides an additional feedback amplitude at the input, as shown in the figure. The gain without feedback is $A_0 = b_{12}b_{23}/b_{13}$. Since there is an additional factor of $r_{23}r_{21}$ at the output there must be an additional $f = r_{23}r_{21}/A_0$ at the input. We substitute these results into the feedback formula [Eq. (7–10)] and obtain

$$G = |A|^2 = \frac{(1 - |r_{23}|^2)(1 - |r_{12}|^2)}{|1 - r_{23}r_{21}|^2(1 - |r_{13}|^2)}$$

which is the result obtained previously [Eq. (7–7)].

We can now turn to the problem of using S-parameters in the design of an amplifier.

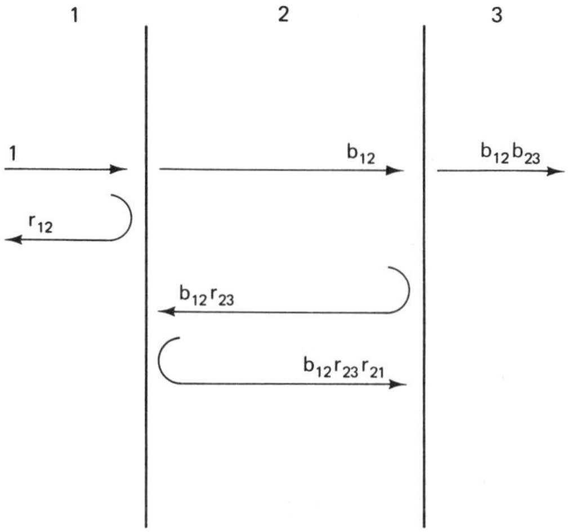

Figure 7–15 Transmitted, reflected, and feedback wave amplitudes in a two-port passive network.

7–9 500-MHz AMPLIFIER DESIGN

We shall use a 500-MHz amplifier design to demonstrate S-parameter and microstrip techniques at UHF/microwave frequencies. The specifications are as follows:

Use an HXTR 3101 transistor with $V_{CE} = 10$ V, $I_C = 10$ mA.
Design for maximum gain at 500 MHz.
Match to 50-Ω input and output.
Use the common-emitter configuration.

We shall calculate the maximum transducer gain, then design input and output matching networks, and finally design the bias circuit. From Eq. (7–9) we calculate a gain of

$$G = \frac{1}{1 - 0.565^2} \times 10.116^2 \times \frac{1}{1 - 0.499^2} = 200 \quad \text{or} \quad 23.0\,\text{dB}$$

as was shown in Example 7–3.

There are several ways to design matching networks. We shall use Smith charts because they help us to visualize the problem. To find the input impedance of the transistor we plot the scattering parameter s_{11} on the chart, and this is done in the following manner:

1. Adjust a pencil compass to the magnitude of s_{11} (0.565 from Fig. 7–12) by using the reflection coefficient scale at the bottom of the chart, and draw a VSWR circle.

2. Use a straightedge to draw a radius from the center of the circle to the angle $-139°$ at the edge of the chart; the intersection of the radius and the VSWR circle is s_{11}, shown as point A in Fig. 7–16.
3. Read the normalized impedance coordinates of point A, $0.320 - j0.340$; the input impedance of the transistor is $50 \times (0.320 - j0.340) = 16.0 - j17\ \Omega$.

To match the input impedance to a 50-Ω source we may either:

1. Rotate from the load (transistor input, point A) toward the source to the nearest resistive point (point B on the chart), then use a quarter-wave section to transform from the resistance at point B to 50 Ω at the source.
2. Find a transmission line of length and characteristic impedance that will transform directly from point A to 50 Ω.
3. Use a matching stub to cancel the transistor input reactance, then use a quarter-wave section to match the remaining resistance to the source impedance of 50 Ω.

For the 500-MHz input section we shall use the first method. We may rotate point A on the chart to either of the resistive points, B or C, but we must rotate in a clockwise direction toward the generator since we are starting at the load. At point B the resistance is $0.280 \times 50 = 14\ \Omega$, and from this point we can match to the 50-Ω source with a quarter-wave line of characteristic impedance $Z = \sqrt{Z_1 Z_2} = \sqrt{14 \times 50} = 26.5\ \Omega$. From the wavelength scale at the edge of the chart we see that the initial rotation will require a line of length 0.056λ. If we choose to rotate to point C, the resistance is $3.6 \times 50 = 180\ \Omega$, and the quarter-wave line would have a characteristic impedance of 94.9 Ω. Since the required line would be longer, we shall remain with the first choice. Before considering the type of line and materials to be used, we shall design the output network.

For the output we will design by the second method: match with a single line. As before, we start by plotting s_{22} on a Smith chart, point A in Fig. 7–17. The output impedance is $(1.55 - j1.30) \times 50 = 77.5 - j65.0\ \Omega$. We now must find the characteristic impedance and length of a transmission line that will transform directly from $77.5 - j65.0$ to 50 Ω. This technique is more difficult than what was used for the input network, but it results in a simpler matching network. We proceed by trial and error, and our first guess for the required characteristic impedance of the matching line is $2 \times 50 = 100\ \Omega$. If the impedance of the transistor had been less than 50 Ω, we should have taken $Z_c = \frac{1}{2} \times 50 = 25\ \Omega$ as our first guess. The remaining procedure is: Renormalize the impedance at point A to a characteristic impedance of 100 Ω, $z = (77.5 - j65.0)/100 = 0.775 - j0.650$; Plot the renormalized impedance at point B on the Smith chart; Draw a VSWR circle through point B; it intersects the real axis at point C, which corresponds to $0.46 \times 100 = 46\ \Omega$.

We see that the 100-Ω line transforms the transistor impedance to 46 Ω, not

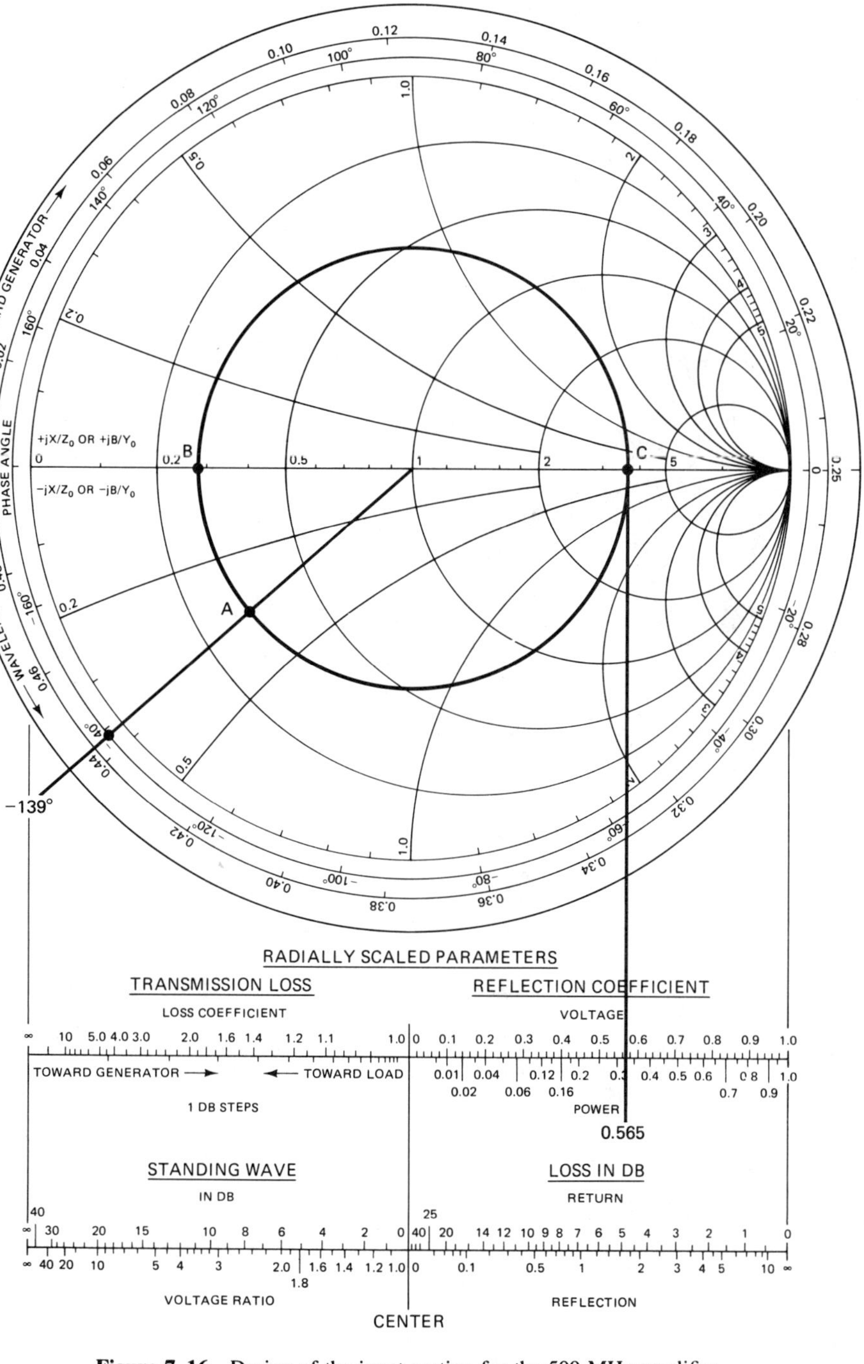

Figure 7–16 Design of the input section for the 500-MHz amplifier.

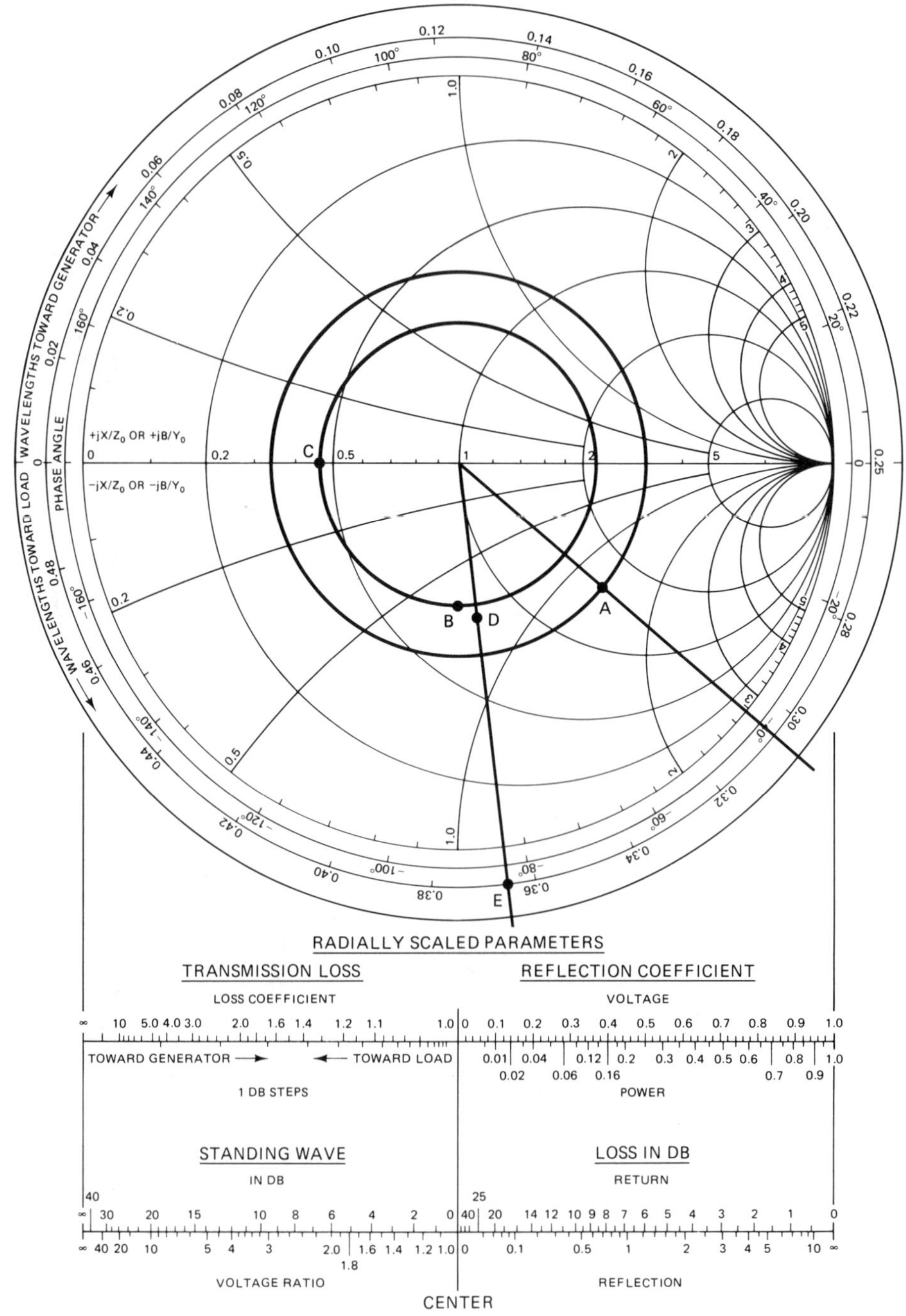

Figure 7–17 Design of the output section for the 500-MHz amplifier.

the 50 Ω we wanted. This almost is close enough, but to demonstrate the method we shall go on to a second approximation.

The transformation would have been perfect if the characteristic impedance of the line had been 50/0.46 = 108.7 Ω. For a second guess we try Z_c = 108.7 Ω. Renormalize the impedance to 108.7 Ω, $z = (77.5 - j65.0)/108.7 = 0.713 - j0.598$, and plot the result at point *D* in Fig. 7.17. A new VSWR circle through *D* now intersects the real axis at 0.44, which corresponds to a resistance of 0.44 × 108.7 = 47.8 Ω, which is close enough.

Draw a radius from the center, through point *D*, to the edge of the circle, point *E*. Since we must rotate from the transistor output impedance, point *D*, *toward the load* to point *C*, the required length of 108.7-Ω line is 0.36λ. This completes the design. We now need to provide a bias circuit and consider construction details.

7–10 500-MHz AMPLIFIER CONSTRUCTION

The matching networks can be transmission lines, either strip line, microstrip, or coaxial. For commercial use, coaxial line with a coaxially packaged transistor is practical, but for the individual experimenter strip line is easier. The required machine shop work to produce coaxial sections with the required impedances is time consuming. Strip-line implementations can be built in an hour.

We shall fabricate the strip lines from aluminum-clad polystyrene insulation board, which is available from building material suppliers in 4 by 8 ft sheets of ½-, 1-, and 2-in. thicknesses. Polystyrene foam has a permittivity almost equal to

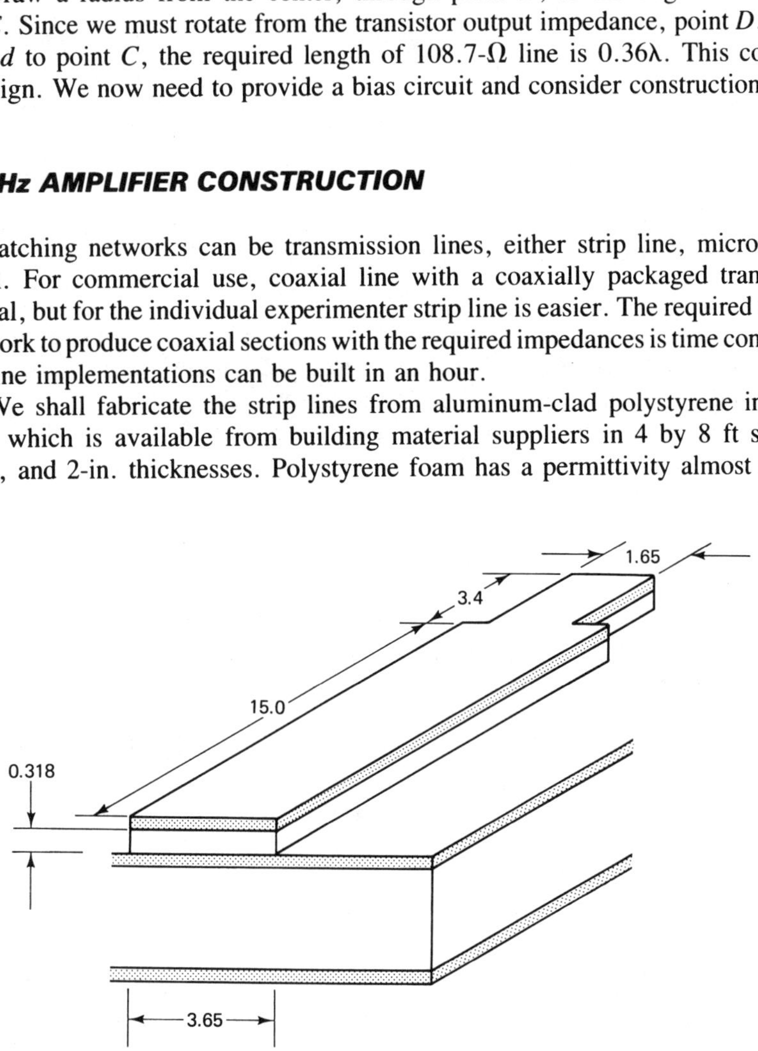

Figure 7–18 Input matching section for the 500-MHz amplifier, not to scale, dimensions in centimeters.

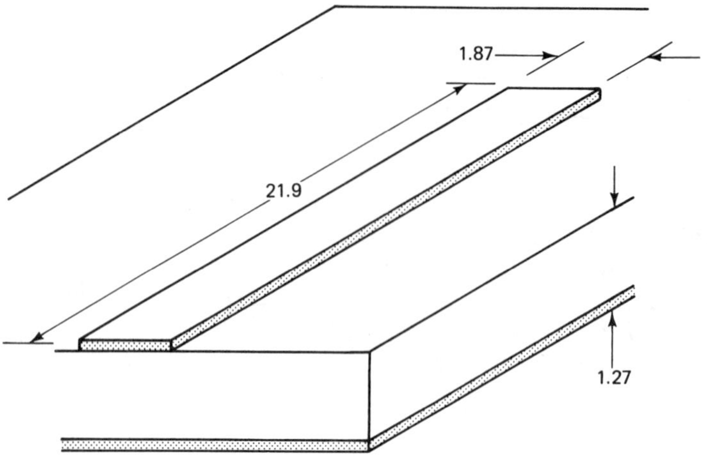

Figure 7–19 Output matching network for the 500-MHz amplifier, not to scale, dimensions in centimeters.

that of free space; thus the wavelength in the transmission lines is the same as the free-space wavelength. The characteristic impedance of strip line for various conductor widths and insulation thicknesses is shown in Fig. 4–6.

For the input network we need a length of 0.056λ of 50-Ω line and 0.250λ of 26.5-Ω line. The operating wavelength is $\lambda = c/f = 3 \times 10^{10}/500 \times 10^6 = 60$ cm, and the required line lengths are:

$$0.056 \times 60 = 3.36 \text{ cm of } 50\text{-}\Omega \text{ line}$$
$$0.25 \times 60 = 15.0 \text{ cm of } 26.5\text{-}\Omega \text{ line}$$

From Fig. 4–6 we find that for 50-Ω line we need a ratio of $W/T = 5.2$, and for the 26.5-Ω line we require $W/T = 11.5$. A suitable set of dimensions for polystyrene foam insulation board is shown in Fig. 7–18.

For the output network we require a transmission line of length 0.365λ = 21.9 cm. From Fig. 4–6 we find that the width-to-thickness ratio should be 1.47. If we use ½-in. insulation board the conductor width will be $1.47T = 1.87$ cm. A sketch of the output line is shown in Fig. 7–19.

7–11 500-MHz AMPLIFIER BIAS CIRCUIT

To complete the amplifier we need a biasing circuit. From the manufacturer's data we see that h_{FE} may vary from 50 to 180. We shall use a mean value of $\sqrt{50 \times 180} = 95$. A supply voltage of $V_{CC} = 20$ V will give $V_{CE} = 10$ V and $I_C = 10$ mA if we provide a 1-kΩ collector resistor.

The base current will be $I_C/h_{FE} = 10 \text{ mA}/95 = 0.105$ mA. If we use a voltage-

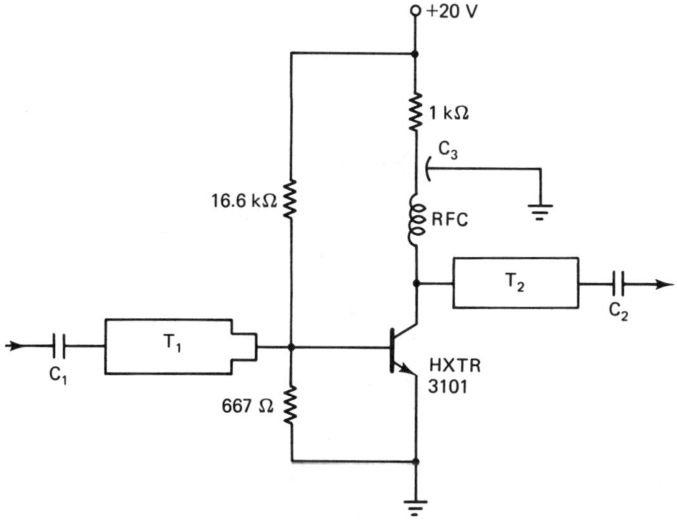

Figure 7–20 Schematic representation of the 500-MHz amplifier.

divider network for the base bias, the bias network current should be about 10 times the base current, or 1.05 mA. The required biasing resistors can be found from Ohm's law:

$$R_2 = \frac{0.7}{1.05 \times 10^{-3}} = 667\ \Omega$$

$$R_1 = \frac{20 - 0.7}{1.16 \times 10^{-3}} = 16.6\ \text{k}\Omega$$

where R_1 and R_2 are shown in a sketch of the complete design (Fig. 7–20).

7–12 2-GHz AMPLIFIER DESIGN

For a second design example we choose a 2-GHz amplifier to be constructed on printed circuit board. The specifications are as follows:

Transistor, HXTR 3101 (data from Fig. 7–12)
$V_{CE} = 10$ V $I_C = 10$ mA
$h_{FE} = 95$
$s_{11} = 0.591\angle 161°$
$s_{12} = 0.080\angle 48°$
$s_{21} = 2.774\angle 49°$
$s_{22} = 0.392\angle -57°$

The expected transducer gain is

$$G = \frac{1}{1 - |s_{11}|^2} \times |s_{21}|^2 \times \frac{1}{1 - |s_{22}|^2}$$

$$= \frac{1}{1 - 0.591^2} \times 2.744^2 \times \frac{1}{1 - 0.392^2} = 14.0 \quad \text{or} \quad 11.5 \text{ dB}$$

For the input network design we shall use the single-line method. As before, we plot s_{11} on a Smith chart and find the input impedance of the transistor, (0.257 + j0.160) × 50 = 12.85 + j8.0 Ω (point A, Fig. 7–21). Since this impedance is less than 50 Ω, our first trial for the characteristic impedance of a matching line will be Z_c = ½ × 50 = 25 Ω.

We proceed by trial and error. Renormalize the impedance to 25 Ω (12.85 + j8.0)/25 = 0.514 + j0.32. Plot the renormalized impedance at point B on the Smith chart and draw a new VSWR circle. Rotate toward the generator to point C, where the resistance is 2.2 × 25 = 55 Ω.

The match would be better if the characteristic impedance had been 50/2.2 = 22.73 Ω. Renormalize to 22.73 Ω, (12.85 + j8.0)/22.73 = 0.565 + j0.352, and plot this as point D on the chart. A new VSWR circle through point D intersects the real axis at point E, which gives 2.10 × 22.73 = 47.7 Ω. This is close enough, but a characteristic impedance of 50/2.1 = 23.8 Ω would be better. Draw a radius through point D to determine the required line length (point F) of 0.250λ − 0.072λ = 0.178λ.

For the output section we shall demonstrate a third matching technique. We shall use a short-circuited stub at the output of the transistor to cancel its output reactance, then use a quarter-wave section to match the remaining resistance to the 50-Ω load. The procedure is as follows. Plot s_{22} at point A in Fig. 7–22. The corresponding admittance is (0.48 + j0.44)/50 = (9.6 + j8.8) × 10^{-3} S. Use a stub line with a characteristic impedance of 114 Ω to neutralize the transistor reactance. Short-circuit the far end of the stub with a large capacitor, a 100-pF chip capacitor. The reactance of the capacitor is $X_c = 1/2\pi fC = 1/2\pi(2 \times 10^9)(100 \times 10^{-12}) = 0.8\ \Omega$. Normalize X_c to the line impedance and plot the result, j0.018, at point B. Rotate from point B, in a direction toward the generator, to point C to obtain the positive reactance, j114 Ω, that will neutralize the capacitive reactance of the transistor. From the scale at the edge of the chart note that the required length of 114 Ω-line to effect the rotation is (0.003 + 0.125)λ = 0.128λ. Provide a quarter-wave section to match the remaining resistance to the load. The required characteristic impedance is $Z_c = \sqrt{104 \times 50} = 72.2\ \Omega$, and the length, of course, is 0.250λ.

It is common practice to not use a Smith chart in the stub-line procedure above but simply to use the following rules:

1. A short-circuited 1/8-wavelength stub provides Z_c ohms of positive reactance at its open end, where Z_c is the characteristic impedance of the stub.
2. An open-circuited 1/8-wavelength stub provides Z_c ohms of negative reactance at its other end.

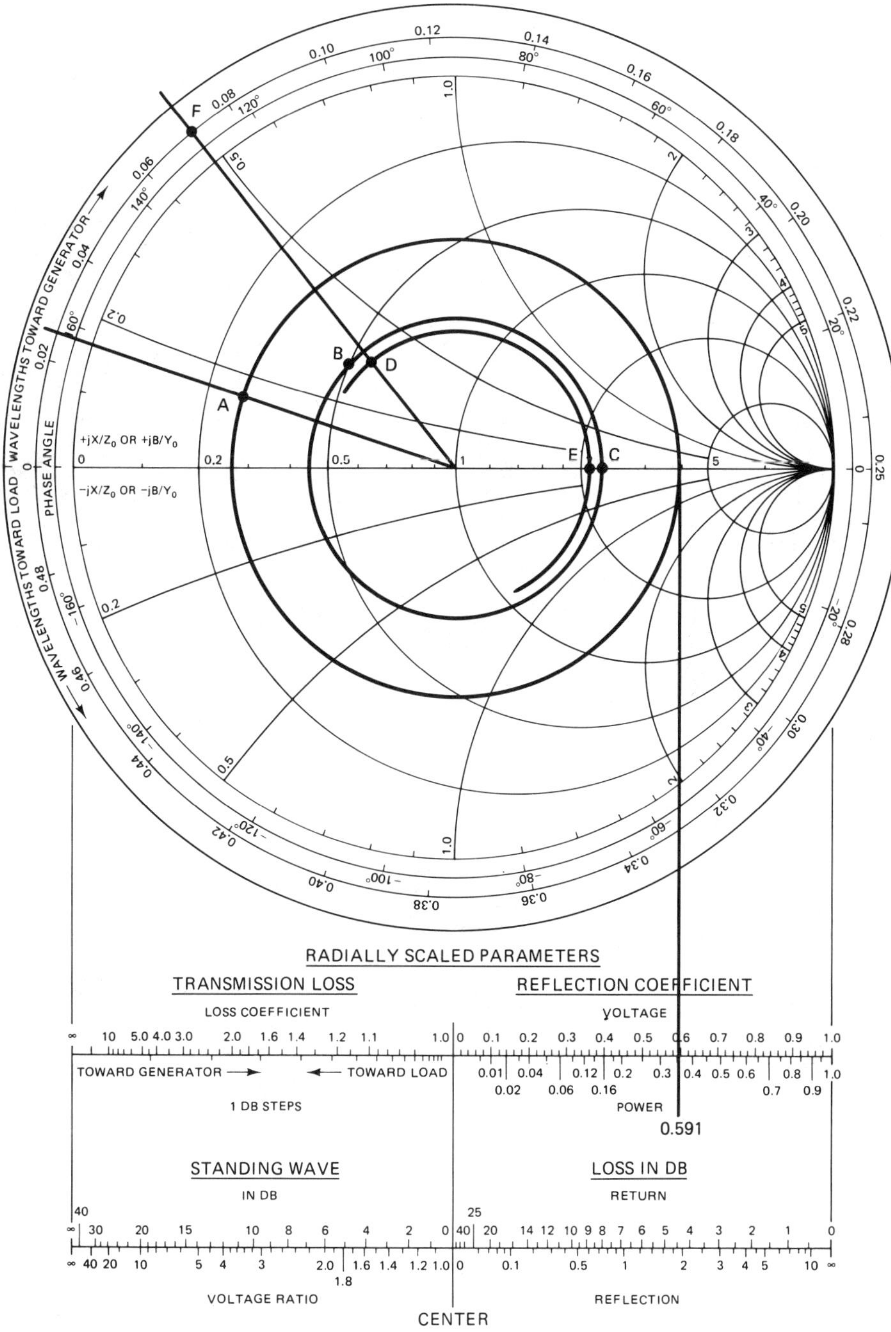

Figure 7–21 Design of the input matching section for the 2-GHz amplifier.

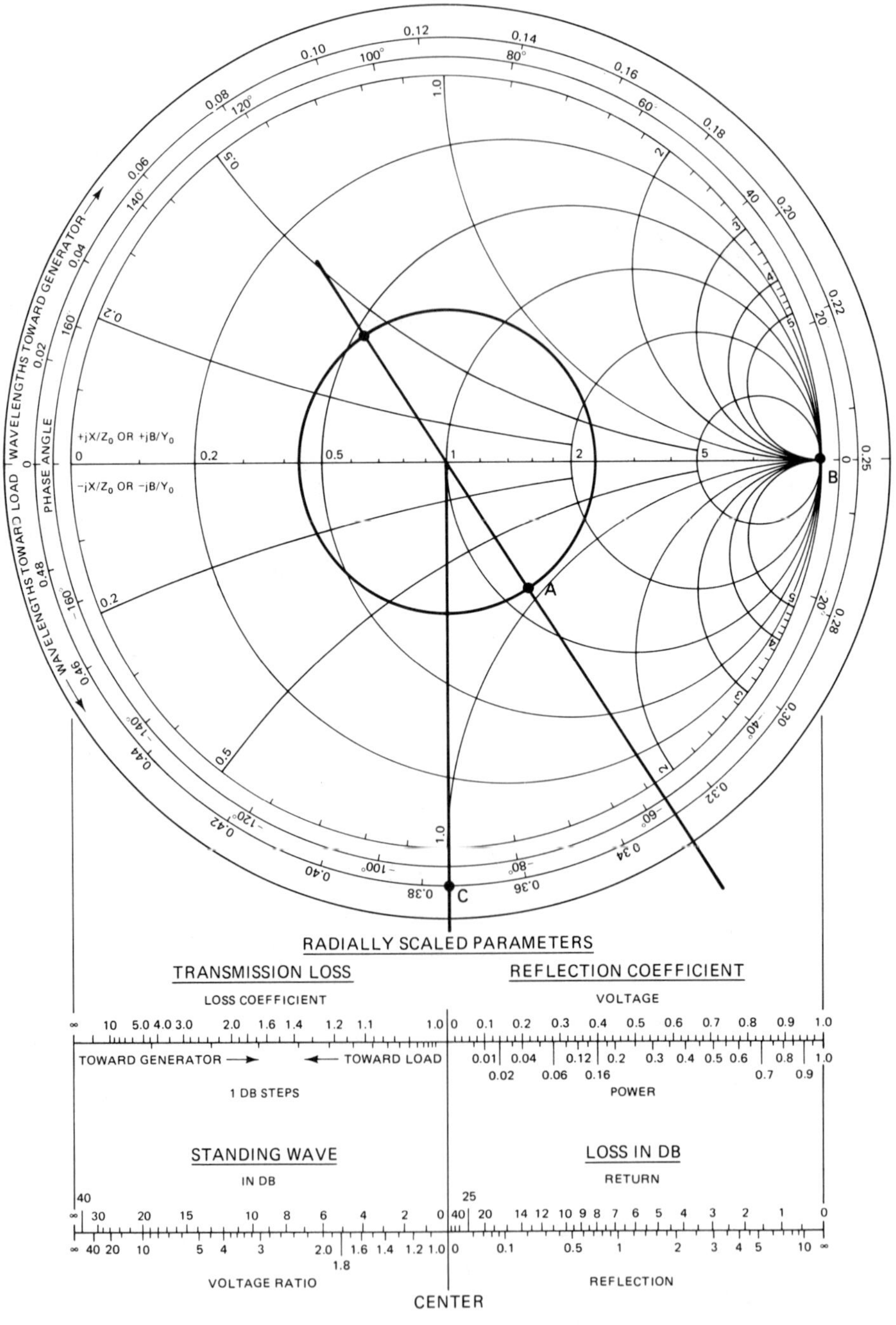

Figure 7–22 Design of the output matching network for the 2-GHz amplifier.

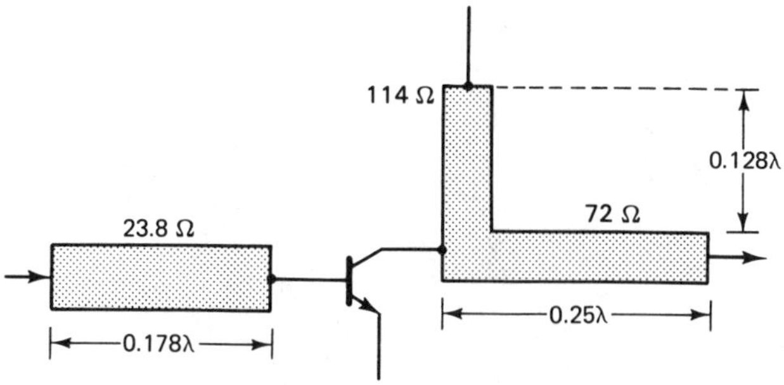

Figure 7–23 Strip-line layout for the 2-GHz amplifier.

It is well, however, to know the Smith chart procedure. There are times when it is not convenient to build a line with a characteristic impedance equal to the magnitude of the required reactance. A line of different characteristic impedance can be used, although its length no longer will be one-eighth of a wavelength. The transmission-line requirements for matching the amplifier are summarized in Fig. 7–23.

7–13 2-GHz AMPLIFIER CONSTRUCTION

For strip-line construction we will use 1/32-in., double-copper-clad, Duroid printed circuit board. The nominal relative dielectric constant is 2.6. The characteristic impedance of strip line for various conductor width-to-insulation thickness ratios is shown in Fig. 4–6. By comparing the required impedances with the W/T ratios shown in the figure, we find that for the input section we need 0.178 wavelength of 23.8-Ω line, $W/T = 7.5$, $W = 7.5 \times T = 7.5 \times 0.0794 = 0.595$ cm. For the output section we require:

For the stub, 0.128 wavelength of 114-Ω line, $W/T = 0.57$, $W = 0.57 \times T = 0.57 \times 0.0794 = 0.045$ cm

For the quarter-wave section, 0.250 wavelength of 72-Ω line, $W/T = 1.50$, $W = 1.50 \times 0.0794 = 0.119$ cm

The line lengths would be given by factors of λ, where $\lambda = \lambda_0/\sqrt{\epsilon_r}$, if all of the electromagnetic field were within the dielectric, but because of fringing effects the effective wavelength is somewhat greater. A graph of the ratio of effective to free-space wavelength for $\epsilon_r = 2.6$ is shown in Fig. 4–5. At 2 GHz the free-space wavelength is $\lambda_0 = 3 \times 10^{10}/2 \times 10^9 = 15$ cm. From the graph we find that the length of the input line is $0.178\lambda_e = 0.178 \times 15 \times 0.661 = 1.76$ cm, where 0.661 is the wavelength ratio when $W/T = 7.5$. For the stub, we have $\lambda_e/\lambda_0 = 0.717$ from the graph when $W/T = 0.57$, and the required line length is $0.128 \times 15 \times 0.717 = 1.38$ cm. For the quarter-wave line we find from the graph that $\lambda_e/\lambda_0 = 0.694$ when the W/T ratio is 1.50, and the line length is $0.250 \times 15 \times$

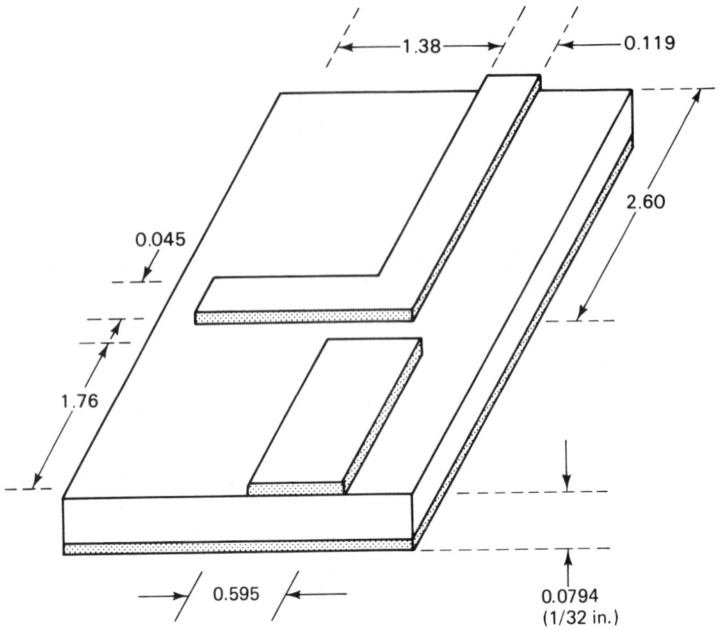

Figure 7–24 Isometric projection of the input and output strip-line sections for the 2-GHz amplifier, not to scale, dimensions in centimeters.

0.694 = 2.60 cm. We summarize all the dimensions in a sketch of the input and output transmission line sections in Fig. 7–24.

7–14 2-GHz AMPLIFIER BIAS CIRCUIT

Since we have designed for the same dc conditions as in the 500-MHz amplifier, we can use the same bias circuit. The ac-grounded end of the stub will be a convenient place to connect to V_{CC}.

7–15 LUMPED-PARAMETER TECHNIQUES

In the preceding examples we used transmission lines for matching networks because they are typical of microwave technique, and Smith charts because they are instructional as well as convenient. It is possible, of course, to use the lumped-parameter matching methods described in Chapter 3; also, we can use the corresponding formulas to compute the required characteristic impedance of a matching line, rather than find it by trial and error on a Smith chart.

EXAMPLE 7–6

Calculate the characteristic impedance of a transmission line to match 77.5 − j65.0 Ω to 50 Ω. Compare the result with the value found by trial and error for the output section of the 500-MHz amplifier.

Solution From Section 3–8 we have

$$Z_c = \sqrt{50 \times 77.5}\sqrt{1 - \frac{65.0^2}{50 \times 77.5 - 77.5^2}} = 107.5\ \Omega$$

From the Smith chart we had 108.7 Ω, an error of 1.1%.

EXAMPLE 7–7

Calculate the characteristic impedance of a transmission line to match 12.85 + j8.0 Ω to 50 Ω. Compare the result with that found by trial and error for the input line to the 2-GHz amplifier described previously.

Solution From Section 3–8 we have

$$Z_c = \sqrt{50 \times 12.85}\sqrt{1 - \frac{8.0^2}{50 \times 12.85 - 12.85^2}} = 23.59\ \Omega$$

From the Smith chart, by trial and error, we had Z_c = 23.8 ohms, which gives an error of 0.9%.

As an example of lumped-parameter technique we shall design the 2-GHz amplifier input network using the circuit shown in Fig. 7–25. Our problem is to match a 50-Ω source to an impedance of $z_1 = r_1 + jX_1 = 12.85 + j8.0\ \Omega$. Let the reactance in the transformer be $X_L' = X_L - X_1$. Since the load reactance now is canceled by X_1, we can use the equation for matching resistances: from Eq. (3–14) we have

$$X_L = \sqrt{R_1R_2 - R_2^2} = \sqrt{50 \times 12.85 - 12.85^2} = 21.85\ \Omega$$

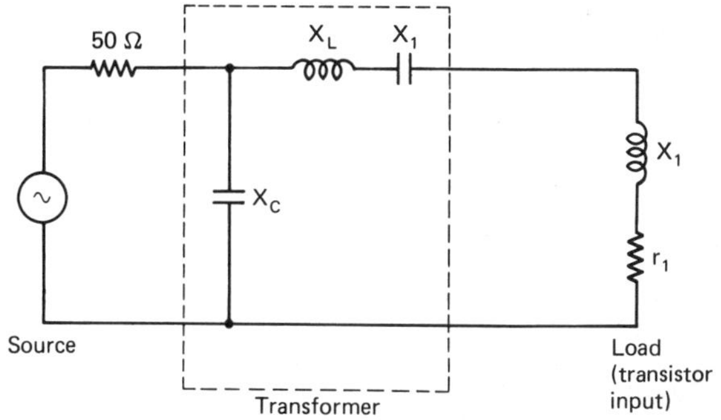

Figure 7–25 Lumped-parameter matching network for a 2-GHz amplifier.

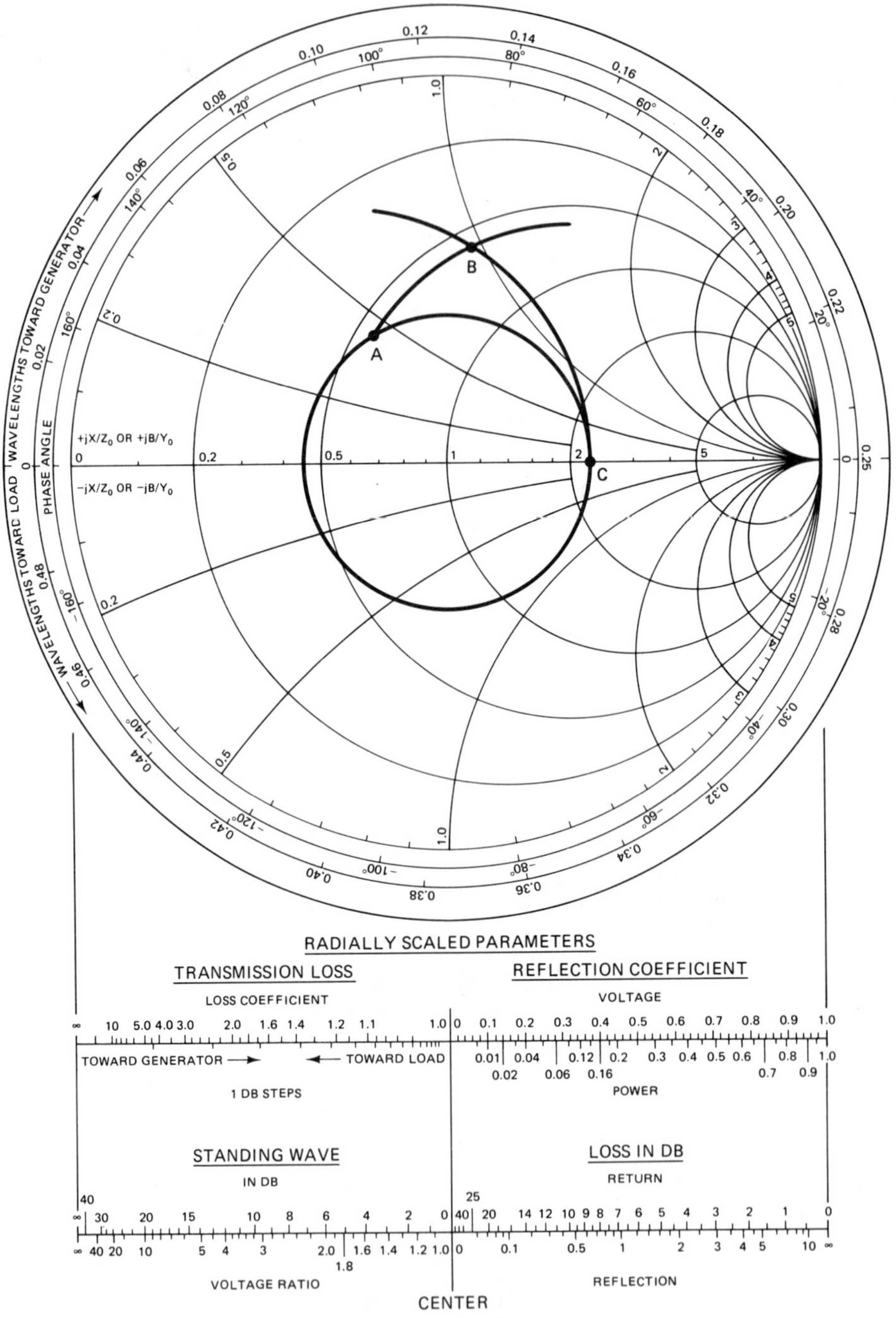

Figure 7–26 Smith chart solution for a lumped-parameter matching network.

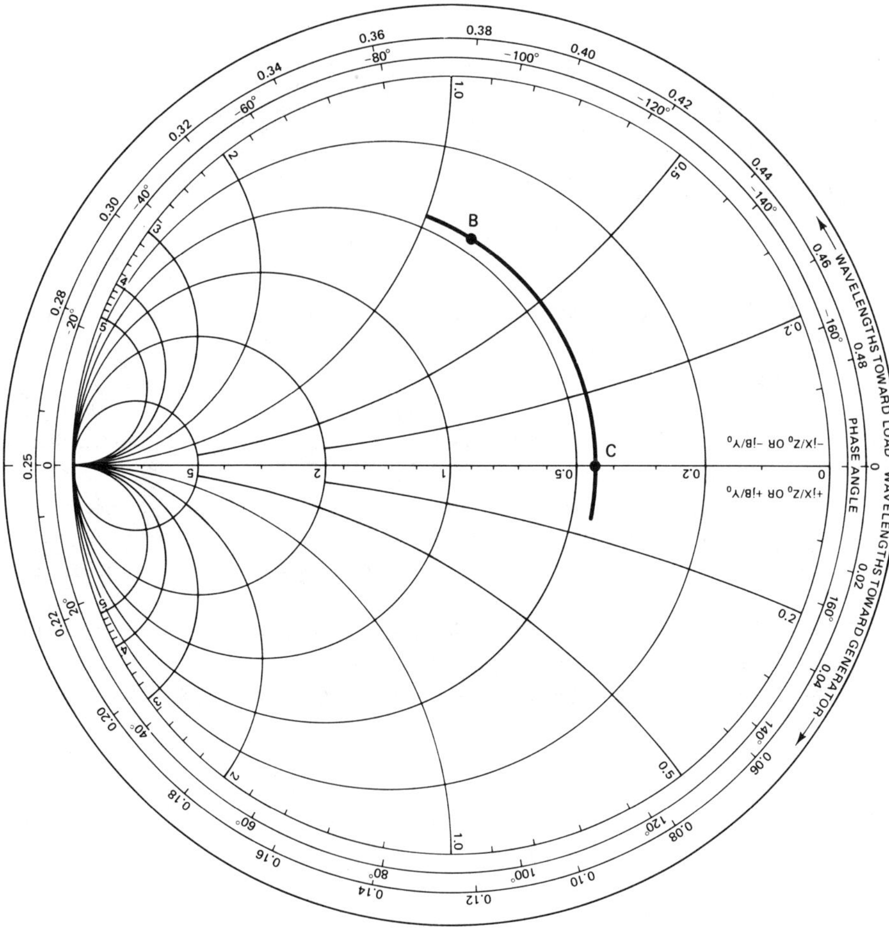

Figure 7–27 Admittance chart used in the lumped-parameter matching network problem shown in Fig. 7–26.

The inductance in the transformer is

$$X_L' = X_L - X_2 = 21.85 - 8.0 = 13.9\ \Omega$$

Now we can compute the required capacitive reactance from Eq. (3–8):

$$X_C = \frac{R_1R_2}{X_L} = \frac{50 \times 12.85}{21.85} = 29.41\ \Omega$$

The capacitance required is $C = 1/2\pi f X_C = 1/2\pi \times (2 \times 10^9) \times 29.41 = 2.7$ pF, and the inductance required is $L' = 1/2\pi f X_L' = 1/2\pi \times (2 \times 10^9) \times 13.9 =$ 5.7 pH. Since this is the inductance of a few centimeters of wire, accurate construction would be difficult. Lumped-parameter elements are used more easily at lower frequencies.

Lumped-parameter circuit elements for matching networks also can be determined by using Smith charts. As an example of the technique we shall repeat the previous problem. We start as we did in Fig. 7–21. Normalize the impedance 12.85 + j8.0 to a characteristic impedance of 23.59 Ω (12.85 + j8.0)/23.59 = 0.545 + j0.339. Plot the result at point A in Fig. 7–26. (A rotation from A to point C is the transmission-line solution, which was previously shown as points D and C in Fig. 7–21.)

Find the required series inductive reactance by rotating from point A along a circle of constant resistance. The stopping point, B, is not yet determined. Plot the objective, the 50-Ω resistance, at point C on an admittance chart (Fig. 7–27). The journey from B to C, along a circle of constant conductance on the admittance chart, will determine the required shunt capacitance or susceptance. Determine the length of the journey by placing the admittance chart on top of the impedance chart and by pushing a pin through the intersection of the constant resistance and constant conductance journeys. This determines the point B, and the journey from A to C is completed.

Read the required inductive reactance on the impedance chart, (0.97 − 0.40) × 23.59 = 13.4 Ω. Find the required capacitance from the admittance chart, X_C = 23.59/0.78 = 30.2 Ω.

There are several other methods for designing matching networks, including methods in which the Q of the network is used to limit or increase the bandwidth of an amplifier. Some of these methods are discussed in Chapter 3. A transmission-line technique for matching the output of an oscillator to a load, while controlling the feedback to its input, is described in Chapter 8.

APPLICATION 7–1: Microwave Transistor Amplifiers

In this application we build, troubleshoot, and test a UHF or microwave amplifier.

Equipment and Materials

model maker's band saw (optional)
electrician's knife
sandpaper
building insulation, aluminum-clad polyurethane or polystyrene foam; polystyrene preferred
printed circuit board, 1/32-in., double copper clad
model maker's knife or razor blade
PC board etching kit (optional)
UHF/microwave signal generator with calibration, attenuation, and modulation capabilities (optional)
UHF/microwave spectrum analyzer (optional)
VSWR tuned-amplifier/meter with detector probe (optional)
Hewlett-Packard HXTR 3101 transistor, or equivalent
assorted 1/8-W resistors and chip capacitors
dc power supply, 0 −30 V, or ±15 V, or equivalent batteries
cellophane tape

Remarks. An inexpensive model maker's band saw is helpful for quickly cutting building insulation to the length, width, and thickness required for strip lines. If a saw is not available, the foam may be cut roughly with a knife and smoothed to size with sandpaper.

PC board etching kits are available from retail electronics outlets. If time is limited, a razor blade can be used to quickly cut strip lines on PC board. The individual experimenter should not use industrial etching materials; the chemicals are corrosive and hazardous.

If laboratory test equipment is not available, build one of the amplifiers shown here and a corresponding oscillator described in Chapter 8; the oscillator can be used to test the amplifier.

Microwave solid-state devices can be destroyed by static electricity. For this reason manufacturers often ship them in a lossy foam material. If static electricity is a problem, the devices should be unpacked, and work done, on a ground plane. The ground plane need not be elaborate. Ordinary aluminum foil wrap can be spread on the workbench. The foil should be connected to an earth ground with a clip lead and the worker's arms should rest on the foil.

If microwave transistors are not available, a fast-switching transistor, such as the 2N 2222A, can be substituted, but the amplifier will have to be redesigned for operation at about 250 MHz.

PART 1: 500-MHz AMPLIFIER

Build the amplifier shown in Figs. 7–18 to 7–20 and 7–28 using building insulation to form the strip lines. If component leads are cleaned carefully, electrical connections can be made by pressing the parts together with cellophane tape, and this will provide sufficient heat sinking for the transistor if it is operated at the low power levels we have designed for. If a more permanent amplifier is desired, elect to build on PC board, as described in Part 2.

If a microwave generator and detector are available, the amplifier can be tested simply by connecting it between them. Connect outer conductors of the coaxial lines from the test equipment to the ground plane of the strip lines, and connect the inner conductors of the coaxial lines to the strip lines. Measure the amplifier gain at the design frequency and at several nearby frequencies. Include gain and bandwidth information in your report. Bandwidth may be determined with a spectrum analyzer, if one is available.

If troubleshooting is necessary, remember that a microwave amplifier, like any other, can be saturated by too large an input signal. Connect the generator directly to the detector to be sure that the test equipment is working properly. Check the dc voltages on the transistor terminals. The transistor will not work if the bias voltages are wrong; replace the detector with a 50-Ω load at the amplifier output. Connect a small coil across the detector input coax and use it as a probe to signal trace the amplifier; increase V_{CC} while watching for output on the detector. We have designed for low power and occasionally a transistor may require more power to turn on.

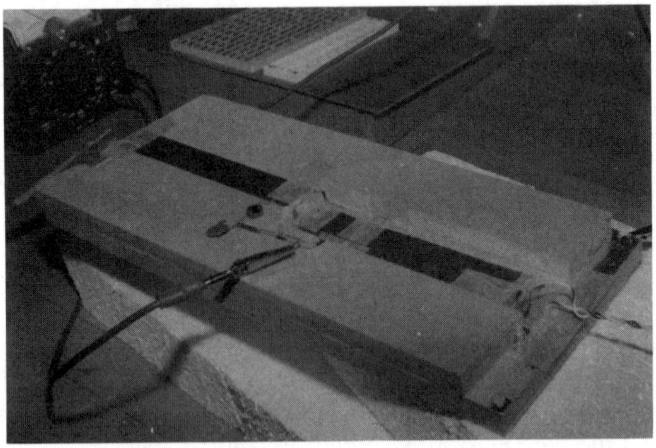

Figure 7–28 Photograph of a 500-MHz amplifier constructed from aluminum-clad polystyrene foam insulation board.

PART 2: 2-GHz AMPLIFIER

Build, on PC board, the 2-GHz amplifier described in the text. The input and output lines are sketched in Fig. 7–24. The bias circuit is similar to that shown in Fig. 7–20. For an RFC use five turns of fine wire wound on a plastic tube, 1.6 mm diameter, 5 mm long. Use either an etching kit or a razor blade to form the strip lines on PC board. Use a heat sink when soldering the transistor terminals to the strip lines. Troubleshoot and test, following the procedures listed in Part 1.

SUMMARY

Amplification can be achieved by introducing an active parameter into a circuit. The parameter may be any component, device, or physical process capable of accepting energy in two or more ways. In electronics, the name "parametric amplifier" is usually reserved for the special case of an active reactance. The varactor is a commonly used active reactance. It can accept energy from an electromagnetic wave, as an ordinary capacitor, or from having its capacitance decreased while it is charged. Since varactor paramps often are one-port amplifiers, a circulator may be required to separate input and output signals.

One form of traveling-wave amplifier comprises varactors distributed along a distributed- or lumped-inductance transmission line. As the signal wave propagates down the line its amplitude grows at the expense of the pump wave.

If the active device is a medium—gas, liquid, or solid—the amplifier is called a maser or laser, depending on its operating frequency. The medium accepts energy in the form of EM standing and traveling waves, and by atomic or molecular excitation.

The transistor is an example of an amplifier in which the active parameter is a resistance. To prepare for the amplifier projects at the end of the chapter, we discuss transistor amplifier theory and design in detail, starting with a review of S-parameters.

S-parameters are defined and measured with the aid of the equations

$$S_{11} = \left.\frac{b_1}{a_1}\right|_{a_2=0} \qquad S_{21} = \left.\frac{b_2}{a_1}\right|_{a_2=0}$$

$$S_{12} = \left.\frac{b_1}{a_2}\right|_{a_1=0} \qquad S_{22} = \left.\frac{b_2}{a_2}\right|_{a_1=0}$$

Measured values of S-parameters for transistors are provided by manufacturers to specify transistor performance at UHF and microwave frequencies.

Transistor amplifier efficiency is improved if matching networks are used between source, amplifier, and load. If multiple reflections occur in a network or medium, the resultant transmitted intensity is

$$|t_L|^2 = \frac{(1 - |r_{12}|^2)(1 - |r_{23}|^2)}{|1 - r_{23}r_{21}|^2}$$

The gain realized from a matching network is

$$G = \frac{(1 - |r_{23}|^2)(1 - |r_{12}|^2)}{|1 - r_{23}r_{21}|^2(1 - |r_{13}|^2)}$$

$$G_{\text{max}} = \frac{1}{1 - |r_{13}|^2}$$

If a transistor amplifier is embedded in matching networks, the maximum (unilateral) transducer gain is

$$G = \frac{1}{|1 - S_{11}|^2}|S_{21}|^2\frac{1}{|1 - S_{22}|^2}$$

Since multiple reflections are a form of feedback, the equation for loop gain can also be derived from the feedback equation

$$A = \frac{A_0}{1 - fA_0}$$

We use Smith chart techniques to design matching networks for two examples of transistor amplifiers and provide construction details.

PROBLEMS

7–1. Find the input and output impedances of the HXTR 3101 transistor when the operating frequency is 4 GHz. Assume that $V_{CE} = 10$ V, $I_C = 10$ mA, and use the data shown in Fig. 7–12.

7–2. If the HXTR 3101 is properly matched, find the unilateral transducer gain at 4 GHz. Assume that $V_{CE} = 10$ V, $I_C = 10$ mA.

7–3. For the HXTR 3101 plot s_{11} on a Smith chart for operating frequencies from 500 MHz to 2 GHz. Assume that $V_{CE} = 10$ V, $I_C = 10$ mA. Does the curve follow a circle of constant resistance?

7–4. A transmission-line matching section is to be used to match the output of the HXTR 3101 transistor to a 50-Ω load. The operating frequency is 800 MHz. Assume the dc operating conditions to be $V_{CE} = 10$ V, $I_C = 10$ mA. The matching section is to be a ⅛-wavelength stub and a quarter-wave section. The strip-line material is 1/32-in. Duroid, $\epsilon_r = 2.6$. Find the characteristic impedances of the strip lines, in ohms, and their lengths, in millimeters.

7–5. Find the conductor widths of the strip lines in Problem 7–4.

7–6. Molecular energy levels in an active medium are 5.2 and 3.2 μV above the normal ground state. What pumping and signal frequencies should be used if the medium is used in a maser? Assume that Planck's constant is 0.667×10^{-15} V s.

7–7. A sound wave in a cylinder of gas can be amplified by varying an applied pressure. What characteristic of the gas is being used as an active parameter?

7–8. You are asked to design a swing for a playground. Assume that the average person can comfortably reach the ropes at a height of 1 to 1½ m above the seat and that the swing is operated in the degenerate mode. What should be the height of the swing, measured from seat to fulcrum?

7–9. In the text the statement is made that varactors are convenient for microwave paramps because the capacitance is in the picofarad range. Explain.

7–10. If a suitable pump frequency is applied to the collector, would it be possible to construct a transistor amplifier without the use of a dc power supply? Explain your answer.

8

Oscillators

8–1 INTRODUCTION

In an electronic oscillator the energy source may supply a small amount of energy throughout each cycle or a small amount of energy only at the beginning of each cycle, and in either case the external circuitry determines the character of the oscillations. Alternatively, the energy source itself may determine the frequency, amplitude, and other characteristics of the oscillations through its internal nonlinearities, and the auxiliary circuitry is little more than a feedback network. We shall consider examples of all three possibilities.

Since microwave oscillators are more easily understood in terms of wave rather than circuit theory, we shall explain oscillator action by building a bridge from our understanding of circuit theory to a wave interpretation.

We shall start with the well-known Colpitts oscillator and use it to demonstrate the Barkhausen conditions. We shall design two UHF/microwave oscillators suitable for construction and testing in a few hours of laboratory time. After discussing some of the important solid-state microwave/millimeter oscillators we shall conclude with the mathematical theory of linear oscillators. Later we shall find linear oscillator theory useful in the study of antennas and optics.

8–2 OSCILLATOR DEFINITION

An oscillator is a device capable of storing energy in two or more forms. A child's swing and a pendulum alternately store energy in the form of gravitational potential energy and kinetic energy or energy of motion. A parallel *LC* or *tank* circuit alternately stores energy in the form of an electric field, when the capacitor is charged, and a magnetic field, which surrounds the coil when a current flows through it.

DEMONSTRATION 8–1

Loop several rubber bands together and loop the last band through a weight or bob. Hold one end of the string of bands and allow the bob to oscillate up and down with small amplitude. This is called damped motion because the amplitude slowly decreases owing to friction. A damped oscillator will not generate a pure sine wave because its amplitude is not constant.

By a slight motion of the hand the oscillations can be sustained at small amplitude. The motion now satisfies Hooke's law: force (and acceleration) are proportional to stretch or displacement. This type of motion is capable of generating a pure sine wave and so is called *simple harmonic motion*. Because of the importance of SHM in oscillator design and wave propagation it will be discussed later in this chapter.

Increase the amplitude of the oscillations until the bands become slack at the top of the motion. This is an example of *saturation*. Lower the bob and increase the amplitude until the bob hits the floor at the bottom of each oscillation; this is *cutoff*.

Notice in Demonstration 8–1 that a person is able naturally, without precise instruction, to provide a small amount of additional energy at the right time to sustain SHM. SHM is desirable in an oscillator if it is to be used to generate waves in a narrow bandwidth. Nonharmonic motion will generate distorted waveforms with a wide spectrum containing harmonics of the fundamental frequency. In the demonstration above, nonharmonic motion was caused by excessive amplitude or incorrect biasing (holding the bob too close to the floor). SHM was achieved by a subtle control of the oscillator. One of our objectives is to learn how to control a transistor or other active device when it is used as part of an oscillator.

8–3 COLPITTS OSCILLATOR

Self-sustained oscillations occur when a portion of the output energy of an amplifier is sent back to the input. If the feedback energy is derived from a capacitive voltage divider (Fig. 8–1), the circuit is called a *Colpitts oscillator*. Capacitors C_1 and C_2 form the voltage divider; also, the total capacitance of C_1 and C_2 is in parallel with inductor L and forms a tank circuit. The tank circuit, of course, determines the resonant frequency of the oscillator.

We can consider the power supply to be an ac ground if the circuit is operated at low frequencies; however, at high frequencies it is necessary to ensure a good ac ground by including a large shunt capacitance at the top of the tank, C_G in Fig. 8–1. The output is coupled to the load, R_L', through the coupling capacitor, C_L.

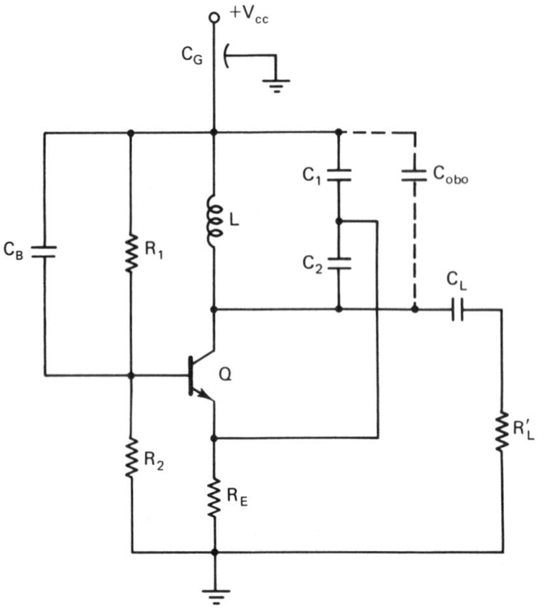

Figure 8–1 Colpitts oscillator: feedback is derived from the split capacitance in the tank circuit.

Since the collector voltage of the transistor amplifier is obtained through the coil L in series with the dc supply, this circuit is called a "series-fed" Colpitts oscillator. More significantly, the large capacitance C_B shunts the base of the transistor to ground; thus the transistor is in the common-base (CB) configuration.

In Fig. 8–2 we have drawn the essential ac circuit, omitting the dc bias network R_1R_2 and some of the capacitors. We see that the tank circuit can be regarded as one cell of a short-circuited, lumped-parameter transmission line. The open end of the transmission line is loaded partly by the transistor input resistance r_{be} and partly by R_L'. The resistance R_L' is the total resistance of the load resistance R_L in parallel with the equivalent parallel coil resistance.

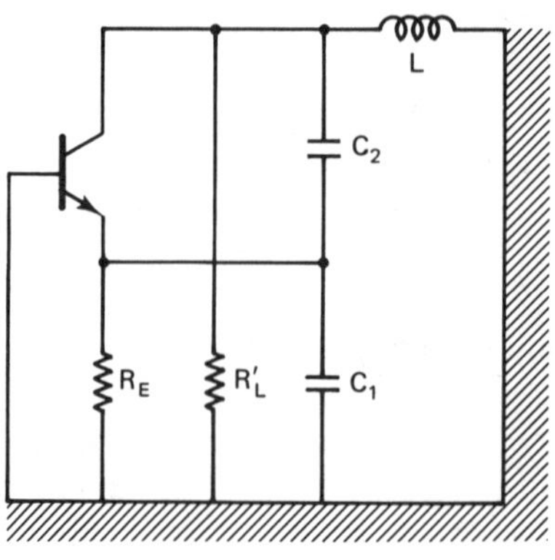

Figure 8–2 Colpitts circuit redrawn as a short-circuited, lumped-parameter transmission line.

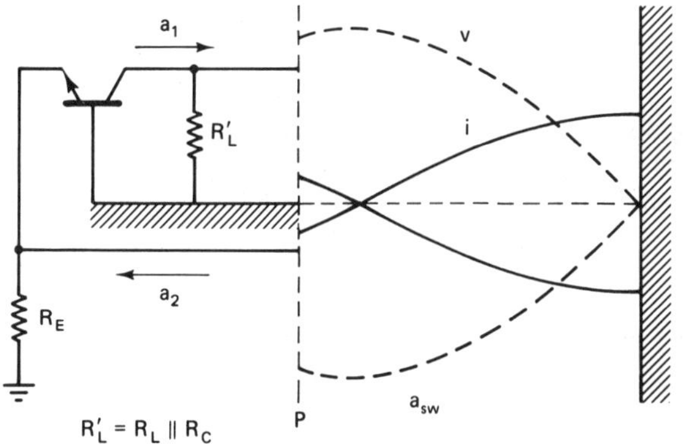

Figure 8–3 Colpitts circuit redrawn to show the standing waves in the tank circuit.

In Fig. 8–3 we have redrawn the circuit again, this time showing the standing wave on the short-circuited transmission line. The output of the transistor, represented by a traveling wave of amplitude a_1, is dissipated in the resistance R_L'; that is, part of its energy is delivered to the load and the other part sustains a standing wave. The part that enters the tank travels an electrical distance of a quarter-wavelength to the short. After reflection, which causes an additional 180° phase change, it travels another quarter-wavelength back to the entrance of the transmission line, plane P; thus the total phase delay is approximately 360°. A traveling wave of amplitude a_2 continues its journey to the input resistance r_{be}, where it is dissipated and produces a feedback voltage at the input of the transistor. The emitter resistance R_E is much larger than r_{be} and can be ignored.

We see that the wave ultimately returned to the transistor input is in phase with the original wave, there being no phase change through a common-base amplifier. This is an example of the Barkhausen phase condition: the loop phase change in an oscillator is an integral multiple of 360°:

$$f = |f|\angle n \times 360^\circ \qquad n = 1, 2, 3, \ldots$$

where f is the fraction of the output fed back to the input.

The loop travel time is only a half period, corresponding to a 180° phase delay. In a half period the feedback voltage should change from positive to negative to sustain the oscillations; this requires an extra 180° phase change, which is supplied by the reflection. The Barkhausen phase condition might be stated better as a temporal phase delay of a multiple of 180° plus a spatial change of 180°, $n \times 180^\circ + 180^\circ$, $n = 1, 2, 3, \ldots$. If feedback were always positive, $n = 0$, the circuit would not oscillate, the transistor would be driven to a final extreme state and held there, the circuit would be a *latch*.

In addition to the phase condition we have to ensure sufficient amplitude in feedback to overcome losses and sustain the oscillations. If f is the fraction of the output voltage v_o that appears at the input to the transistor, the input voltage is $v_i = fv_o$. Since voltage gain is $A_v = v_o/v_i$, we can combine the two equations to obtain $fA = 1$. The *Barkhausen amplitude condition* is $fA \geq 1$. The product fA

must be greater than unity to ensure starting and maintaining oscillations under varying conditions.

In our first design we shall attempt to limit the feedback just enough to overcome losses. The transistor is conducting throughout the cycle, and its amplitude swing is confined to the nearly linear region of its characteristic curves. From an examination of technical literature we conclude that some sine-wave oscillator designs are attempted in this manner.

EXAMPLE 8–1

A common-base transistor oscillator is being designed for an operating frequency of 10 MHz. The following parameters have been tentatively assigned for the circuit shown in Fig. 8–1:

$$V_{cc} = 10 \text{ V}$$
$$R_E = 5 \text{ k}\Omega$$
$$R_L' = 2.5 \text{ k}\Omega$$
$$Q = 10$$

Estimate the remainder of the design for feasibility: calculate A_v, f, C_1, C_2, and L.

Solution In a CB configuration the collector current is approximately equal to the emitter current and the voltage gain is $A_v = v_o/v_i = i_c R_L'/i_e r_{be}$, or

$$A_v = \frac{R_L'}{r_{be}} \tag{8–1}$$

where r_{be} is the input resistance of the transistor. We can estimate the input resistance from the empirical formula for the resistance of a forward-biased diode,

$$r_{be} = \frac{27 \text{ mV}}{I_E} \tag{8–2}$$

and we see that we can find the voltage gain after we have determined the dc emitter current.

We assume that half of V_{cc} is dropped across the transistor and the remaining 5 V across the emitter resistor; then from Ohm's law we have

$$I_E = \frac{\frac{1}{2}V_{cc}}{R_E}$$

$$= \frac{\frac{1}{2}\ 10}{5 \text{ k}\Omega} = 1 \text{ mA}$$

From Eq. (8–2) we find the input resistance,

$$r_{be} = \frac{27\ \text{mV}}{1\ \text{mA}} = 27\ \Omega$$

and from Eq. (8–1) we find the voltage gain,

$$A_v = \frac{2.5 \times 10^3}{27} = 92.6$$

The specified quality determines the total capacitance. From the definition of quality as the ratio of stored to dissipated energy per radian cycle we have $Q = \omega CR_L'$, or

$$C = \frac{Q}{\omega R_L'}$$

$$= \frac{10}{2\pi \times 10^7 \times (2.5 \times 10^3)}$$

$$= 63.7\ \text{pF}$$

Here we have neglected the small loading of the tank circuit by the input resistance of the transistor.

To determine the capacitance ratio for adequate feedback we assume the Barkhausen amplitude condition to be satisfied by an excess feedback of 10%,

$$fA = 1.1 \geq 1$$

and use the voltage-divider equation,

$$f = \frac{v_{C1}}{v_C} = \frac{C}{C_1}$$

Solving for C_1 we have $C_1 = C/f$, or

$$C_1 = \frac{AC}{1.1}$$

$$= \frac{92.6 \times (63.7 \times 10^{-12})}{1.1}$$

$$= 5360\ \text{pF}$$

Check: From the voltage-divider equation we have

$$f = \frac{63.7 \times 10^{-12}}{5360 \times 10^{-12}}$$

$$= 0.0119$$

and

$$fA = 0.0119 \times 92.6$$
$$= 1.10$$

We can find the value of the other capacitor from the equation for the total capacitance, $1/C = 1/C_1 + 1/C_2$,

$$\frac{1}{C_2} = \frac{1}{C} - \frac{1}{C_1}$$

$$= \frac{1}{63.7 \times 10^{-12}} - \frac{1}{5360 \times 10^{-12}}$$

$$C_2 = 64.5 \text{ pF}$$

Finally, we find the required inductance from the approximate equation for the resonant frequency $\omega^2 = 1/LC$,

$$L = \frac{1}{\omega^2 C}$$

$$= \frac{1}{(2\pi \times 10^7)^2 \times (63.7 \times 10^{-12})}$$

$$= 3.98 \ \mu\text{H}$$

The design procedure given in Example 8–1 is sufficient to build a prototype to see why it does not work. A common difficulty is "motorboating" or "squegging," caused by excessive or stray feedback. Excessive feedback drives the transistor to cutoff, and a large $R_E C_1$ time constant holds it there long enough to amplitude modulate the oscillations. The modulation is manifested in unwanted side bands. This difficulty can be avoided by:

Readjusting the capacitor, to reduce feedback

Partially bypassing the emitter resistor, to reduce the effective transistor input resistance (Fig. 8–4)

Rebiasing the transistor at a larger emitter current, to reduce R_E

Inserting an inductance in the emitter circuit between the emitter and feedback network

Adopting an alternative approach

An alternative design procedure is to use extremely excessive feedback. The transistor is driven to cutoff and remains there throughout most of the cycle. Like a child being pushed on a swing, the oscillator receives a small "push" at the end of each cycle. Since the character of the oscillation is determined by the tank circuit, not the transistor, a nearly pure sine wave can be achieved.

Extreme feedback can be achieved by connecting the feedback line to the smaller capacitor. In a particularly simple example we make the two tank capacitors equal in value. Not only is the feedback increased, the time constant $R_E C_1$ is greatly

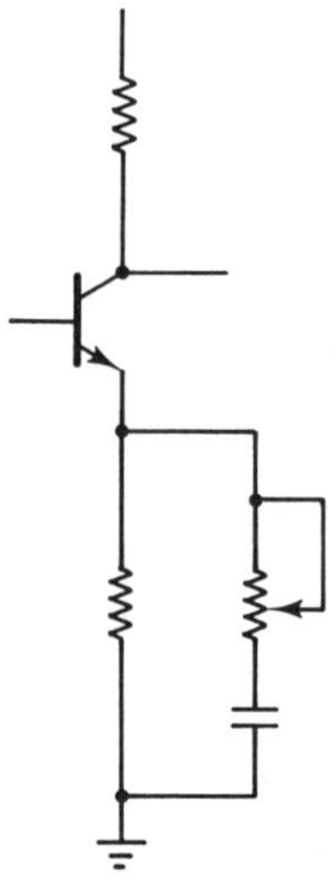

Figure 8–4 A partially bypassed emitter resistance is one way to control the feedback ratio in an oscillator circuit.

reduced. The large negative swings of the feedback voltage negatively charge capacitor C_B (Fig. 8–1). If the time constant $R_1 C_B$ is sufficiently large, the transistor is held near cutoff throughout most of a cycle. Only a small "push" is given to the tank circuit at the beginning of each cycle. The circuit values and performance for a 60-MHz example are as follows:

$$V_{cc} = 10\text{ V}$$
$$I_E = 10\text{ mA}$$
$$R_E = 500\ \Omega$$
$$C_1 = C_2 = 27\text{ pF}$$
$$C = \frac{C_1}{2 + C_{obo}}$$
$$= \frac{27}{2} + 8 = 21.5\text{ pF}$$
$$L = 0.334\ \mu\text{H}$$

Transistor, 2N2222A
Second harmonic, -18 dB
Third harmonic, < -20 dB

At this frequency it is necessary to include in the design the internal capacitance of the transistor, C_{obo} (Fig. 8–1).

8–4 SHUNT-FED COLPITTS OSCILLATOR

The collector voltage of the Colpitts oscillator shown in Fig. 8–5 is obtained through a collector resistor in parallel with the tank, and for this reason it is called "shunt fed." A more significant change is the reversal of the sign of the feedback voltage,

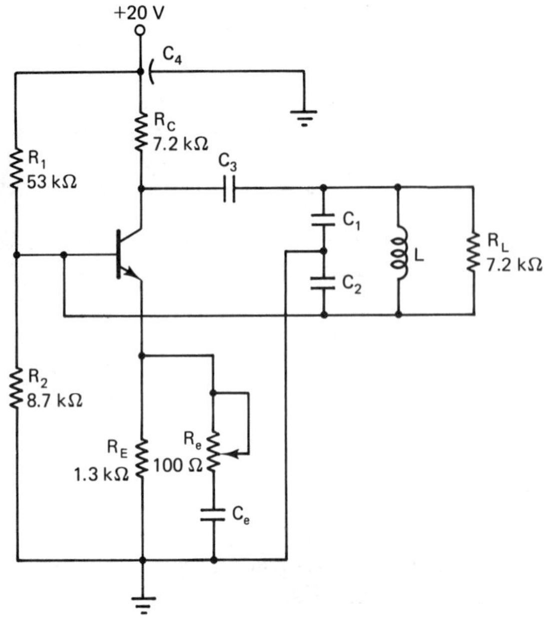

Figure 8–5 CE series-fed Colpitts oscillator. Component values are for a 100-kHz operating frequency.

which must now be connected to the base of the transistor. Input and output now have the emitter in common and the transistor is in CE configuration. This configuration is not as useful as the CB at high frequencies because the higher input resistance is shunted by the base capacitance. In the 2N2222A transistor, for example, the common-base capacitance is $C_{ibo} = 25$ pF. The CB configuration has an additional advantage of greater stability.

Design procedure is much the same as in the preceding examples. Component values and specifications for a 100-kHz oscillator are

$$Q = 15$$
$$C_1 = 0.007\ \mu\text{F}$$
$$C_2 = 0.063\ \mu\text{F}$$
$$L = 0.382\ \text{mH}$$
$$A_v = 15$$
$$f = 0.1$$

and the remaining parameters are shown in Fig. 8–5.

8–5 HIGH-FREQUENCY OSCILLATORS

At high frequencies the transistor parameters are complex, and we cannot use the simple design procedures given in the previous examples. Further, the lumped-parameter tank circuit becomes impractical and must be replaced by a cavity or mismatched length of transmission line. A mismatched transmission line stores electric and magnetic energy in the form of standing waves and so satisfies our definition of an oscillator. The Q of the oscillator is more difficult to calculate; we shall have to consider the multiple reflections of waves to find the energy stored and lost in the mismatched line.

We have already discussed the advantages of the CB configuration at high frequencies. An additional advantage is the low input impedance of the CB configuration, which makes it easy to mismatch the feedback line. Two design examples are given in the following sections.

8–6 500-MHz OSCILLATOR DESIGN

A suitable driver at UHF/microwave frequencies is the HXTR 3101 transistor in the CB configuration. A suitable oscillator is a half-wavelength coaxial line, loaded by the input impedance of the transistor at one end and the transistor's output impedance at the other. A symbolic representation of the circuit is shown in Fig. 8–6. In the figure we also show a $\lambda/8$ stub; its purpose is to cancel the output reactance of the transistor. We also show a quarter-wave line which will match the remaining transistor resistance to a 50-Ω load. The general plan resembles the 0.5-GHz amplifier; however, the design is made more complicated by the CB configuration, the calculation of mismatch in the feedback line, and the Barkhausen requirements.

From the manufacturer's catalog, for the HXTR 3101 transistor with V_{CE} = 10 V and I_C = 10 mA, we have

$$s_{11} = 0.565\angle{-139^\circ}$$
$$s_{21} = 10.116\angle{101^\circ}$$
$$s_{12} = 0.036\angle{41^\circ}$$
$$s_{22} = 0.499\angle{-40^\circ}$$

In addition, we shall need the following normalized impedances which can be calculated from the S-parameters or obtained by plotting the S-parameters on a Smith chart:

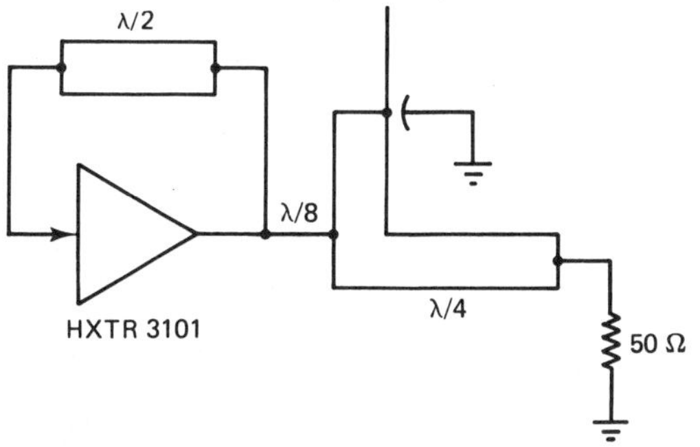

Figure 8–6 Symbolic representation of an 0.5-GHz oscillator.

$$z_{1e} = 0.31 - j0.34 = 0.46\angle{-47.6^\circ}$$
$$z_{2e} = 1.55 - j1.33 = 2.042\angle{-40.6^\circ}$$

We have added a subscript e to the input and output impedances to remind ourselves that the data are for the CE configuration. We shall have to derive equivalent data for the CB configuration.

8–7 TRANSITION TO THE COMMON-BASE CONFIGURATION

To make the transition to the CB configuration, we refer to an approximate equivalent circuit for high-frequency transistors shown in Fig. 8–7. The CB output impedance is measured by replacing the load resistor, shown at the top of the figure, by a generator. Since the reactance of capacitor C_e is much smaller than that of C_c, the measured reactance will be approximately that of C_c. The same statement can be made when the generator is moved to the CE terminals and the output impedance is measured in the CE configuration. We can conclude that Im $z_{2b} \approx$ Im z_{2e}.

In the CE configuration the measurement of the input impedance is referred to the base current, but in the CB configuration it is referred to the emitter current. As a result, the input impedance is higher for the CE configuration by a factor of the current ratio, $z_{1b} \approx z_{1e}/|\beta|$. Since the impedances are determined from S-parameters, when the transistor is matched to 50 Ω, a better estimate of the common-base input impedance is

$$z_{1b} \approx \frac{z_{1e}}{|s_{21}|} \tag{8–3}$$

A corresponding relation for the real part of the output resistance is

$$\text{Re } z_{2b} \approx \text{Re } z_{2e}|s_{21}| \tag{8–4}$$

With these approximations and the preceding data we have

$$z_{1b} = \frac{z_{1e}}{|s_{21}|} = \frac{0.31 - j0.34}{10.116} = 0.0306 - j0.0336,$$

$$Z_{1b} = 1.53 - j1.68\ \Omega$$
$$z_{2b} = \text{Re } z_{1e}|s_{21}| \parallel \text{Im } z_{ie} = 1.55 \times 10.116 \parallel -j1.33 = 15.17 \parallel -j1.33,$$
$$Z_{2b} = 784 \parallel -j66.5\ \Omega$$

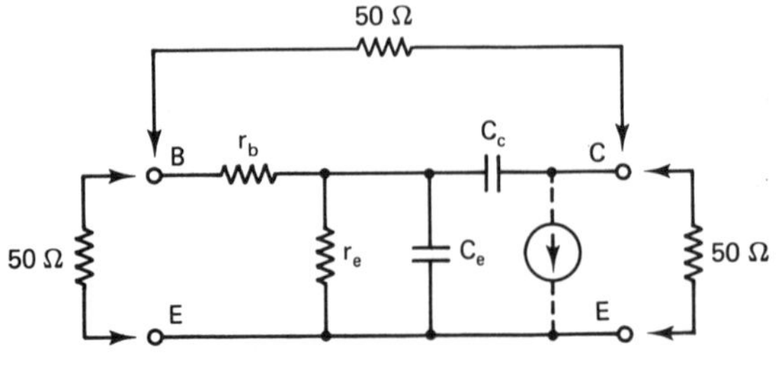

Figure 8–7 Equivalent circuit for a high-frequency transistor. B, C, E, base, collector, and emitter terminals; r_b, base spreading resistance, 1 to 5 Ω; r_e, forward-biased junction resistance, 10 to 100 Ω; C_e, emitter capacitance, about 100 pF; C_c, collector–base capacitance, about 3 pF.

The reverse transfer amplitude s_{12} has been ignored because it is small.

From the above results we see that to cancel the output reactance of the transistor we shall need a short-circuited, 118-λ stub with a characteristic impedance of 66.5 Ω. The remaining resistance, 784 Ω, can be transformed to the 50-Ω load with a quarter-wave line of characteristic impedance $Z_c = \sqrt{50 \times 784} = 198\ \Omega$.

8–8 500-MHz OSCILLATOR BARKHAUSEN CONDITIONS

We now turn to designing the oscillator itself, and we tentatively assume that a half-wavelength of transmission line connecting the transistor output to the input will satisfy the Barkhausen phase condition.

To see if the Barkhausen amplitude condition is satisfied we must calculate the feedback product, fA, where f is the feedback factor and A the amplitude gain. The feedback factor is the magnitude of the transmission coefficient of the 50-Ω coaxial feedback line. From Eq. (7–5) we have

$$|t| = \frac{|t_{12}|\,|t_{23}|}{|1 - r_{23}r_{21}|}$$

where t_{12} is the transmission coefficient through the interface between the transistor output and the feedback line, and t_{23} the coefficient between the output of the feedback line and the input to the transistor. The reflection coefficients r_{21} and r_{23} are for the same interfaces. These coefficients can be calculated from the impedance data with Eq. (2–11) or with a Smith chart:

$$r_{21} = \frac{784 - 50}{784 + 50} = 0.880\underline{/0^\circ}$$

$$r_{23} = \frac{1.53 - j1.68 - 50}{1.53 - j1.68 + 50} = 0.942\underline{/-176.5^\circ}$$

$$|t_{12}|^2 = 1 - |r_{12}|^2 = 1 - |r_{21}|^2 = 1 - 0.880^2 = 0.225$$
$$|t_{23}|^2 = 1 - |r_{23}|^2 = 1 - 0.942^2 = 0.1152$$

and the evaluation of Eq. (7–5) is

$$|t| = \frac{(0.225 \times 0.1152)^{1/2}}{|1 - 0.942\underline{/-176.5^\circ} \times 0.880\underline{/0^\circ}|} = \frac{0.161}{|1.828\underline{/1.6^\circ}|} = 0.0881$$

The current gain of a transistor in CB configuration is approximately unity, and the effective voltage gain is approximately $A_v = A_i r_2/r_1 = 1 \times 50/1.53 = 32.7$. The feedback product is

$$fA_v = 0.0881 \times 32.7 = 2.88 \geq 1$$

and we see the Barkhausen amplitude condition is satisfied.

We turn now to deciding whether or not the half-wavelength feedback line satisfies the Barkhausen phase condition. The half-wavelength, mismatched line can be regarded as a resonant tank circuit, sometimes called "antiresonant." The standing waves on the line play the role of the circulating currents and fields in an *LC* tank circuit. At resonance there is a phase delay of 90° in the transmitted wave. There is a phase delay of 180° due to propagation delay through the half-wavelength line. Since the operating frequency will be somewhat higher than the resonant frequency, there will be an additional phase delay of almost 90°. The total phase delay is nearly 360°, and it looks as if the Barkhausen conditions may be satisfied. Actually, the oscillator is more complicated than is apparent. As the oscillator self-adjusts its operating frequency—trading phase error against feedback amplitude—the phases of the coefficients in Eq. (7–5) change with the changing wavelength and electrical path length of the feedback line. The Barkhausen phase and amplitude conditions are interdependent. We shall proceed with the design, assuming that we have done enough to ensure that it probably will work. A more thorough analysis will be given in Section 8–15.

8–9 500-MHz OSCILLATOR QUALITY

To find the Q of the proposed oscillator we rewrite Eq. (5–52) as

$$Q = \frac{2\pi(1 + |r_{23}|^2)}{|t_{23}|^2/z_3 + |t_{21}|^2|r_{23}|^2/z_1}$$

where $z_3 = Z_3/Z_2$ and $z_1 = Z_1/Z_2$. With the conservation equations $|t_{23}|^2/z_3 = 1 - |r_{23}|^2$ and $|t_{21}|^2/z_1 = 1 - |r_{21}|^2$, the equation for the quality simplifies to

$$Q = \frac{2\pi(1 + |r_{23}|^2)}{1 - |r_{23}|^2 + (1 - |r_{21}|^2)|r_{23}|^2}$$

or

$$Q = \frac{2\pi(1 + |r_{23}|^2)}{1 - |r_{21}|^2|r_{23}|^2} \tag{8–5}$$

We recall that the reflection coefficients of the feedback line were calculated in Section 8–8:

$$r_{21} = 0.880\underline{/0^\circ} \qquad |r_{21}|^2 = 0.774$$
$$r_{23} = 0.942\underline{/-176.5^\circ} \qquad |r_{23}|^2 = 0.887$$

and with these values in Eq. (8–5) we have

$$Q = \frac{2\pi(1 + 0.887)}{1 - 0.774 \times 0.887}$$
$$= 37.8$$

which is quite satisfactory.

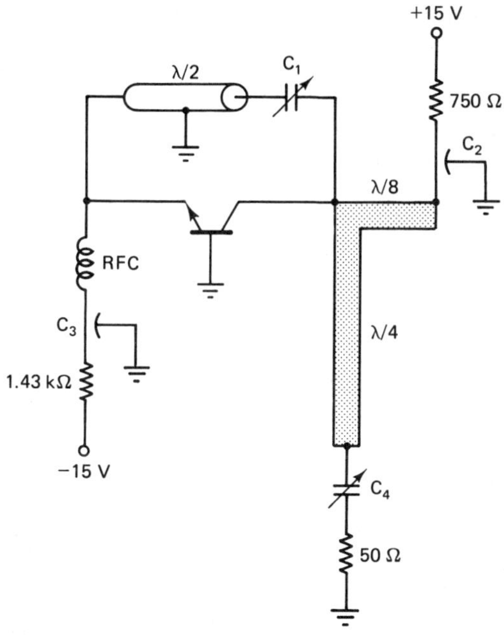

Figure 8–8 Schematic diagram of the 500-MHz oscillator.

8–10 500-MHz OSCILLATOR CONSTRUCTION

For the feedback line, any type of 50-Ω coaxial cable is acceptable, but miniature (small-diameter) line is preferred. Since the operating wavelength is 60 cm, the required length of coax is $\frac{1}{2}\lambda/\epsilon_r^{1/2} = \frac{1}{2} \times 60/(2.25)^{1/2} = 20$ cm. Here we have used $\epsilon_r = 2.25$ for Teflon insulation.

If we use ½-in. polystyrene foam insulation board for strip-line construction of the stub and quarter-wave section, the lengths will be 7.5 cm and 15.0 cm. The conductor widths, for characteristic impedances of 66.5 and 198 Ω, will be 4.45 cm and 0.400 cm, from Fig. 4–6.

A schematic diagram of the oscillator is shown in Fig. 8–8. The 1- to 10-pF tuning capacitor also serves as a blocking capacitor. RF ground is established with a bypass capacitor C_2 at the end of the stub and C_3 at the radio-frequency choke. Output capacitor C_4 blocks the dc bias from the load and provides some small control of matching. The RFC can be six turns of fine wire wound on a plastic tube or straw.

8–11 2-GHz OSCILLATOR

In a second example of oscillator design at microwave frequencies we shall again use the CB configuration and strip-line construction. At this frequency PC board construction will be necessary. We again use the HXTR 3101 transistor and the manufacturer's scattering parameter data:

$$s_{11} = 0.591\angle 161^\circ$$
$$s_{21} = 2.774\angle 49^\circ$$
$$s_{12} = 0.80\angle 48^\circ$$
$$s_{22} = 0.392\angle -57^\circ$$

where the bias conditions are

$$V_{CE} = 10 \text{ V} \qquad I_C = 10 \text{ mA}$$

We convert these data to the common-base configuration:

$$\begin{aligned} Z_{1b} &= (\text{Re } z_{1e}/|s_{21}| + \text{Im } z_{1e}) \times 50 = (0.0973 + j0.155) \times 50 \\ &= 4.87 + j7.75\ \Omega \\ Z_{2b} &= (\text{Re } z_{2e}|s_{21}| \parallel \text{Im } z_{2e}) \times 50 = (3.25 \parallel -j0.900) \times 50 \\ &= 162 \parallel -j45\ \Omega \end{aligned}$$

With the information above we see that we can reduce the transistor output impedance to a pure resistance if we use a one-eighth wavelength, short-circuited stub with a characteristic impedance of 45 Ω. The remaining output resistance, 162 Ω, can be matched to a 50-Ω load by a quarter-wave line with a characteristic impedance of $Z_c = \sqrt{Z_1 Z_2} = \sqrt{162 \times 50} = 90\ \Omega$.

For the feedback network we shall again try a mismatched half-wave transmission line. In checking the feedback product we note that the voltage gain is approximately $A_i r_2/r_1 = 1 \times 50/4.87 = 10.27$. The feedback factor is the magnitude of the transmission coefficient calculated from Eq. (7–5):

$$|t| = \frac{0.849 \times 0.569}{|1 - 0.528\angle 0^\circ \times 0.822\angle 162^\circ|} = 0.340$$

The feedback product is

$$|t|A = 0.340 \times 10.27 = 3.49 \geq 1$$

and we see that the Barkhausen amplitude condition is probably satisfied. The values of the coefficients used in the calculation above were obtained as they were for the 500-MHz oscillator with Eq. (2–11) and the results are

$$t_{12} = 0.849 \qquad r_{23} = 0.528\angle 0^\circ$$
$$t_{23} = 0.569 \qquad r_{21} = 0.822\angle 162^\circ$$

To find the Q of the oscillator we substitute the values of the reflection coefficients into Eq. (8–5):

$$Q = \frac{2\pi(1 + 0.528^2)}{1 - 0.528^2 \times 0.822^2}$$
$$= 9.90$$

which is satisfactory. The interdependence of the Barkhausen conditions will be considered in Section 8–15.

8–12 2-GHz OSCILLATOR CONSTRUCTION

The 2-GHz oscillator design is similar to the 500-MHz oscillator; we need only to implement it in PC board rather than polystyrene foam.

For the $\lambda/8$ matching stub we require a characteristic impedance of 45 Ω. If we use a PC board with a relative dielectric constant of 2.6, we find from Fig. 4–6 that the ratio of conductor width to insulation thickness should be $W/T = 3.25$. If the board is $\frac{1}{32}$ in. thick, the conductor width will be $3.25 \times \frac{1}{32} \times 2.54 = 0.258$ cm.

We require an impedance of 90 Ω for the quarter-wave matching line. From Fig. 4–6 we find the required ratio of width to thickness to be $W/T = 0.98$, which gives a conductor width of $0.98 \times \frac{1}{32} \times 2.54 = 0.078$ cm.

The lengths of the lines above depend on the effective wavelength in the dielectric, and we can find this from Fig. 4–5. For the $\lambda/8$ line the ratio of conductor width to insulation thickness is $W/T = 3.25$, which corresponds to a ratio of effective wavelength to free-space wavelength of $\lambda_e/\lambda_0 = 0.678$. Since the free-space operating wavelength is $\lambda_0 = 15$ cm, the effective wavelength will be $\lambda_e = 0.678 \times 15 = 10.17$ cm. The stub length is one-eighth of the effective wavelength, or $10.17/8 = 1.27$ cm. For the quarter-wave line the ratio $W/T = 0.98$ corresponds to an effective-to-free-space wavelength ratio of 0.706, and the length of the line is $0.706 \times 15/4 = 2.65$ cm.

The construction details are now complete except for the bias resistors, blocking capacitors, and the radio-frequency choke. The bias circuit is identical to the one we used for the 500-MHz oscillator. The blocking capacitors and RFC need only to be scaled to the new operating frequency.

8–13 OTHER TYPES OF OSCILLATORS

It is evident from the two preceding designs that at higher frequencies Q may become too small and at lower frequencies the transmission lines too long. The long-line problem can be avoided by using lumped-parameter components, and at higher frequencies Q can be increased by coupling the feedback from a short-circuited line or cavity (Fig. 8–9). The feedback factor can be controlled by the degree of coupling between the cavity and the line. Since the cavity is loaded only at one end, its Q can be higher than the Q of the 2-GHz design.

To see exactly how such an oscillator works, let us assume that the coupling is magnetic. With magnetic coupling there is no feedback at the maximum or minimum of a sinusoidal waveform, that is, when the change of current with change in time is zero. With no feedback at a current maximum, the transistor will start to return

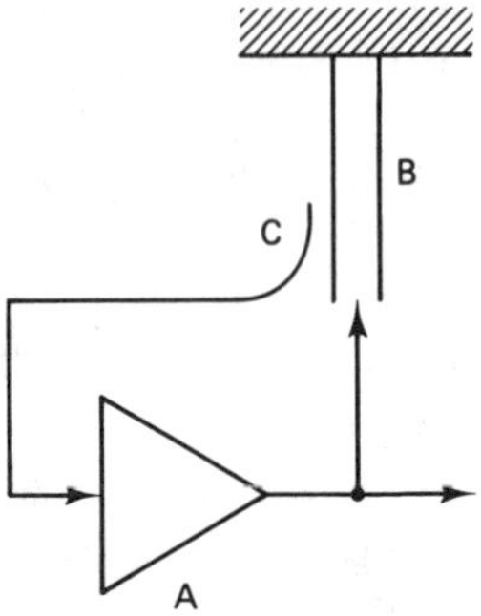

Figure 8–9 High-frequency oscillator: A, amplifier; B, short-circuited line; C, directional coupler.

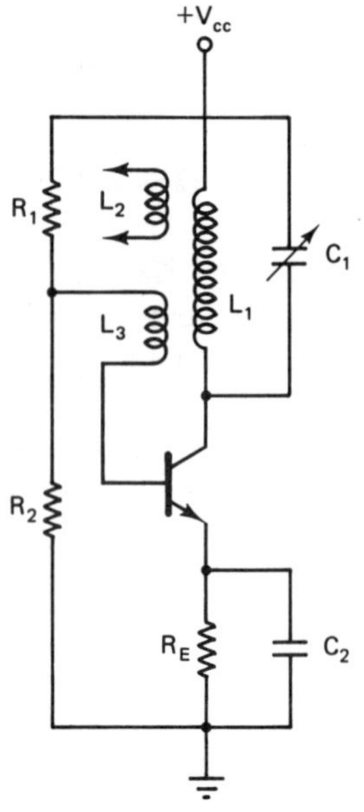

Figure 8–10 Schematic diagram of an Armstrong oscillator: L_1, C_1, tank oscillator; L_2, output coil; L_3, feedback or "tickler" coil; R_1, R_2, voltage-divider bias network; C_2, bypass capacitor.

to its quiescent bias point. As the transistor returns toward quiescence a wave with reversed phase is generated. After reflection from the short-circuited end of the quarter-wave line, the wave with reversed phase is coupled to the input of the transistor; thus the transistor is driven toward cutoff. When the amplitude approaches a minimum, the process is repeated in reverse and the cycle moves to completion.

A lumped-parameter version of the foregoing type of oscillator is called an *Armstrong oscillator* (Fig. 8–10). A transformer or "tickler" coil provides feedback from the tank circuit to the input of the transistor. The Barkhausen condition is satisfied by a phase change of 180° through the amplifier and an additional 180° through the transformer. The feedback product is controlled by the turns ratio of the transformer.

The oscillations of the Armstrong circuit can be described as in the microwave version above. Assume a positive voltage to be present at the input of the transistor. The amplified output is fed back positively and the transistor is driven toward saturation. When the change in current with time is zero, feedback ceases: transformers do not transform dc. The transistor now drifts back toward its quiescent bias condition, and this action generates a wave with opposite phase. This negative-going output is fed back to the input and the transistor is driven toward cutoff. When there is again a steady-state or dc condition, feedback ceases, the transistor drifts toward its Q point, and the cycle is brought to completion.

We have discussed oscillators in which the active device is a transistor operating as a linear or nonlinear driver. In general, any active device having a negative resistance can be combined with a high-Q tank or cavity to produce SHM. We shall next consider negative resistance devices capable of generating appreciable power at frequencies as high as 100 GHz.

8–14 NEGATIVE-RESISTANCE DEVICES

By *negative resistance* we mean a dynamic resistance with a negative slope: as voltage decreases, current increases. The transistor amplifier in the common-emitter connection can be regarded as a negative-resistance circuit. In Fig. 8–11 we show the ac load line of a CE transistor circuit superimposed on the transistor I–V curves. When a signal is impressed on the base, the operating or Q point moves along the load line, and an increase in collector current corresponds to a decrease in collector voltage. The negative-resistance region of a device usually is bounded by regions of positive resistance, and this means that the device can by itself be an oscillator. Let us examine the action of the CE circuit in more detail.

Consider the input to the CE transistor to be excessive positive feedback that drives the transistor to saturation (point A in Fig. 8–11). Saturation occurs as the base–collector bias drops to zero. As the device is driven further into saturation, the base–collector junction becomes forward biased. The resulting reverse current reduces slightly the net collector current, while the emitter current increases. We

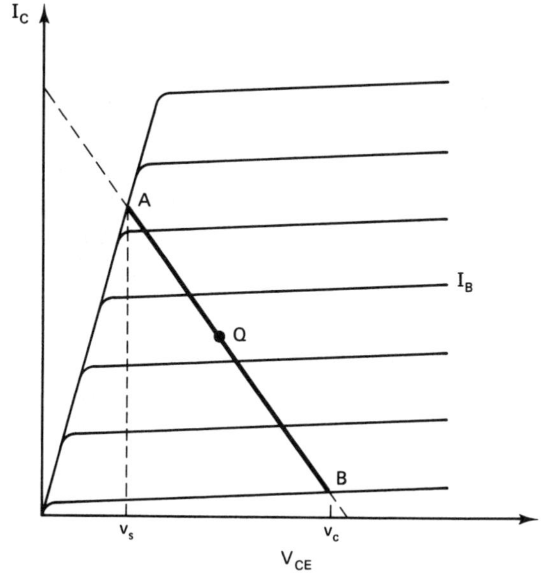

Figure 8–11 Characteristic curves and load line for a CE transistor circuit: ABQ, *ac* load line; v_s, v_c, saturation and cutoff voltages; Q, quiescent point.

now have a positive resistance: decreasing collector current with decreasing collector-to-emitter voltage. A similar change from negative to positive resistance occurs when the device is driven beyond cutoff.

A change from negative to positive resistance is equivalent to changing the feedback from positive to negative, a phase change of 180°. The Barkhausen condition is satisfied, there is now a temporary negative feedback in the positive resistance region and the direction of the oscillation is reversed. Like the child swinging itself, driver and oscillator are one.

Although the transistor can be used as a negative resistance device, it is stable because of the isolation of its input and output circuits. Some of the output energy must be coupled to the input through a feedback network before the device becomes unstable and oscillates. There are other solid-state devices, called *diodes*, because they have only two terminals, which also show negative-resistance regions. Such devices are inherently unstable because the input and output ports are not isolated. When such devices are biased to operate in a region of negative resistance, the slightest decrease in voltage will correspond to an increase in current. The increase in current results in an increase in voltage drop across the bias circuit resistor, thus further reducing the bias voltage. There is automatic positive feedback until the device reaches the limit of the negative-resistance region. Stability does not follow. The slightest increase in bias voltage now reverses the process and the diode is driven to the other extreme; thus the device is a self-oscillator. Such devices can be used to generate energy at frequencies as high as 100 GHz. One of the earliest of these was the tunnel diode.

Heavily doped *n*- and *p*-type semiconductors can be used to form diodes with very thin depletion layers of the order of 10^{-6} cm. To put this in perspective, this is a distance of about 100 hydrogen atom diameters. Evidently with distances this small we can expect quantum effects to be significant. Before the theory, let us describe what happens in a tunnel diode.

As the diode is forward biased, the current rises rapidly, as is shown from A to B on the characteristic curve in Fig. 8–12. If the forward-bias voltage is increased, there is a region of negative resistance from B to C. With a further increase in forward-bias voltage the diode shows a normal characteristic from C to D. We have a device that has a negative resistance between two regions of positive resistance. Once again, as in the CE transistor circuit discussed previously, we can create an oscillator. With positive feedback the device will oscillate; the amplitude of the oscillation is limited by a change from positive to negative feedback as the resistance changes from negative to positive. The oscillations occur at rather low voltage and current levels, and the output is rather small, 10 to 100 μW.

We shall explain the tunnel diode action in terms of the *Heisenberg uncertainty principle*, not just because it provides a simple explanation, but because we shall later find the uncertainty principle useful in antenna theory. The statement of the principle is

$$\Delta E \, \Delta t \geq \hbar \tag{8–6}$$

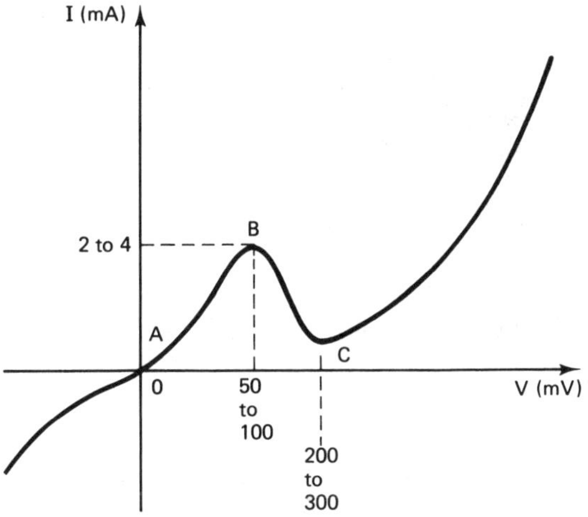

Figure 8–12 Tunnel diode characteristic curve: *AB*, tunnel region; *BC*, negative-resistance region; *CD*, part of the normal diode characteristic curve.

and is interpreted as follows: the product of the uncertainty is energy, ΔE, and the uncertainty in time, Δt, is greater than, or at least equal to, a universal constant, $\hbar = h/2\pi$. The universal or fundamental constant is called *Planck's constant*, and it has a value of 1.06×10^{-27} erg s.

The expression (8–6) is not convenient in electronics because we prefer to measure energy per unit charge (voltage) rather than energy in ergs. We obtain a more convenient form of the uncertainty relation if we divide it by the charge of an electron and convert ergs to joules; in addition, we multiply by the speed of light to convert uncertainty in time to uncertainty in distance, $c\ \Delta t = \Delta s$. The result is

$$\Delta V\ \Delta s \geq 2.0 \times 10^{-5}\ \text{V cm} \qquad (8\text{–}7)$$

Expression (8–7) is interpreted as follows. The product of the uncertainty in the energy of an electron, measured in volts, and the uncertainty in its location, measured in centimeters, is greater than or equal to 2.0×10^{-5} V cm. Consider, for example, the uncertainty in the energy of an electron if its uncertainty in location is 6.5×10^{-5} cm. From expression (8–7) we have

$$\Delta V \times (6.5 \times 10^{-5}) \geq 2.0 \times 10^{-5}$$

or

$$V \geq 0.3\ V$$

In the example above we chose an uncertainty in electron location equal to the width of the depletion layer in a tunnel diode, and we found that the corresponding uncertainty in electron energy was 0.3 V. Since this uncertainty in energy is greater than the potential difference or *barrier* separating the *p*- and *n*-type semiconductors, an electron in the conduction band of the *n*-type material can appear in the valence band of the *p*-type material (Fig. 8–13). It is equally probable that an electron in a

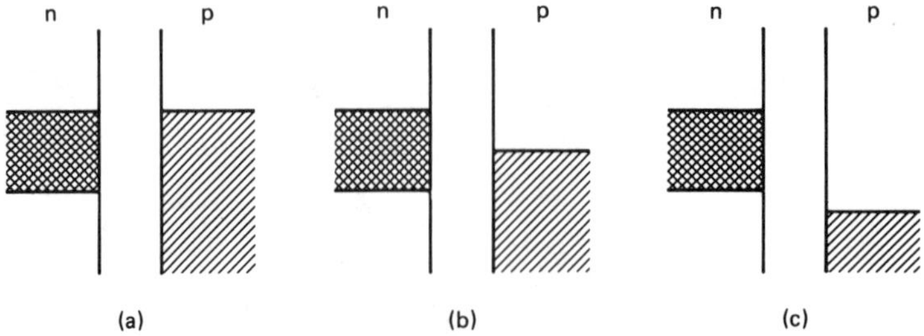

Figure 8–13 Energy levels associated with the tunnel effect: (a) equal and opposite currents flow between *n*- and *p*-type materials; (b) with external bias there is a net flow of current from *n* to *p*; (c) with greater bias the net flow decreases as the depletion layer widens.

valence band can appear on the other side, in the conduction band of the *n*-type semiconductor. The two electron currents cancel, and there is no net current flow.

If an external bias voltage is applied, the current flow from *n*- to *p*-type material is favored and there is a net current flow, as shown from *A* to *B* in Fig. 8–12 and at *B* in Fig. 8–13. As the external bias is increased, the depletion layer increases in width. It follows from expression (8–7) that as Δs increases, the uncertainty in energy per unit charge must decrease. A decrease in energy per unit charge or voltage, ΔV, means that it is less probable that an electron can appear on the other side of the potential barrier. The net current flow decreases, as shown from *B* to *C* in Fig. 8–12. As the external bias is further increased, the potential barrier and the tunnel effect disappear, and the normal diode characteristic curve is shown (*C* to *D* in Fig. 8–12).

Uncertainty in the position of an electron enables it to appear on the other side of a potential barrier. This is likened to the ability of a person to walk through a tunnel, rather than climb a mountain, to get to the other side. No such tunnel exists in a tunnel diode. ''Tunnel'' is a colloquialism, a convenient figure of speech that does not really explain uncertainty in location and energy.

EXAMPLE 8–2

If the potential barrier of a diode is 0.2 V and the depletion layer width is 6.5×10^{-5} cm, what operating frequency can be achieved if the diode is used as an oscillator?

Solution We need some additional information to solve this problem; namely, if an electron acquires 0.2 V of energy, it has a speed of 2.7×10^7 cm/s, about one-thousandth the speed of light.

If the diode is used as an oscillator, the electrons must travel across the depletion layer twice in every cycle, a distance of $2\ \Delta s$. The time required to execute a cycle is $\tau = 2\ \Delta s/v = 2 \times (6.5 \times 10^{-5})/2.7 \times 10^7 = 4.8 \times 10^{-12}$ s, which corresponds to a frequency of $f = 1/\tau = 200$ GHz.

Tunnel diodes have been used in oscillators with operating frequencies as high as 100 GHz, which suggests that the effective width of the depletion layer, after biasing, is somewhat greater than what we assumed in Example 8–2. The low- and high-frequency tunnel diode circuits in Fig. 8–14 show some of the techniques used to build negative resistance oscillators. In each example the Barkhausen condition is satisfied by the sum of phase delay in propagation time and phase reversal, owing to negative- and positive-resistance changes.

The Gunn and IMPATT diodes were developed after the tunnel diode. In the Gunn device negative resistance occurs when electrons are energetic enough to be transferred to a higher energy level in which they are less mobile; accordingly, there is a decrease in current with increasing voltage. This diode is also called a *transferred electron device* (TED).

The IMPATT (*imp*act *a*valanche *t*ransit *t*ime) device is made to withstand operation under avalanche conditions. When avalanche occurs, the resulting current pulse develops 90° behind the original voltage pulse; in addition, the width of the device is designed to provide an additional 90° phase delay due to propagation time. A maximum in the current pulse arrives at the cathode when the voltage is a minimum; in other words, there is negative resistance. These devices are capable of generating 10 W to 50 mW of power at frequencies of 10 to 200 GHz.

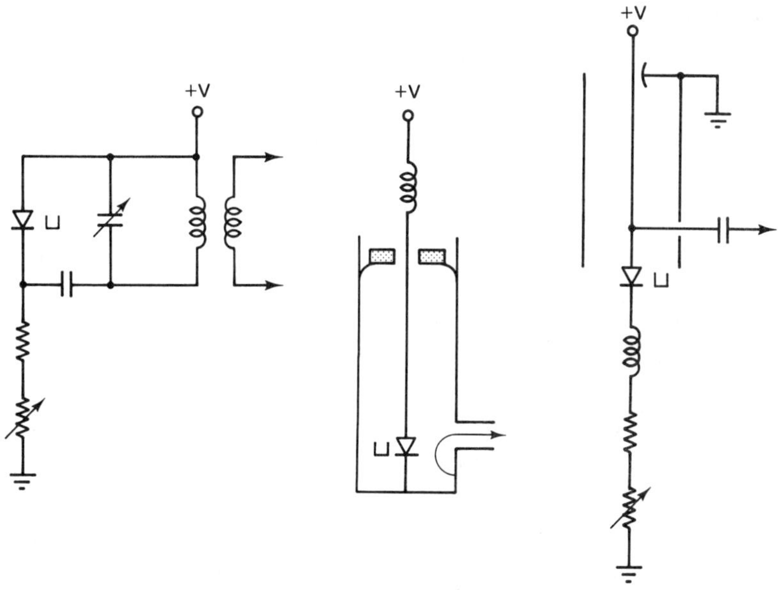

Figure 8–14 Low- and high-frequency oscillator circuits: utilization of the tunnel diode as a negative resistance device.

8–15 LINEAR OSCILLATOR THEORY

At the beginning of this chapter we demonstrated linear oscillations with a bob attached to a rubber band. For small amplitudes the motion conforms to Hooke's law, force is proportional to stretch or displacement, $f = \pm kx$, where k is the spring constant and x the displacement. The right side of the equation is positive if f is the applied force, negative if f is the force exerted by the spring on the bob. Since the Hooke's law equation is linear, we have a *linear oscillator*.

The relation between a linear oscillation and *simple harmonic motion* (SHM) is shown in Fig. 8–15. Let P be a point moving on a circle of radius r with uniform angular velocity $\omega = \phi/t$, where the phase ϕ is measured from the horizontal baseline. As point P moves upward, its projection, point M, moves rapidly at first, decelerates to a halt at the top, then accelerates downward. Maximum velocity on the vertical line occurs as point M crosses the baseline, maximum acceleration occurs at the extremities of the vertical line. The maximum acceleration of point M is $\omega^2 r$, and the equivalent spring constant is $m\omega^2$, where m is the mass of the bob.

In summary, the projection of uniform angular velocity onto a straight line gives Hooke's law motion; further, circular motion can be displayed in rectangular coordinates as a sine wave, as shown in Fig. 8–15. Ideally, when a perfect sine wave at one fixed frequency is generated, the motion is called "simple harmonic."

The mathematical theory of optical refraction and dispersion in terms of mechanical oscillators of atomic size was completed in the nineteenth century. By the beginning of the twentieth century the theory had been refined in electromagnetic theory, and a result of importance to microwave electronics is the amplitude of the electric field scattered by an electric dipole oscillator,

$$\frac{E_s}{E_0} = \frac{j\sigma f Q}{1 + jQ(f - 1/f)} \qquad (8\text{–}8)$$

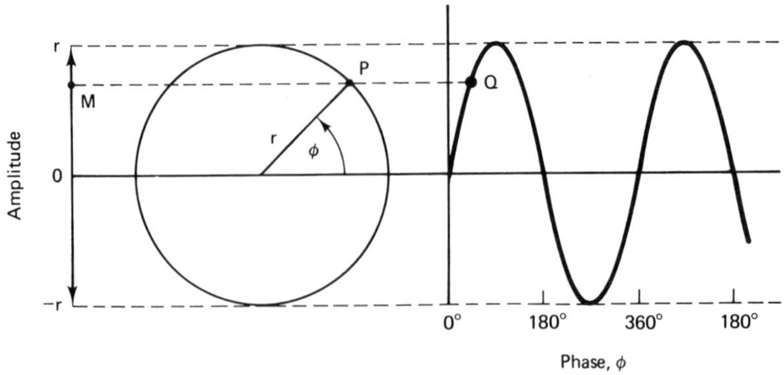

Figure 8–15 Simple harmonic motion: center, point P moves with uniform angular velocity on a circle; right, the motion is displayed in rectangular coordinates as a sine wave; left, the motion projected on a vertical line is that of a linear oscillator of amplitude r.

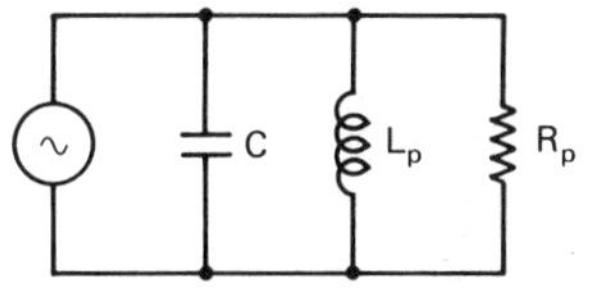

Figure 8–16 Parallel resonant tank circuit.

where E_s/E_0 is the ratio of scattered to incident electric field strength, σ is a scattering cross section, Q is the quality, and f is the ratio of the frequency to the resonant frequency, ω/ω_p.

By the middle of the twentieth century the concepts of impedance and circuit theory were well developed, and today we can present linear oscillations as the theory of a resonant "tank" circuit.

We assume, for the tank circuit shown in Fig. 8–16, that the load is principally parallel, that the series resistance of the coil is small and negligible. The admittance of the circuit is

$$Y = \frac{1}{R_p} + j\left(\omega C - \frac{1}{\omega L_p}\right)$$
$$= \frac{1}{R_p}\left[1 + j\frac{R_p}{\omega L_p}(\omega^2 L_p C - 1)\right]$$

This equation can be simplified with the substitutions:

Parallel resonant angular frequency, $\omega_p = (L_pC)^{-1/2}$
Relative frequency, $f_p = \omega/\omega_p$
Quality, $Q_p = R_p/\omega_p L_p$
Characteristic impedance, $Z_c = (L_p/C)^{1/2}$

and the admittance becomes

$$Y = \frac{1}{R_p}\left[1 + j\frac{R_p}{f\omega_p L_p}(f^2\omega_p^2 L_p C - 1)\right]$$
$$= \frac{1}{QZ_c}\left[1 + \frac{jQ(f^2 - 1)}{f}\right]$$

and the normalized impedance is

$$z = \frac{Z}{Z_c} = \frac{Q}{1 + jQ(f - 1/f)} \tag{8–9}$$

Except for a factor of σf and a phase difference of 90°, we recognize the right side of Eq. (8–9) as the same as that of Eq. (8–8). The phase reference in Eq. (8–8) is with respect to the radiation field, whereas in circuit theory the reference is with respect to the induction field; in other words, there is a 90° phase difference between a traveling wave and a standing wave.

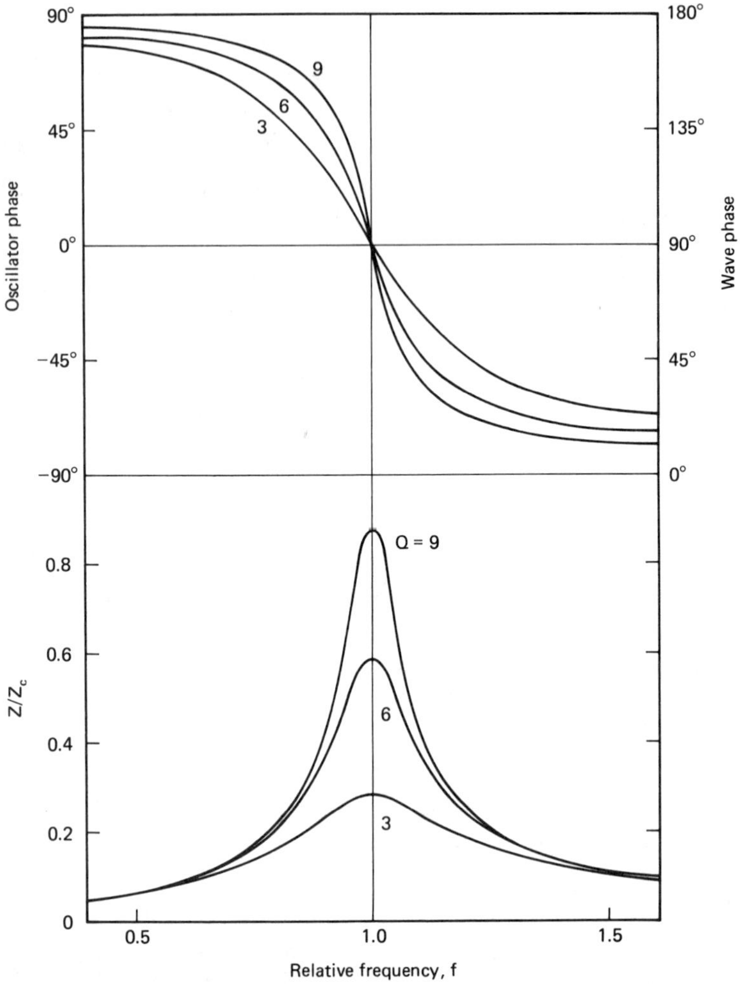

Figure 8–17 Amplitude and phase curves for parallel resonant tank circuits.

A graph of Eq. (8–9) for various values of Q is shown in Fig. 8–17. We see that for the smaller values of Q the amplitude is less and the phase change more gradual. These results are appropriate for circuit oscillators, such as those described earlier in this chapter.

To apply oscillator theory to the problem of the mismatched feedback line used in the 500-MHz and 2-GHz oscillators, we consider the problem of a parallel LC admittance embedded in transmission lines, (Fig. 8–18), where we assume that the generator is matched to the first transmission line. The admittance of the tank circuit and load is

$$Y = \frac{1}{R} + j\omega C - \frac{j}{\omega L}$$

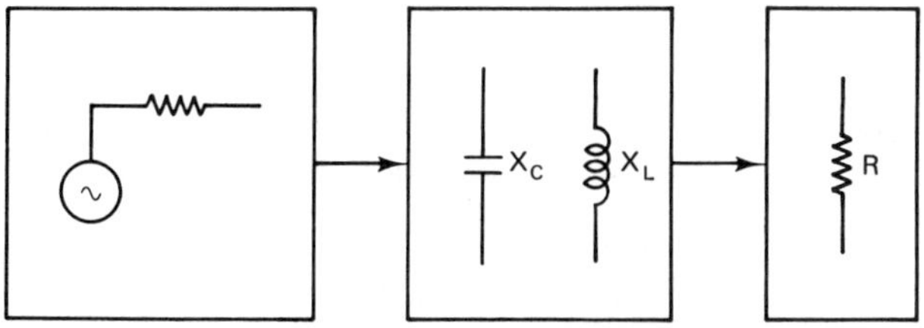

Figure 8–18 Tank circuit as an admittance embedded in transmission lines.

and after normalization this becomes

$$y = YR = 1 + j\frac{R}{\omega L}(\omega^2 LC - 1)$$

$$= 1 + j\frac{\omega_0}{\omega}\frac{R}{\omega_0 L}\left(\frac{\omega^2}{\omega_0^2}\omega_0^2 LC - 1\right)$$

$$= 1 + \frac{jQ(f^2 - 1)}{f}$$

where the resonant frequency is $\omega_0 = (LC)^{-1/2}$, the relative frequency is $f = \omega/\omega_0$, and the quality is $Q = R/\omega_0 L$. The transmission coefficient into the tank circuit is, from Eq. (5–45),

$$t = \frac{2}{1 + y} = \frac{2}{1 + 1 + jQ(f^2 - 1)/f}$$

$$= \frac{1}{1 + j(1/2)Q(1 - f^2)/f}$$

$$= \frac{1 - j(1/2)Q(1 - f^2)/f}{1 + [\tfrac{1}{2}Q(1 - f^2)/f]^2}$$

The phase of the transmission coefficient is determined by the ratio of the imaginary and real components,

$$\delta = \tan^{-1}\frac{-\tfrac{1}{2}Q(1 - f^2)}{f} \qquad (8\text{–}10)$$

and the magnitude is

$$|t| = \frac{1}{\sqrt{1 + [-\tfrac{1}{2}Q(1 - f^2)/f]^2}}$$

$$= \frac{1}{\sqrt{1 + \tan^2\delta}} = \cos\delta$$

If we combine this result with Eq. (8–10), we obtain a transmission coefficient of

$$t = \cos\delta\,\underline{/\delta} \qquad (8\text{–}11)$$

EXAMPLE 8–3

Find the transmission coefficient through a tank circuit when the operating frequency is 10% above the resonant frequency. Assume that the quality of the circuit is $Q = 10$.

Solution At 10% above resonance the relative operating frequency is $f = \omega/\omega_0 = 1.1$. We find the phase angle from Eq. (8–10):

$$\delta = \tan^{-1} \frac{10/2 \times (1 - 1.1^2)}{1.1} = 43.67°$$

and from Eq. (8–11) we now find

$$t = \cos(43.67°)\underline{/43.67°} = 0.723\underline{/43.67°}$$

We must make some additional assumptions before we can use Eq. (8–11) to calculate the loop phase delay in the 500-MHz oscillator. In a loop there is no terminating load as shown in Fig. 8–18. The load in turn becomes a source and the wave circulates endlessly. The transistor is not the source of the waves; it only amplifies. We shall assume that Eq. (8–11) is valid not only for a tank circuit but for any linear oscillator and that the load resistance is the radiation resistance of the oscillator.

It has been known for more than two centuries that the amplitude of a radiated or traveling wave is proportional to the current in the source but lags the current by 90°; accordingly, we shall modify Eq. (8–11) to include an additional 90° phase lag,

$$t = \cos\delta\underline{/\delta + 90°} \tag{8–12}$$

where t is now the transmission coefficient into R, and where R may represent the external load of the tank circuit or the characteristic impedance of the medium following the tank circuit.

EXAMPLE 8–4

Calculate the feedback amplitude and total loop phase delay in the 500-MHz oscillator feedback network when $f = 1.09$. Assume that $Q = 37.7$.

Solution From Eq. (8–10) we have

$$\delta = \tan^{-1} \frac{-1/2 \times 37.7(1 - 1.09^2)}{1.09}$$

$$= 72.9°$$

and from Eq. (8–11)

$$|t| = \cos 72.9° = 0.294$$

The total loop phase delay is

$$\begin{aligned}\phi &= 180° \times 1.09 + \delta + 90° \\ &= 180° \times 1.09 + 72.9° + 90° \\ &= 359.1°\end{aligned}$$

We see that the Barkhausen phase condition is satisfied at a frequency of about $f = 1.09$; however, this is not necessarily the operating frequency of the oscillator. The oscillator will behave as if it had a mind of its own and seek that frequency at which maximum real power is fed back to the load, the load being

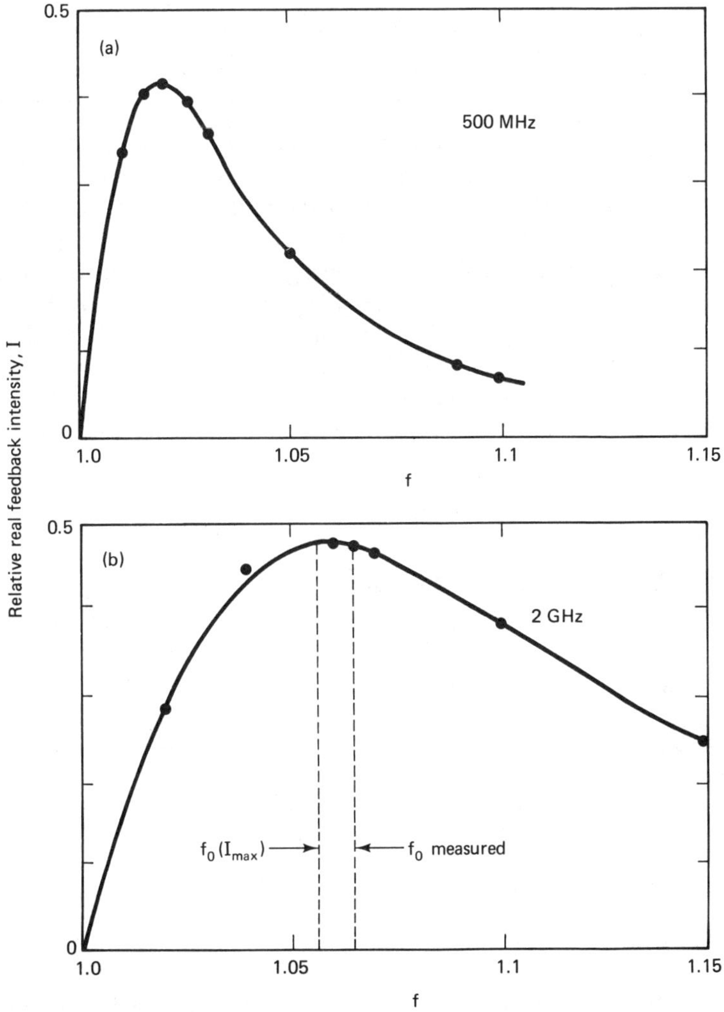

Figure 8–19 Relative real intensity fed back to the transistor input of (a) the 500-MHz oscillator and (b) the 2-GHz oscillator.

the input impedance of the transistor. We can find the relative real intensity to the load from

$$I = P_{del}|z| = |t|^2 \cos \phi \qquad (8\text{–}13)$$

where $\cos \phi$ is the power factor.

EXAMPLE 8–5

Calculate the relative real intensity delivered to the input of the transistor in the 500-MHz oscillator if the operating frequency is $f = 1.09$.

Solution Using Eq. (8–13) we have

$$I = 0.294^2 \cos (359.13°) = 0.086$$

By repeating the numerical calculations in Examples 8–4 and 8–5 for various values of f we can produce a graph of relative intensity versus relative frequency (Fig. 8–19a). We see that the maximum value of I occurs at $f = 1.02$. The operating frequency of the 500-MHz oscillator will be 2% above the design frequency, at 510 MHz. Since this shift in frequency is small enough to be compensated by the variable capacitor in the feedback line, no design change is necessary.

We repeat the numerical calculations for the 2-GHz oscillator and obtain the curve shown in Fig. 8–19b. We see that the lower Q of the feedback network causes a greater drift from the design frequency, which can be corrected with a variable capacitor or by shortening the feedback line about 6%.

APPLICATION 8–1: Oscillators

In this application we shall use strip-line techniques and build, troubleshoot, and test either a UHF or microwave oscillator.

Equipment and Materials

The supplies and tools required are the same as those described in Application 7–1. If none of the optional test equipment is available, the oscillator can be tested with the amplifier built in Application 7–1.

Procedure

Build and test either the 500-MHz or the 2-GHz oscillator described in the text and shown in Fig. 8–20. When testing or troubleshooting, remember that the oscillator

Figure 8–20 A 500-MHz oscillator can be constructed from polystyrene foam and metal foil. This construction is satisfactory for immediate testing, unsatisfactory for permanent use.

may be generating several hundred milliwatts of power; *always keep your body, especially your head, at least a foot from the oscillator*. Use a UHF/microwave spectrum analyzer to measure the operating frequency, bandwidth, and harmonics generated by the oscillator. Calibrate the spectrum analyzer, if necessary, with a microwave signal generator and measure the power output of the oscillator. Your report should include a description of the measurement procedures and a summary of the results, as well as a description and drawing of the oscillator. Some spectrum analyzers are restricted to very small input signal levels. You may have to load the oscillator with a 50-Ω step attenuator to reduce the signal.

If no test equipment is available, build two dipole antennas. Each antenna should be a conducting cylinder or rod about 0.48 wavelength in length; for example, for a 500-MHz operating frequency the dipole length is 28.8 cm. The diameter of the cylinder should be about 1/50 of its length: 0.3 to 0.6 cm or 1/8 to 1/4 in. A 60-Ω feed line for a dipole can be made by twisting together two lengths of insulated hookup wire. Make the feed line a quarter-wavelength long and solder one end to the dipole as shown in Fig. 8–21.

Connect one vertical dipole to the oscillator output and the other to the amplifier input. Connect the output of the amplifier to a microwave receiver. Separate the transmitting and receiving sites by a distance of at least 5 or 10 m. Tune the receiver to measure the operating frequency and bandwidth of the transmitter. The bandwidth measured is approximately the bandwidth of the oscillator, since the Q of the dipole is small. To estimate the gain of the amplifier you built for Application 7–1, compare the signal level on the receiver to the signal level after you have removed the amplifier and connected the dipole directly to the receiver.

If a receiver is not available, you may be able to tell whether or not the oscillator is functioning by receiving its modulated signal on a television receiver. Modulate the oscillator by loading its output to the dipole with a shunt PIN diode and apply a small AF signal to the diode through an RF choke.

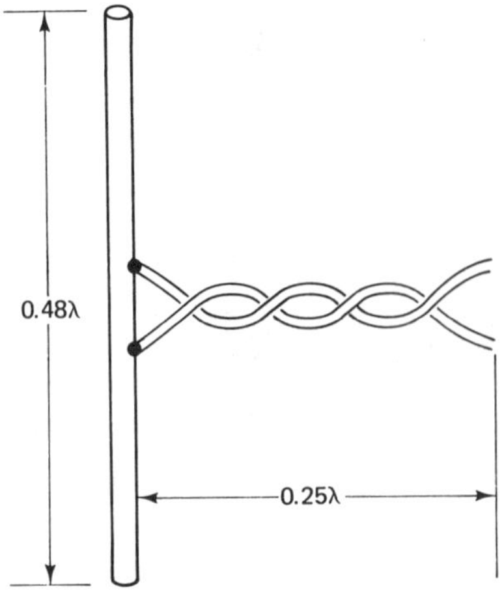

Figure 8–21 500-MHz, half-wave, dipole antenna for testing the 500-MHz oscillator and amplifier.

SUMMARY

In designing a transistor oscillator it is convenient to start with the tank circuit. Use the desired Q to determine the total capacitance required; use the desired operating frequency to determine the inductance required from the expression $L = 1/\omega^2 C$.

Bias the circuit and determine the voltage gain of the amplifier, then determine the required feedback ratio to satisfy the Barkhausen condition $fA_v \geq 1$.

At high frequencies the CB configuration is preferred. The lumped-parameter tank circuit is replaced by a cavity or mismatched transmission line. The Q of the oscillator can be calculated from a consideration of multiple-reflected waves [Eq. (8–5)]. Because the transistor parameters are complex, the Barkhausen phase condition must be calculated using complex transmission and reflection coefficients derived from S-parameters; also, matching stubs and quarter-wave transformers may be needed to load the oscillator properly.

At very high frequencies the Barkhausen phase condition may be satisfied by phase changes in the active device itself, as, for example, in negative-resistance devices. The tunnel diode is an example of a negative-resistance device, and its operation is properly explained by the Heisenberg uncertainty relation [Eq. (8–6)]. Additional examples are the Gunn and IMPATT diodes.

A linear oscillator generates a pure sine wave. The classical theory of optical refraction and dispersion is based on the theory of linear oscillators, which in turn can be expressed in the equations for a resonant series or parallel circuit.

PROBLEMS

8–1. Calculate the transmission coefficient at an interface between a 300-Ω line and a load impedance of $50 + j100\ \Omega$.

8–2. Calculate the transmission coefficient at the interface of a 300-Ω line and a load impedance of 50 Ω in parallel with $-j75$ Ω.

8–3. If the feedback product is 1.2 and the feedback loop phase is 310°, will the circuit oscillate?

8–4. Calculate the loaded Q of a half-wavelength, 50-Ω coaxial line. The input is loaded with 200 Ω and the output load is $5 - j5$ Ω.

8–5. Design a tank circuit for a Colpitts oscillator to meet the following specifications:
Feedback factor, 0.05
$Q = 10$
Tank load, 1 kΩ
Operating frequency, 25 MHz
Determine the required capacitances and coil inductance.

8–6. The output impedance of a transistor at 1 GHz is $125 \parallel -j75$ Ω. Design a strip-line stub and quarter-wave section to match the transistor output impedance to a 50-Ω load. Use 1/32-in. Duroid, $\epsilon_r = 2.6$.

8–7. Redesign the oscillator in Example 8–1 for a Q of 15.

8–8. Redesign the oscillator in Example 8–1 more accurately by taking into account the loading of the tank circuit by the transistor input resistance. *Suggestion*: From the definition of Q as the ratio of stored to dissipated energy per radian cycle we have

$$Q = \frac{\tfrac{1}{2}Cv_0^2}{\tfrac{1}{2}v_i^2/r_{be} + \tfrac{1}{2}v_0^2/R_L'}$$

$$= \frac{1}{f^2/Q_2 + 1/Q_1} = Q_1Q_2/(f^2Q_1 + Q_2)$$

8–9. Redesign the 10-MHz oscillator given in the text (Example 8–1) using a 2N2222A transistor in the CE configuration.

8–10. Calculate the time constant C_1R_E for the 60-MHz oscillator given in the text and show that it is comparable to the period of the oscillator.

8–11. For a 2N2222A transistor the CB input capacitance is $C_{ibo} = 25$ pF. At approximately what frequency would the transistor cease to be useful in the *CB* configuration? In the CE configuration? Assume that $h_{ie} = 250$ Ω, $h_{fe} = 75$.

8–12. If feedback voltage is derived from a tapped inductor rather than a split capacitance, the circuit is called a *Hartley oscillator*. Use the 10-MHz oscillator example in the text to design a 10-MHz Hartley oscillator: calculate the total number of turns in the coil and the number of turns on each side of the tap. Assume that the coil is made of AWG No. 30 bare wire wound on a plastic soda straw of diameter 0.6 cm. A useful approximation for the inductance of an air core coil is

$$L = \frac{n^2A\mu_0}{x} \quad \text{H}$$

where n is the number of turns, A and x are the cross-sectional area and length of the coil, and the permeability of free space is $\mu_o = 4\pi \times 10^{-7}$ H/m.

Parabolic horn antennas. Excitation is at the apex of a conical horn, reflection from an offset section of a paraboloidal reflector focuses and directs the beam in a horizontal direction. The apertures are covered with protective radomes.

9

Antennas and Propagation

9–1 INTRODUCTION

The generation of waves is an everyday occurrence. A disturbance on the surface of a pond generates waves that propagate away from the source. This is a commonplace observation and does not seem mysterious; however, the radiation of radio waves by an antenna is more difficult to comprehend. We cannot see or hear radio waves, and their electric and magnetic components form a wave structure more complicated than that of a wave on a pond.

An antenna receives energy from an electronic circuit and transforms it into waves traveling through space. It is a bilateral device, a two-way street. The same antenna can intercept waves traveling in space and deliver their energy to a receiver. By the *principle of reversibility* or *reciprocity*, the gain of an antenna is the same for transmitting as for receiving.

The most-used and simplest of antennas is the oscillating dipole. A *pole* is a center of force. At one instant a half-wave dipole appears as two centers of force, positive and negative charges, at either end of a half-wavelength-long conductor. A quarter-cycle later the poles and associated electric field have disappeared, replaced by a current element, the source of a magnetic field (Fig. 9–1). The action is similar to that of a resonant circuit: at one instant energy is stored in the form of an electric field in a capacitor; a quarter-cycle later the energy appears in the form of a magnetic field surrounding a conductor. In a resonant circuit dissipation results from the ohmic resistance of the coil. In a dipole the principal loss is from the radiation of waves and is measured by radiation resistance.

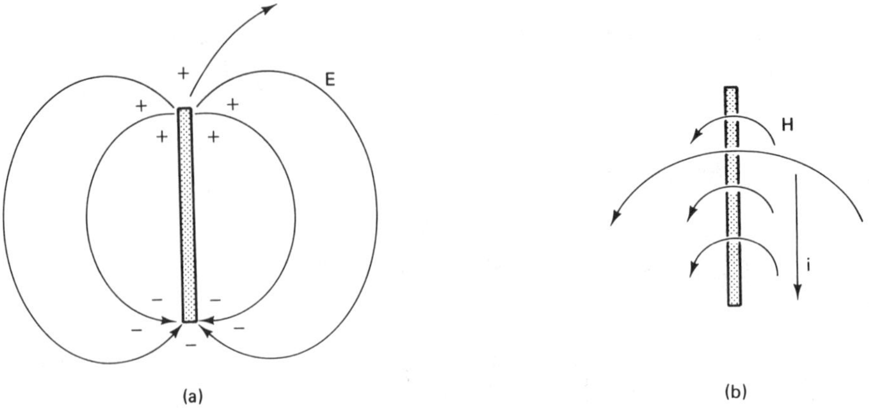

Figure 9–1 Elementary oscillating dipole or "doublet": (a) electric part of the cycle; (b) magnetic part.

In our study of the dipole we shall demonstrate the several techniques that can also be used to analyze, design, and measure other types of antennas.

There are several lengthy derivations in this chapter which may be omitted on a first reading. The reader may wish to concentrate on the results and their applications. The derivations are included for completeness, to check the correctness of our results, and because some of the material is difficult to find in current literature. As a further reassurance to the reader, we have checked some of our results against published experimental data.

9–2 DIPOLE RESONANCE

We can discover some of the properties of a dipole antenna by referring to the characteristics of a resonant circuit. We can discover why, for example, the dipole is resonant when its length is a half-wavelength.

The angular resonant frequency of an LC circuit is

$$\omega_0 = (LC)^{-1/2} \tag{9–1}$$

where L and C are the inductance and capacitance of the coil and capacitor. Now consider L and C to be the self-inductance and capacitance of one arm of a dipole of length s, as suggested in Fig. 9–2. If we multiply Eq. (9–1) by s we have

$$\omega_0 s = \left(\frac{L}{s}\frac{C}{s}\right)^{-1/2} = (L'C')^{-1/2}$$

or, from Eq. (1–7),

$$\omega_0 s = v \tag{9–2}$$

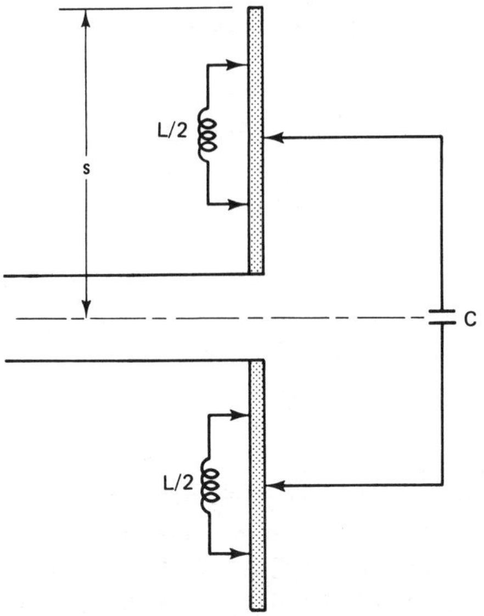

Figure 9–2 Schematic representation of a dipole as an equivalent *LC* circuit.

If we compare this result with the equation for the speed of light, $\omega_0\lambda = v$, we see that if

$$s = \lambda \tag{9–3}$$

the antenna is resonant at frequency ω_0. Since λ is an effective quarter-wavelength, the dipole is resonant when its total length is a half-wavelength.

Effective quarter-wavelength is a form of the average value of a sinusoidal voltage or current. A quarter-wavelength of a unit-amplitude, sinusoidal voltage is shown in Fig. 9–3. The area under the curve also is unity, $A = 1$. The average

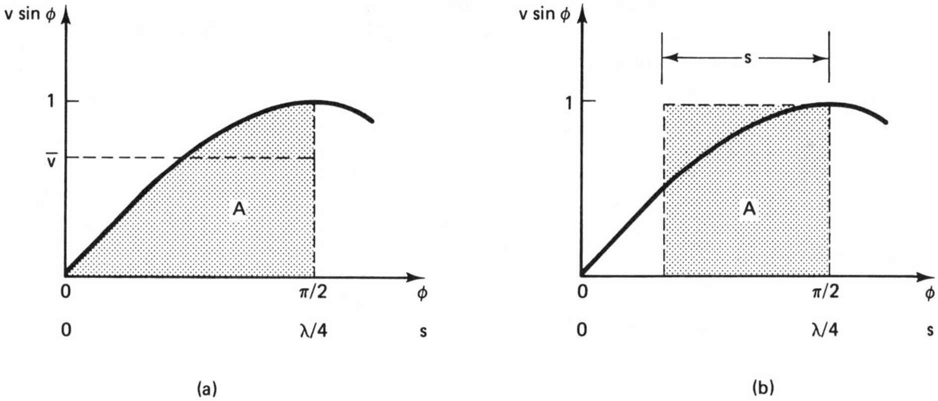

Figure 9–3 Two ways of expressing the average value of a sinusoid: (a) average amplitude; (b) average or effective wavelength.

voltage or current is defined as the multiplier of the baseline that gives the same area: $\bar{V} \times \pi/2 = A = 1$, or

$$\bar{V} = \frac{2}{\pi} = 0.6366 \tag{9–4}$$

a numerical result familiar to us as the average current available from a full-wave rectifier.

To express the result as an average length rather than an average voltage, we choose a length that when multiplied by the peak voltage, gives the same area. From Fig. 9–3b we see that $s \times 1 = A = \bar{V} \times \lambda/4$, or

$$s = \frac{2}{\pi}\left(\frac{\lambda}{4}\right) = \bar{\lambda} \tag{9–3}$$

The angular wavelength $\bar{\lambda}$ appears frequently in antenna theory, wave propagation, and optics as an effective quarter-wavelength of a sine wave.

The dipole half-wavelength resonance is easy to visualize as a transmission-line effect. Suppose that a dipole is to be made by opening the end of a twin-wire transmission line (Fig. 9–4a). At the ends of the dipole arms the open circuit condition requires that the current vanish and the voltage be a maximum (Fig. 9–4b). A half-wave of current and voltage is the simplest standing wave that satisfies the boundary conditions at both ends.

Higher-order modes that satisfy the boundary conditions are shown in Fig. 9–5. Figures 9–5a and b show dipoles called *full-wave dipoles*, which occasionally causes confusion. Unless we say otherwise, the antenna with the symmetric current distribution (Fig. 9–5a) is a full-wave dipole. The dipole shown in Fig. 9–5b is seldom used at microwave frequencies and is used at low frequencies more as a matter of necessity than choice. At low frequencies it is impractical to erect a vertical antenna of great height. Since the antisymmetric full-wave dipole radiates in directions inclined toward the axis of the dipole, it is useful as a low-frequency horizontal antenna. When it is long enough to contain several harmonics, it is called a "long-wire" antenna.

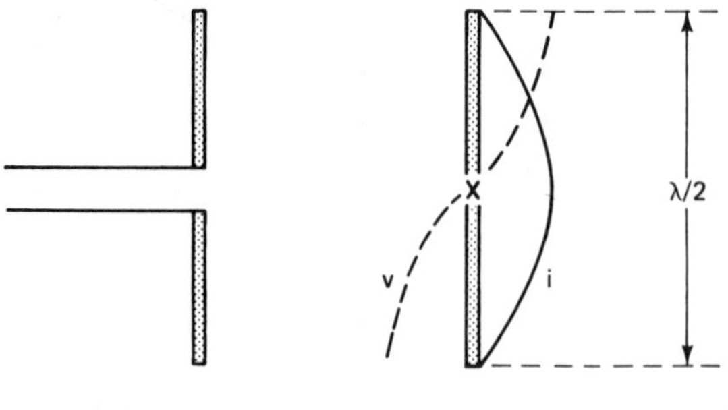

(a) (b)

Figure 9–4 Half-wave dipole as an open transmission line.

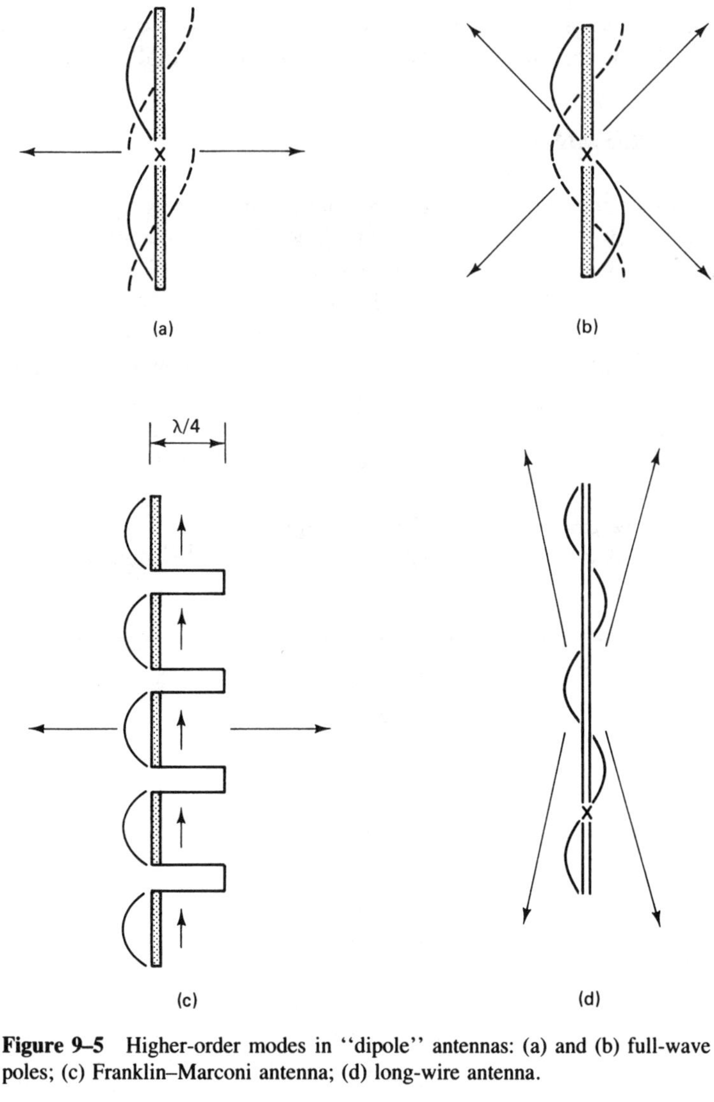

Figure 9–5 Higher-order modes in "dipole" antennas: (a) and (b) full-wave dipoles; (c) Franklin–Marconi antenna; (d) long-wire antenna.

The multiple-harmonic form of the symmetric dipole is called a *Franklin–Marconi* antenna, and it radiates a beam in the broadside direction (Fig. 9–5c). It is a series of quarter-wave elements connected by short-circuited, quarter-wave stubs.

At the resonant frequency the input impedance of an *LC* circuit is purely resistive; similarly, the input impedance of a resonant dipole is a pure resistance. From a transmission-line point of view, the impedance at a standing-wave maximum or minimum is purely resistive. A short dipole, a dipole shorter than a half-wavelength, is called a *Hertz* dipole. Since its input impedance is capacitively reactive, it has to be matched to a resistive source with an inductive reactance, called a "loading coil." It has the added disadvantage of having a small input resistance.

9–3 MONOPOLE ANTENNA

The *monopole* antenna is a dipole antenna in which the missing half is supplied by an image in a conducting ground plane. It is also called a *unipole* antenna and, especially if made of a flexible conductor, a "whip" antenna.

The method of images follows from the fact that a charge induces an equal but opposite charge in a conductor. Consider a positive charge placed above a conducting plane (Fig. 9–6). The charge attracts negative carriers in the conductor, gathering them together on the surface below. When the negative charge density is great enough to repel additional negative carriers, the system is in equilibrium. A sketch of the equilibrium surface charge density is shown directly below the ground plane in Fig. 9–6a. The electric field lines from the charge terminate on the negative surface charges and form a field configuration identical to what would be formed by an equal but opposite charge placed an equal distance below the ground plane (Fig. 9–6b).

EXAMPLE 9–1

Calculate the force on a charge of +100 μC placed 10 cm above a ground plane.

Solution We replace the ground plane by a charge of −100 μC placed at a distance of 20 cm from the original positive charge. From Coulomb's force law we find

$$f = \frac{1}{4\pi\epsilon_0}\frac{q_1 q_2}{r^2}$$

$$= \frac{(100 \times 10^{-6}) \times (100 \times 10^{-6})}{4\pi \times (8.85 \times 10^{-12}) \times (0.2)^2}$$

$$= 2248 \text{ N}$$

$$= \frac{2248 \text{ N}}{4.45 \text{ N/1b}}$$

$$= 505 \text{ lb wt}$$

Since this is a ¼-ton force, a substantial support should be provided.

Consider a charge moving upward with a velocity v on a vertical conducting cylinder above a horizontal ground plane (Fig. 9–7a). We replace the ground plane by a negative charge moving downward, as shown. The current caused by a negative

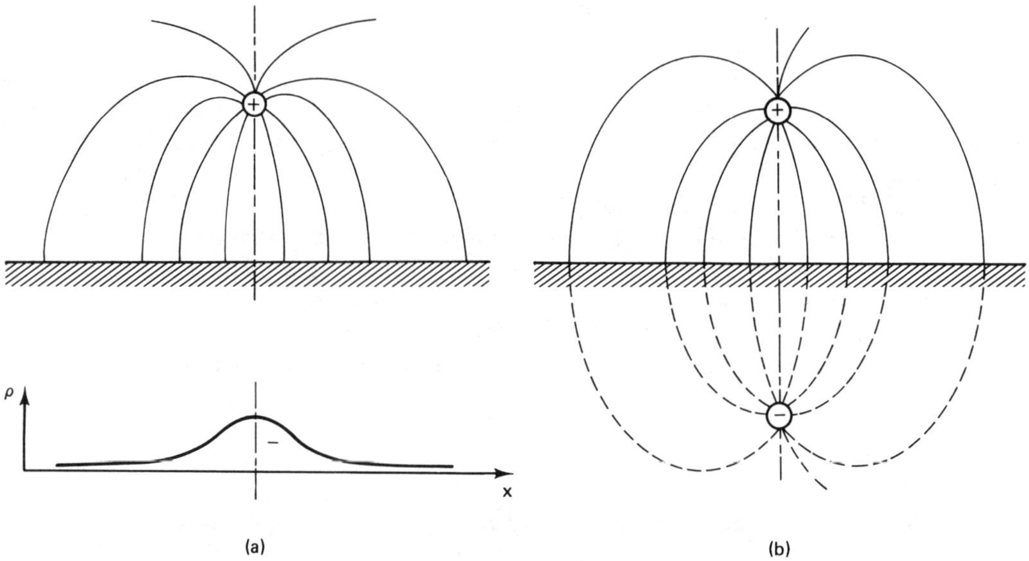

Figure 9–6 Charge image in a conducting plane: (a) charge and surface charge density induced by the charge; (b) charge and its image.

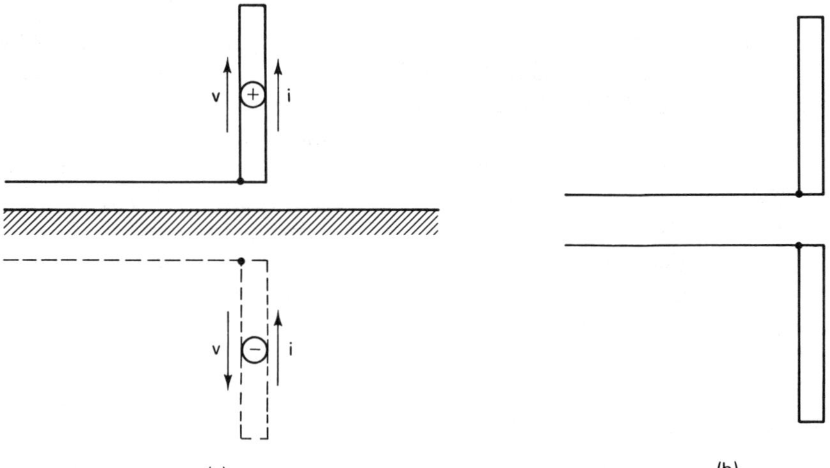

Figure 9–7 Method of images applied to a monopole antenna: (a) monopole and its image; (b) equivalent dipole.

charge moving downward is the same as that produced by a positive charge moving upward: an outflow of cash is equivalent to an inflow of monthly bills. An appreciation of this fact would eliminate a lot of discussion about whether or not electricity flows from positive to negative. We see that a quarter-wave monopole and its image are equivalent to a half-wave vertical dipole (Fig. 9–7b); however, since the radiated field is only half that of a complete half-wave dipole, the input resistance also is half, 32.35 Ω instead of 64.69 Ω.

A horizontal dipole near a ground plane induces an equal but opposite current in the image; thus the two currents tend to cancel one another. For this reason a horizontal dipole near a ground plane is a poor radiator.

The monopole antenna is used frequently in communications as a "whip," in test measurements as a probe, and as a transition from coaxial line to waveguide.

9–4 DIPOLE AS A TRANSMISSION LINE

The power radiated by an antenna can be equated to the input power,

$$P = i^2 R_1 \tag{9–5}$$

where i is the input current and R_1, the input resistance, is called the *radiation resistance*. We assume that the dipole is excited at its midpoint, that is, "center fed," and that the current and resistance are measured at the same point. If the dipole were "end fed," the radiated power would be equated to the input power as

$$P = \frac{v^2}{R_2} \tag{9–6}$$

where v and R_2 are the voltage and input resistance at either end of the dipole.

If the average characteristic impedance of the dipole arms is chosen properly, an arm can be regarded as a quarter-wave transformer, matching R_1 and R_2. We shall see that the proper characteristic impedance is that of free space, 377 Ω.

We show the dipole as a quarter-wave transformer in Fig. 9–8. As a receiving antenna, a signal is developed across the radiation resistance R_2 and matched by the transmission line transformer to a load impedance, R_1. A peculiarity of a receiving dipole is that the radiation resistance simultaneously rebroadcasts part of the received signal. Nearly all of the received signal will be rebroadcast or "scattered" if the antenna is short-circuited. If the antenna is matched to a load, then in accordance

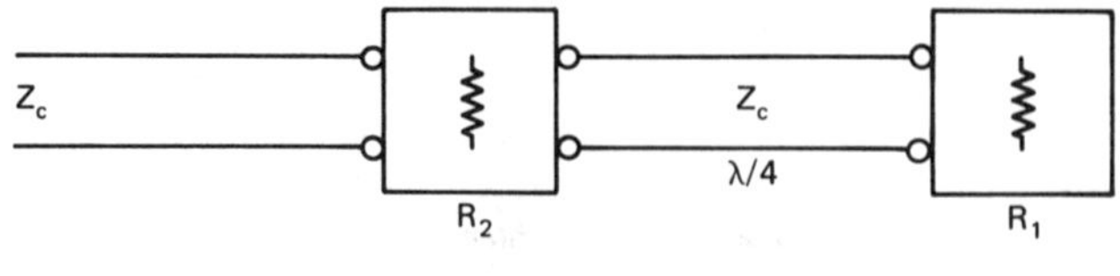

Figure 9–8 Half-wave dipole as a quarter-wave transformer.

with the maximum power transfer theorem, half the energy is delivered to the load and the other half is scattered.

If half of the energy is rebroadcast, the amplitude reflection coefficient is $(2)^{-1/2}$, and the VSWR in the vicinity of the antenna and on the dipole arm is

$$s = \frac{1 + r}{1 - r} = \frac{1 + 2^{-1/2}}{1 - 2^{-1/2}} = 5.8284 \tag{9–7}$$

If we draw the VSWR circle on a Smith chart (Fig. 9–9), we see that it intersects the real axis at 5.8284 and 1/5.8284 = 0.1716, corresponding to unnormalized impedances of

$$R_2 = 5.8284 \times 377 = 2197\ \Omega$$

and

$$R_1 = 0.1716 \times 377 = 64.68\ \Omega \tag{9–8}$$

The value usually quoted for the input impedance of a center-fed dipole is 73 Ω. This is appropriate for very thin dipoles, dipoles made of long wires for use at low frequencies. The result we have obtained, 64.68 Ω, is in agreement with measurements made on UHF and microwave dipoles, which usually are much thicker.

To determine the quality or Q of a dipole we note that the characteristic impedance of the dipole arms at resonance is $Z_c = (L/C)^{1/2} = (\omega_0 L/\omega_0 C)^{1/2} = (\omega_0^2 L^2)^{1/2}$, or

$$Z_c = \omega_0 L \tag{9–9}$$

where ω_0 is the resonant frequency from Eq. (9–1). The unloaded Q of an optimum dipole is $Q = R_2/\omega_0 L$, and combining this with Eqs. (9–8) and (9–9), we have

$$Q = \frac{R_2}{Z_c} = \frac{Z_c}{R_1} = 5.8284 \tag{9–10}$$

If the dipole is matched to a load, the loaded Q is found as the ratio of the reactance to resistance at the point marked Q_s on the Smith chart in Fig. 9–9,

$$Q_s = \frac{Q}{2} = 2.914 \tag{9–11}$$

We shall call Q_s the quality of the system—dipole and generator or load—and reserve Q for the quality of the dipole alone.

The results we have obtained are for an "optimum" dipole, a dipole with a

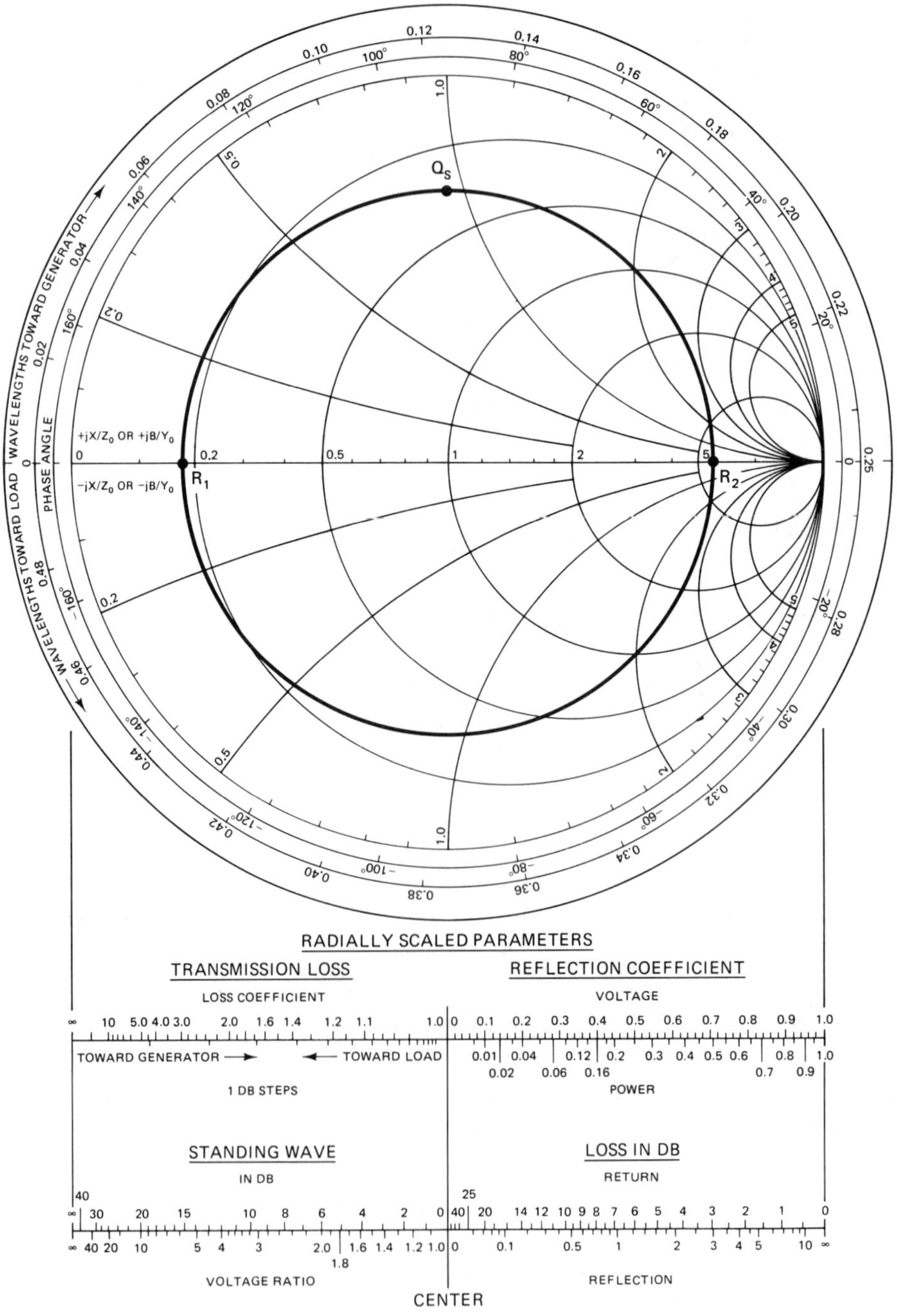

Figure 9–9 VSWR circle for a half-wave dipole.

characteristic impedance of 377 Ω. We can find the physical dimensions of an optimum dipole from the equation for its characteristic impedance,

$$Z_c = 120\left(\ln \frac{2s}{a} - 1\right) \tag{9–12}$$

where s is the length of one arm of the dipole and a is its radius.

EXAMPLE 9–2

Calculate the length-to-diameter ratio of an optimum dipole antenna.

Solution The length-to-diameter ratio is the same as the ratio of the length of one arm to its radius. Solving Eq. (9–12) for s/a, we have

$$\frac{s}{a} = \frac{1}{2} \exp\left(\frac{Z_c}{120} + 1\right) \tag{9–13}$$

$$= \frac{1}{2} \exp\left(\frac{377}{120} + 1\right)$$

$$= 31.5$$

If the dipole were 1 m long, its diameter would be 1/31.5 = 0.0318 m or 1.25 in.

The half-wave dipole is seldom end fed; however, the full-wave dipole can be regarded as two end-fed, half-wave dipoles excited in shunt. We then expect the input impedance of a full-wave dipole to be half that of a single end-fed dipole, or

$$R_\lambda = \frac{R_2}{2} = 1099\ \Omega \tag{9–14}$$

The Q_s of the full-wave dipole is identical to the Q_s of either of the half-wave dipoles from which it is formed; for an optimum dipole, 2.914.

The results we have obtained so far are appropriate for an optimum dipole, a dipole with a characteristic impedance of 377 Ω. If a nonoptimum dipole is to be built, its characteristics can be calculated by starting with Eq. (9–12) and modifying the previous results accordingly. For brevity we shall continue to discuss only the optimum dipole.

9–5 END EFFECT

The lowering of the resonant frequency of an oscillator by loading is well known, but for some reason its occurrence in a dipole is called an "end effect." We shall establish a relationship between the equivalent series and parallel reactances of an inductor and its coil resistance and the Q of an LC resonant circuit. From this we derive an expression for the lowering of the resonant frequency by the resistive component. We then allow the reactances to represent those of a dipole, allow the coil or load resistance to represent the radiation resistance of the dipole, and obtain an expression for the end effect. We justify our procedure by the Smith chart diagram shown in Fig. 9–9. We see that the end-fed dipole is equivalent to a parallel resonant oscillator, the center-fed to a series resonant oscillator.

Equivalent parallel and series inductive reactances are related as

$$X_p = \frac{R_s^2 + X_s^2}{X_s} \tag{9–15}$$

and

$$X_s = \frac{R_p^2 X_p}{R_p^2 + X_p^2} \tag{9–16}$$

where R_s and R_p are equivalent series and parallel resistances in combination with the equivalent series and parallel inductive reactances X_s and X_p. By rearranging Eq. (9–15), we find $X_p/X_s = (1 + X_s^2/R_s^2)/(X_s^2/R_s^2)$, and at the resonant frequency this becomes

$$\frac{X_p}{X_s} = \frac{1 + Q^2}{Q^2} \tag{9–15a}$$

For either the series or parallel circuit the capacitive and inductive reactances are equal at resonance and we have $X_p/X_s = X_C/X_s = 1/\omega_0' C \omega_0' L$. With this result and Eq. (9–15a) we find $(\omega_0')^2 LC = Q^2/(Q^2 + 1)$, or

$$\omega_0' = \omega_0 \left(\frac{Q^2}{Q^2 + 1} \right)^{1/2} \tag{9–17}$$

where ω_0 is the resonant frequency of an unloaded circuit, $\omega_0 = (LC)^{-1/2}$. If now we let X_p, X_C, R_p, R_s, and Q represent the reactances, radiation resistances, and quality of a dipole, then Eq. (9–17) becomes an expression for the so-called "end effect."

EXAMPLE 9–3

Design an optimum half-wave dipole for operation at 150 MHz.

Solution The operating wavelength is $\lambda = c/f = 3 \times 10^8/150 \times 10^6 = 2$ m. If there were no end effect, the dipole would be 1 m in length. For an optimum dipole $Q_s = 2.914$, and from Eq. (9–17) we have

$$\frac{\omega_0'}{\omega_0} = \left(\frac{2.914^2}{2.914^2} + 1 \right)^{1/2}$$

$$= 0.946$$

Since the resonant frequency would be 94.6% of the desired operating frequency, we compensate by shortening the length of the dipole by the same factor:

$$2s' = \frac{\omega_0'}{\omega_0} 2s = 0.946 \times 1 = 0.946 \text{ m} \tag{9–18}$$

For $Z_c = 377\ \Omega$, we find from Eq. (9–13) that $s/a = 31.5$, and the diameter of the tubing should be $2s/a = 0.946/31.5 = 0.030$ m or 1.18 in.

The design of a dipole antenna seems straightforward, but it may be instructive to consider a bad design.

EXAMPLE 9–4

A half-wave dipole, intended for operation in the 150-m band ($f = 2$ MHz) is made of 246 ft of AWG 28 copper wire suspended 30 ft above ground. Evaluate the antenna.

Solution We use the method of images to estimate the ground effect. We assume a good ground and replace it with an image antenna 180° out of phase and 60 ft distant. The operating wavelength is 150 m or 492 ft. If a is the amplitude at the antenna in free space, the amplitude in the presence of the image antenna is proportional to

$$\begin{aligned} a + a\underline{/180° + 360° \times 60/\lambda} &= a + a\underline{/180° + 360° \times 60/492} \\ &= a(1 + 1\underline{/224°}) \\ &= a(0.281 - j0.695) \end{aligned}$$

neglecting multiple reflections. Since the real part of the amplitude is reduced by a factor of 0.281, we estimate that the input resistance or radiation resistance is reduced by the same factor. The radiation resistance is reduced to $0.281 \times 73 = 20.4\ \Omega$. This is a rather low input resistance, and the matching required will have to include cancellation of the $-j0.695$ reactance factor. A height of $\lambda/4 = 123$ ft would be ideal but perhaps prohibitively expensive. If the height could be doubled to 60 ft, the input resistance would be 70 Ω, a considerable improvement.

The characteristic impedance of the dipole can be found from Eq. (9–12). The radius of No. 28 wire is 0.0063 in. and the length of the antenna is 2772 in. We have

$$Z_c = 120 \ln \left(\frac{2772}{0.0063} + 1\right) = 1559\ \Omega$$

and we estimate the Q as $Z_c/R_1 = 1559/20.4 = 76.4$; however, this Q will not be realized. The dc resistance of No. 28 copper wire is 65 Ω/1000 ft. Since the skin depth at this frequency is almost equal to the radius of the wire, no correction for the skin effect is necessary. The ohmic resistance of the antenna is 65(246/1000) = 16 Ω. The antenna efficiency is the useful power divided by the total power, 20.4/(20.4 + 16) = 56%. About half the input power is wasted in ohmic loss.

9–6 DIPOLE IMPEDANCE

We cannot use the Smith chart diagram shown in Fig. 9–9 to find the dipole impedance at off-resonant frequencies because our model does not include the fact that radiation resistance varies with the fourth power of the frequency. Rather than trying to modify the Smith chart presentation we shall find it easier to proceed algebraically.

Since series resonant and parallel resonant circuits are identical, differing only in series or shunt excitation, we can start with either. The input impedance of a series resonant circuit is

$$Z_1 = R_1 + j\left(\omega L - \frac{1}{\omega C}\right)$$

$$= R_1 + jR_1 \frac{\omega_0 L}{R_1}\left(\frac{\omega}{\omega_0} - \frac{1}{\omega_0 L \omega C}\right)$$

We can use Eq. (9–1) to eliminate LC, and we have $Z_1 = R_1 + jR_1 Q(\omega/\omega_0 - (\omega/\omega_0)^{-1})$, or

$$Z_1 = R_1 + jR_1 Q\left(f - \frac{1}{f}\right) \tag{9–19}$$

where $f = \omega/\omega_0$ is the relative frequency. To account for variation of radiation resistance with frequency we replace R_1 by $R_1 f^4$. To account for the end effect we refer f to the unloaded resonant frequency as

$$f' = \frac{\omega}{\omega_0'} = \frac{\omega}{e\omega_0} = \frac{f}{e} \tag{9–20}$$

where e is the end effect factor, $e = Q/(Q^2 + 1)^{1/2}$. With these changes Eq. (9–19) becomes, approximately,

$$Z_1 = R_1 \left(\frac{f}{e}\right)^4 \left[1 + jQ\left(\frac{f}{e} - \frac{e}{f}\right)\right] \tag{9–21}$$

We have evaluated Eq. (9–21) for a loaded optimum dipole, $Q_s = 2.914$, and the results, shown in Fig. 9–10, are in substantial agreement with measurements made at UHF and microwave frequencies.

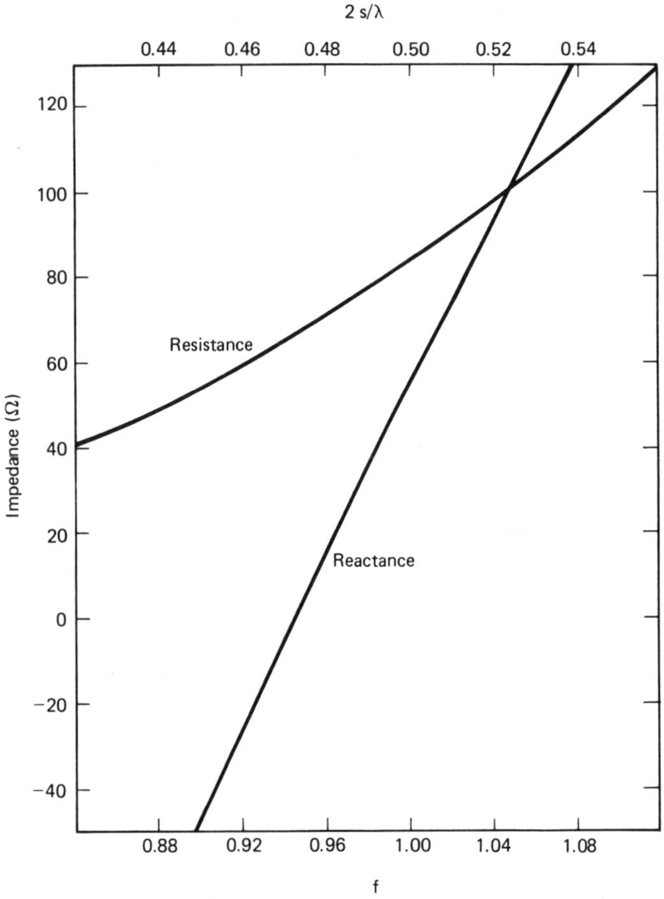

Figure 9–10 Real and reactive components for the input impedance of a half-wave dipole as a function of the relative frequency of operation.

EXAMPLE 9–5

The 150-MHz dipole of Example 9–3 is to be used for reception at 165 MHz. Match it to a 50-Ω load.

Solution The operating frequency has been increased from 150 to 165 MHz, a factor of 165/150 = 1.1. The relative frequency is increased from 0.946 to 0.946 × 1.1 = 1.0406. From Fig. 9–10 we find $Z_1 = 98 + j99\ \Omega$, and the impedance normalized to a 50-Ω line is $z = 1.96 + j1.98$.

We plot the normalized impedance at point A on a Smith chart and draw a diameter and VSWR circle through the point (Fig. 9–11). We see that a rotation of 0.136λ transforms the source admittance at point B to an

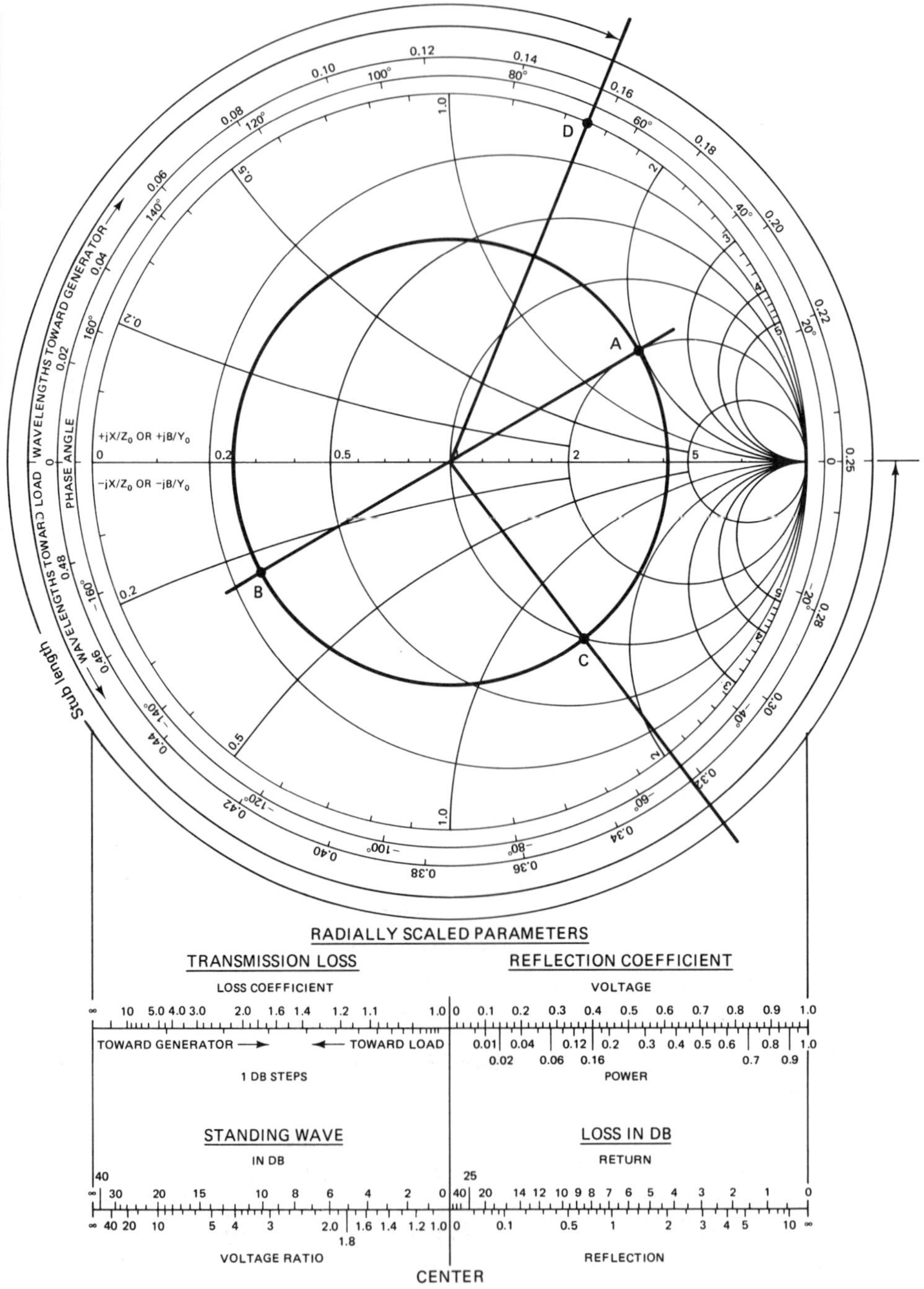

Figure 9–11 Smith chart procedure for matching an off-resonant dipole antenna to a 50-Ω line.

admittance of $1 - j1.52$ at point C. Here the real part of the admittance is matched to the characteristic admittance of the line; we need now a stub to cancel the inductive susceptance of $-j1.52$. We plot $+j1.52$ at point D and find that we need a short-circuited stub of length 0.408λ or an open-circuited stub of length 0.158λ. If the latter were executed in coax with a dielectric constant of $\epsilon_r = 2.25$, its length would be $0.158(\lambda_0/(\epsilon_r)^{1/2}) = 0.158[1.82/(2.25)^{1/2}] \times 39.37 = 7.55$ in. Short-circuited stubs are usually preferred, to avoid radiation loss.

9–7 APERTURE THEORY AND DIPOLE GAIN

We can approach aperture theory from two points of view: either radiation is in the form of waves or is in the form of particles, called "corpuscles" or *photons*. Whether light is wavelike or particlelike was debated in the eighteenth and nineteenth centuries. Scientists on the European continent inclined toward the wave point of view, but Isaac Newton held the opinion that light is corpuscular. When confronted with the evidence that light is polarizable, a wave characteristic, he replied that the particles must "have sides."

The discovery of the photoelectric effect at the beginning of the twentieth century was one of several experiments that indicated that light has a dual nature. It is both wavelike and corpuscular. We can imagine light waves spreading out from a source, like waves on a pond, but we find that light can eject an electron from its binding in an atomic or crystalline structure as though all of the energy of the light waves were concentrated in one place. In this respect light acts like a pellet shot from a gun. If it strikes an electron it transfers a large portion of its energy to the electron.

This puzzling behavior of light is part of what is called the *principle of duality*. Perhaps it is not so puzzling if we say that light propagates like a wave but is emitted or absorbed like a particle. It is perhaps even less puzzling if we say that a photon can be regarded as a small bunch of waves, a "wave packet," which in turn can be regarded as the Fourier wave components of a pulse or quantum of light.

In this section we develop aperture theory from photon theory, since it is simpler than the wave theory of apertures. For our purpose, a photon is a spherical particle of radius $\lambdabar$ (Fig. 9–12a). Its radius is an effective quarter-wavelength. It moves in free space with a velocity c and has an energy of

$$U = hf \tag{9–22}$$

and a momentum of

$$p = \frac{hf}{c} = \frac{h}{\lambda} \tag{9–23}$$

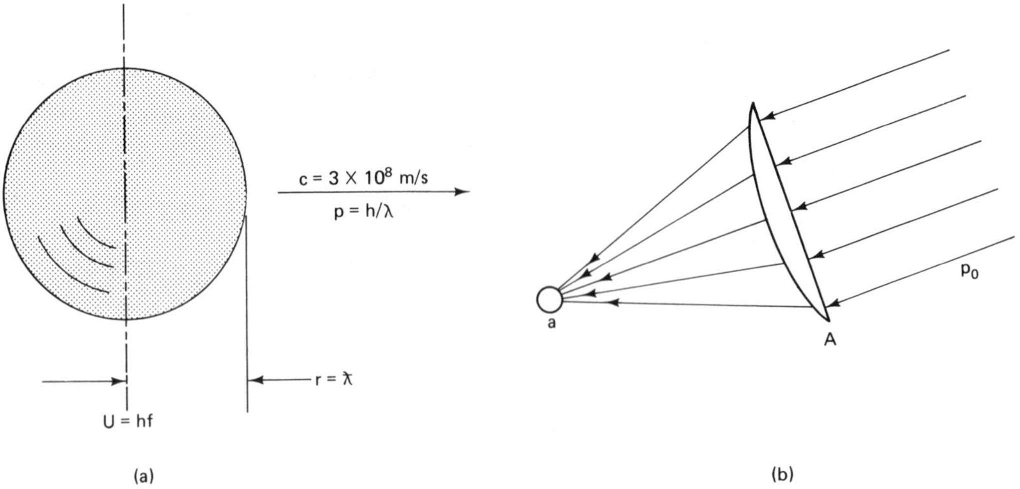

Figure 9–12 (a) Characteristics of a photon; (b) burning glass as a receiving antenna.

where f is frequency, λ is wavelength, and h is a universal constant called *Planck's constant*, $h = 6.622 \times 10^{-27}$ erg-s or 4.139×10^{-15} eV-s.

The gain of a receiving antenna is defined as the ratio of the power delivered to a load to the power that would be delivered without the antenna, that is, to the power that would be delivered by an omnidirectional antenna. Consider, for example, a burning glass—a lens used to concentrate the sun's rays to a focal point (Fig. 9–12b). Let A be the cross-sectional area of the lens, the area of the *aperture*, and let a be the area of the focal point. If p_0 is the energy flux or power density in space, the power delivered to load when the lens is in place is $P_1 = p_0A$. If the lens is not used, the power delivered is $P_2 = p_0a$. The gain is defined as $G = P_1/P_2$, or $G = A/a$. Since the area of the focal point cannot be smaller than the cross-sectional area of a photon, $a = \pi r^2 = \pi ƛ^2$, the gain is $A/\pi ƛ^2$, or

$$G = \frac{4\pi A}{\lambda^2} \tag{9–24}$$

Equation (9–24) is without approximations and is valid for all antennas, from VLF to optical frequencies. By the principle of reciprocity it is valid for transmitting as well as receiving antennas. It is, however, idealized in that it assumes 100% aperture efficiency. Aperture efficiency is discussed in a later section.

EXAMPLE 9–6

Calculate the power gain of a satellite dish. The diameter is 2 m and the received signal has a frequency of 3.25 GHz.

Solution The operating wavelength is $\lambda = c/f = 3 \times 10^8/3.25 \times 10^9 = 9.23 \times 10^{-2}$ m, and the area of the dish is $\pi d^2/4 = \pi \times 2^2/4 = 3.14\ \text{m}^2$. From Eq. (9–24) we have

$$G = 4\pi \times \frac{3.14}{(9.23 \times 10^{-2})^2}$$

$$= 4.63 \times 10^3 \quad \text{or} \quad 36.7\ \text{dB}$$

In Example 9–6 we not only assumed 100% aperture efficiency but also assumed that there are no losses in transferring the energy to a receiving antenna at the focal point and no additional losses in mismatch and in transmission lines. In practice we call power gain the gain that is actually measured under field conditions, and the ideal gain resulting from the directivity of the antenna we call *directive gain* or *directivity*.

We can use Eq. (9–24) to find the gain of a dipole antenna if we can define the equivalent aperture of a dipole. We assume that if a photon hits a dipole dead center it will be absorbed—disappear as a burst of energy in the load resistance. If it is a near-miss, grazes the dipole as it goes by, it has a 50–50 chance of being absorbed. We shall define the effective absorption area just as we would define beamwidth or bandwidth: cut it off at the half-power point. The effective absorption area becomes that shown in Fig. 9–13. We see that the area is composed of two half-circles of area $\pi ƛ^2$ and a rectangle of area $(\lambda/2 - 2ƛ)(2ƛ)$. The total area is $A = \pi ƛ^2 + (\lambda/2 - 2ƛ)(2ƛ) = ƛ^2(3\pi - 4)$, or

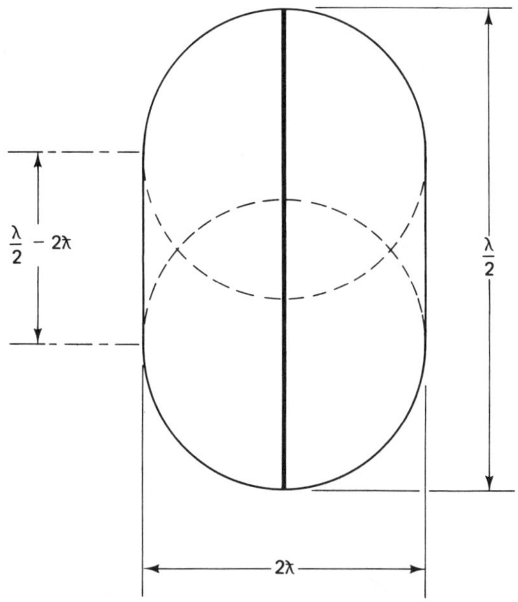

Figure 9–13 Effective absorption area or cross section for a half-wave dipole antenna.

$$A = \frac{\lambda^2(3\pi - 4)}{4\pi^2} \qquad (9\text{–}25)$$

$$= 0.137\lambda^2$$

By substituting A/λ^2 from Eq. (9–24) into Eq. (9–25) we find

$$G = 4\pi(3\pi - 4)4\pi^2$$
$$= 1.73 \qquad (9\text{–}26)$$

This result often is given as the gain of a thin wire or high-Q dipole, but for a microwave or optimum dipole we must shorten the length by a factor of 0.946 because of the end effect. The gain is reduced to

$$G = 0.946 \times 1.73$$
$$= 1.630 \qquad (9\text{–}27)$$

The photon or quantum theory of antennas often is simpler than the wave theory, and we shall return to it again to analyze more difficult problems. It is not often used, possibly because some workers in the first half of the twentieth century thought radio dipoles were somehow different from atomic dipoles, as though the laws of physics changed with change in scale. For the present we shall take a more conventional approach, so that we may become familiar with standard technique.

9–8 DIRECTIVITY AND RADIATION RESISTANCE

We have derived radiation resistance and gain of a dipole separately, but the two are related. To be intuitively aware of the relationship, we should know that the radiation pattern of a dipole is shaped like a doughnut, half of which is sketched in Fig. 9–14.

If the power, P, transmitted by an antenna were radiated uniformly in all directions, that is, if it were an *omnidirectional source*, the power density at a

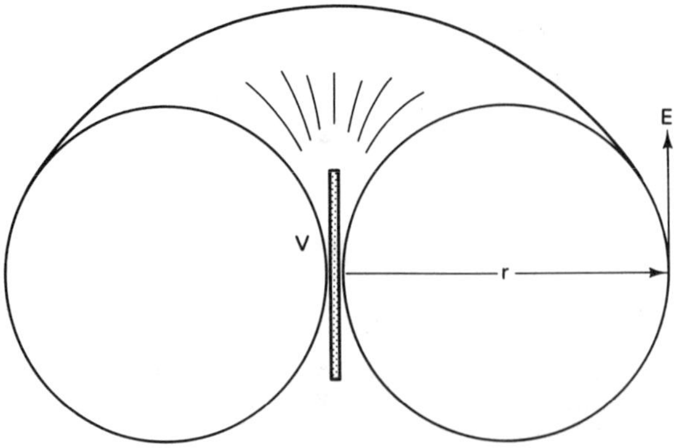

Figure 9–14 Cross section of an elementary dipole radiation pattern.

distance r would be the total power divided by the area of a sphere of radius r: $p = \frac{P}{4\pi r^2}$. If the power is radiated in a preferred direction, for example, concentrated in the broadside direction, as shown in Fig. 9–14, the power density at a distance r in the preferred direction is

$$p = \frac{GP}{4\pi r^2} \tag{9–28}$$

Here G is our definition of gain for a transmitting antenna, but by reciprocity the gain is the same when the same antenna is used for receiving.

So that our description of the radiation pattern of a dipole as "doughnut shaped" be not misleading, we should explain that although the E-plane amplitude pattern actually is circular in cross section, it is usually described as

$$E - E_0 \cos \theta \tag{9–29}$$

where E_0 is the field strength at maximum amplitude, $\theta = 0°$, and θ is the angular coordinate. A polar graph of the dipole amplitude radiation pattern is shown on the right side of Fig. 9–15, and the power pattern

$$P = P_0 \cos^2 \theta \tag{9–30}$$

is shown on the left side. Radiation patterns may be shown on polar or rectangular graphs, and the vertical scale may be linear or in dB.

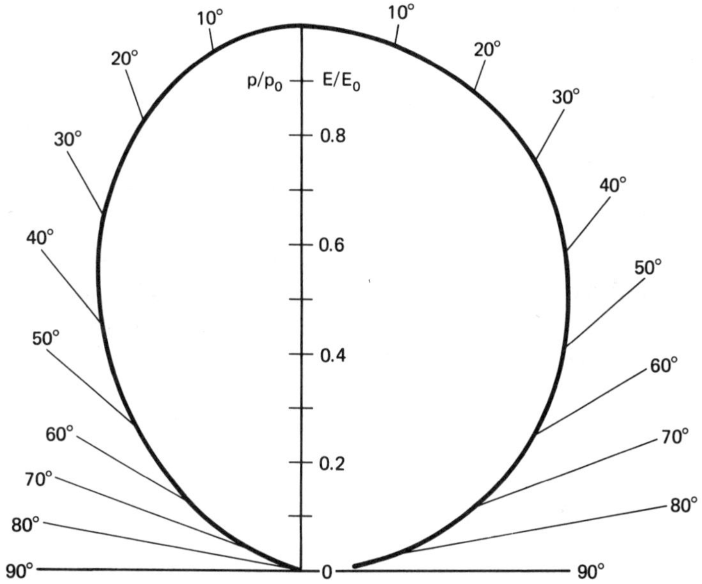

Figure 9–15 Polar diagram of power and amplitude radiation patterns of an elementary dipole antenna.

EXAMPLE 9–7

A vertical, half-wave dipole on a hilltop is used as a radio beacon at 250 MHz. The transmitted power is 200 W. Find the power received by an approaching aircraft if it is ½-mile distant and flying at an elevation 3000 ft above the hilltop. Assume that the aircraft antenna has a power gain of 6 dB.

Solution Our first task is to find the power density at the aircraft. From Fig. 9–16 we find that the distance from beacon to aircraft is $r = (3000^2 + 2640^2)^{1/2} = 4000$ ft or 1218 m, and the angle of approach is $\theta = \tan^{-1}(3000/2640) = 48.4°$. We find that the power density at the distance r is, from Eq. (9–28), $p = G\ 200/4\pi(1218)^2 = 1.073 \times 10^{-5}\ G\ \text{W/m}^2$. The gain of the dipole is 1.63 in the preferred direction, $\theta = 0°$, but from Fig. 9–15 we see that at 48.4° the gain is reduced by a factor of 0.435. The power density at the aircraft is $p = (1.073 \times 10^{-5}) \times 1.63 \times 0.435 = 7.65\ \mu\text{W/m}^2$.

We next need to find the effective aperture area of the aircraft antenna. The operating wavelength is $\lambda = c/f = (3 \times 10^8)/(250 \times 10^6) = 1.2$ m. We solve Eq. (9–24) for the aperture area and find

$$A = \frac{G\lambda^2}{4\pi} = \frac{3.98 \times 1.2^2}{4\pi} = 0.456\ \text{m}^2$$

where the number 3.98 comes from $\log^{-1}(6\ \text{dB} \times 10)$.

The final step is the multiplication of the aperture area by the power density to determine the total power received:

$$7.65\ (\mu\text{W/m}^2) \times 0.465\ (\text{m}^2) = 3.49\ \mu\text{W}$$

This is the power delivered by the aircraft antenna to its receiver and is adequate for modern receivers; however, it is evident that if the aircraft flies directly over the dipole beacon it will enter a "zone of silence."

The power density at a distant receiving site can also be described in terms of the local field parameters as $i_D^2 Z_c$, where i_D is the current density in space at the receiving site and is called the *displacement current* and Z_c is the characteristic impedance of free space. By equating this expression to the power density from Eq. (9–28) we find

$$\frac{G}{4\pi r^2} i_1^2 R_1 = i_D^2 Z_c \tag{9–31}$$

where we have also substituted $i_1^2 R_1$ for the transmitted power, P. The antenna current, i_1, eventually is spread out as a current sheet of density $i_1/2\pi r$ at the receiving site. The factor $2\pi r$ is the circumference of the "doughnut" shown in Fig. 9–14. Equation (9–31) becomes, with $i_D = i_1/2\pi r$,

$$G = \frac{Z_c}{\pi R_1} \tag{9–32}$$

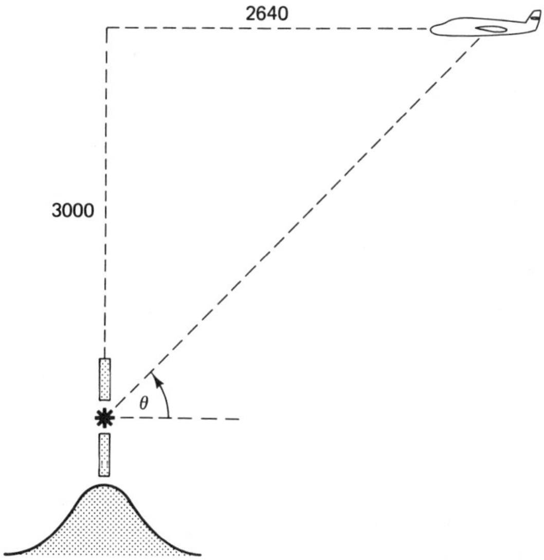

Figure 9–16 Aircraft approaching a dipole antenna beacon, Example 9–7.

For an optimum dipole we have, from Eq. (9–10), $Z_c/R_1 = Q$, and Eq. (9–32) can be expressed as

$$\frac{G}{Q} = \frac{1}{\pi} \tag{9–33}$$

For a loaded dipole this becomes

$$\frac{G}{Q_s} = \frac{2}{\pi} \tag{9–33a}$$

From Eq. (9–33) or (9–33a) we see that gain and bandwidth are related, not independent. The derivation is intuitive in that we are asked to believe a current density of $i_1/2\pi r$ exists at the receiving site; however, a more rigorous mathematical derivation gives the same result. Although Eq. (9–33) often is found in technical literature, it should be corrected for the end effect. If the total current i_1 remains the same, the current density at the receiving site is reduced by a factor 0.946. We substitute $0.946i_D$ for i_D in Eq. (9–31), rederive our equations, and now have

$$\frac{G}{Q} = \frac{0.946^2}{\pi} \tag{9–34}$$

and

$$\frac{G}{Q_s} = \frac{2 \times 0.946^2}{\pi} \tag{9–34a}$$

$$= 0.5697$$

We have used the concept of electromagnetic field variables and circuit theory to derive and correct Eq. (9–34a), the transmission-line concept and SWR to derive Q, and the concept of absorption area to derive G. We now put all of our results into Eq. (9–34a) to check for consistency:

$$\frac{1.630}{2.914} = 0.5697$$

$$0.5594 = 0.5697$$

Equation (9–34) is valid for optimum or nearly optimum dipoles but in general should be regarded as an inequality,

$$\frac{G}{Q} \leq \frac{0.946^2}{\pi} \tag{9–35}$$

since we can always build a thin dipole with a high Q and narrow bandwidth. On the other hand, if we build a very fat dipole to increase bandwidth, we lose mode control at the feed point, and the gain in the desired direction decreases. The optimum dipole is a compromise between extremes.

We now shall repeat our previous derivations in which we introduced the concept of a current density in space—the displacement current, but this time we shall use the concept of an electric field strength. If a transmitting dipole is end fed, we have $P = V_1^2/R_2$ at the feed point and $p = E^2/Z_c$ for the power density at the receiving site, where V_1 is the voltage impressed on the antenna and E is the electric field strength at a distant receiver. With the assumption $V_1/2\pi r = E$, we repeat our previous derivation to obtain

$$G = \frac{0.946^2 R_2}{\pi Z_c} \tag{9–36}$$

which is, since $R_2/Z_c = Q$, the same as Eq. (9–34).

A horizontal dipole near a ground plane may have a low input impedance, as we saw in Example 9–4; similarly, a dipole close to parasitic elements may have a low input impedance. In such circumstances it may be desirable to start with an antenna that has a high input impedance, such as a full-wave dipole or, more compactly, a folded dipole.

A *folded dipole* can be regarded as two end-fed, half-wave dipoles connected in parallel (Fig. 9–17). Connecting both ends of half-wave dipoles in shunt reduces the impedance by a factor of 4, and the resistance becomes, from Eq. (9–8), 2197/4 = 549.3 Ω. To find the impedance at the input we first normalize to 377 Ω and we have 549.3/377 = 1.457. We plot this on a Smith chart, rotate a quarter-wavelength to the feed point, obtain a normalized impedance of 1/1.457 = 0.686, and we conclude that the unnormalized radiation resistance of a folded dipole is 0.686 × 377 = 259 Ω.

In a folded dipole the current at the center of the folded element is reversed

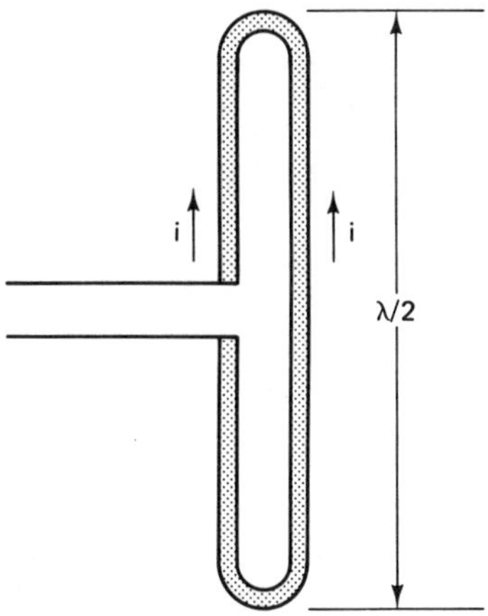

Figure 9–17 Folded dipole antenna.

because it is a half-wavelength removed from the feed point; in addition, it is reversed again by the folding. The result is the current and voltage distributions are much the same as in a single half-wave dipole. This means that the radiation pattern and directive gain of a folded dipole are almost the same as for a single half-wave dipole.

9–9 DISPLACEMENT CURRENT

We introduced the concept of displacement current in Chapter 6 and again in Section 9–8, but we need to review and further develop the concept so that we may better understand the launching of waves into space.

If we connect a capacitor to a battery, current flows and the capacitor is "charged." Actually, the capacitor contains no more charge than before; charge has only been displayed across its terminals. When all the negative and positive charges are added together, the total is still zero. If we reverse the battery terminals, the current is reversed and the capacitor is oppositely charged. Replace the battery by an RF generator and the process continues automatically. An ac ammeter in the circuit indicates that a current is flowing, as if through the capacitor (Fig. 9–18).

The above "as if" explanation of ac current flow through a capacitor is incorrect. The difficulty lies in not considering the magnetic field associated with the circuit. From elementary circuit theory we know that a current loop is the source of a magnetic field, $h = ni/s$, where ni/s is the number of ampere turns per meter. For our one-turn circuit the field at the center of the loop is

$$H = \frac{\frac{1}{2}i}{a} \tag{9–37}$$

where a is the radius of the loop. Equation (9–37) follows from the law of Biot and Savart and ultimately is based on experimental evidence. Since the circuit in Fig. 9–18 is open—no ampere turns—there ought not be a complete magnetic field

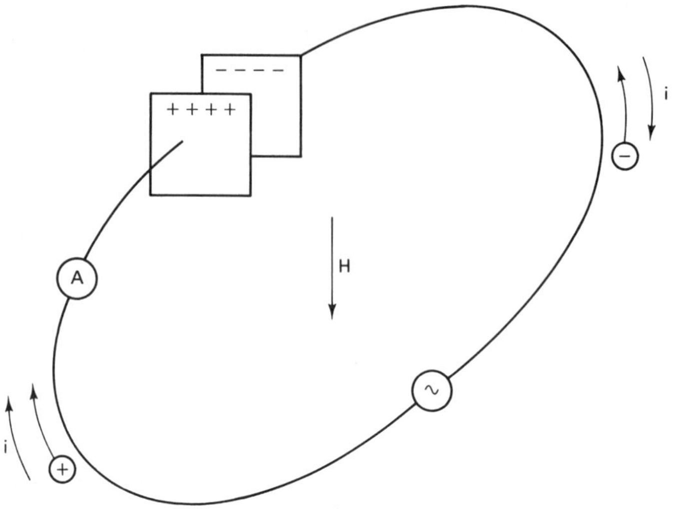

Figure 9–18 Current flow through empty space, the "displacement current."

associated with it, but experience shows that there is. We cannot create a magnetic field with "as if's."

The difficulty was resolved in the second half of the nineteenth century by James Clerk Maxwell, who assumed that there actually is a current flowing through the capacitor, even if the capacitor plates are separated by a vacuum. Maxwell called it a *displacement current*, and set it equal to the time rate of change of the electric displacement in the capacitor,

$$i_D = \frac{\Delta D}{\Delta t} \tag{9–38}$$

The circuit shown in Fig. 9–18 now is completed by a conduction current, $i_c = \Delta q/\Delta t$, through the conductors and a displacement current through the capacitor.

For sinusoidal variations Eq. (9–38) can be expressed as

$$i_D = -j\omega\epsilon E \tag{9–39}$$

where ω is the angular frequency of the sine waves and ϵ the permittivity of the dielectric. By using the relations $c = \lambda\omega$ and $Z_c = (\mu/\epsilon)^{1/2}$ we can also write Eq. (9–39) in the form

$$E = jZ_c\lambda i_D \tag{9–40}$$

In this form the displacement current equation somewhat resembles Ohm's law. Georg Ohm originally formulated his law in terms of field variables as

$$E = \rho i_c \tag{9–41}$$

where E is field strength in volts/meter, ρ is resistivity in ohm-meters, and i_c is the conduction current density—the current flow through a cross section of unit area—in amperes/meter^2.

EXAMPLE 9–8

Assume that the capacitor in Fig. 9–18 is a vacuum capacitor with circular plates of diameter 10 cm and a separation of $d = 2$ mm. Find the displacement current when the circuit is driven by a 10-V peak, 200-MHz signal generator.

Solution The peak electric field strength in the capacitor is $E = V/d = 10/0.002 = 5$ kV/m. The characteristic impedance of the vacuum is 377 Ω, and the operating wavelength is $\lambda = c/\omega = 3 \times 10^8/2\pi \times (200 \times 10^6) = 0.239$ m. From Eq. (9–40) we now have

$$i_D = \frac{-j5 \times 10^3}{377 \times 0.239}$$

$$= 55.5\angle{-90^\circ} \text{ peak A/m}^2$$

Check: Since in a closed circuit the total displacement current must be equal in magnitude to the total conduction current, we can check our result by calculating the conduction current. The capacitance of the parallel-plate capacitor is $C = \epsilon_0 A/d = \epsilon_0 \pi r^2/d = [(8.85 \times 10^{-12}) \times \pi \times 0.05^2]/0.002 = 34.8$ pF. The reactance is $X_C = 1/\omega C = 1/[2\pi \times (200 \times 10^6) \times (34.8 \times 10^{-12})] = 22.9\ \Omega$. We now can find the conduction current from Ohm's law: $i_c = v/X_C = 10/22.9 = 0.437$ A. The magnitude of the total displacement current is, from Example 9–8, $i_D A = i_D \pi r^2 = 55.7\pi \times 0.05^2 = 0.437$ A. For another way to calculate displacement current, see Example 6–2.

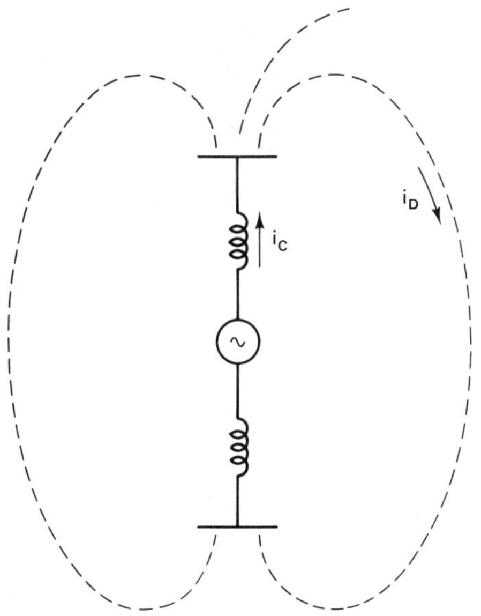

Figure 9–19 Dipole as an inside-out *LC* resonant circuit.

With the concept of displacement current we can understand dipole radiation more fully. Let us build a resonant circuit by adding an inductor to the circuit shown in Fig. 9–18. Now turn the circuit inside out, as suggested in Fig. 9–19. We see that the path length of the displacement current has been greatly increased, and for this reason the principal loss is to radiation rather than to the ohmic resistance of the inductor. The purpose of the dipole structure is to increase the path of the displacement current in space. The fact that the distributed capacitance and inductance cause a resonance is incidental. The resonant frequency is of interest only because it is the frequency at which the input impedance is purely resistive. What remains to be answered is why the energy is launched into space rather than returned to the antenna as in an induction field.

9–10 MAXWELL'S EQUATIONS

With the introduction of the displacement current Maxwell was able to complete a hydrodynamical model of the electromagnetic ether. In the nineteenth century the ether was thought to be an invisible substance existing throughout all space, inside and outside of material objects. Matter and mechanical forces had little or no effect on the ether, but charges and electric and magnetic fields interacted with it strongly. It was the invisible substance that did away with the unthinkable "action at a distance." Electric charges attract or repel one another because of tensions or compressions created in the invisible ethereal gelatin. Several models of the ether were considered in the nineteenth century; what was new in Maxwell's model was the displacement current.

A magnetic field is represented as a vortex. The magnetic field line is the axis of the spinning vortex, and its magnitude or field strength, H, is proportional to the spin velocity of the vortex, as indicated in cross section in Fig. 9–20a. A smoke ring is a good model of the magnetic vortex. The magnetic field is a dynamic concept, being proportional to the velocity of rotation of the ethereal fluid. The electric field is regarded as a tension or compression that stretches or compresses the ethereal gelatin.

Maxwell perceived a problem at the interface between two vortices: the rim velocities are in opposite directions and the motions are in opposition (Fig. 9–20b). He solved the problem much as a mechanic would solve the problem in a gear train. If two gears are to rotate in the same direction, we put an idler gear between them. Since there are no fixed axles in the ether, Maxwell provided an endless supply of idler wheels or "ball bearings." If neighboring vortices rotate at the same speed, the idler wheel spins but has no linear motion. If the rim speed of one vortex is faster than its neighbor's, $v_2 > v_1$, the idler wheels will have a linear speed equal to half the difference,

$$v = \tfrac{1}{2}(v_2 - v_1) \tag{9–42}$$

as shown in Fig. 9–20c. This effect can be observed by rotating a ball bearing wheel: the balls move at half the speed of the rim. The idler wheels are ethereal

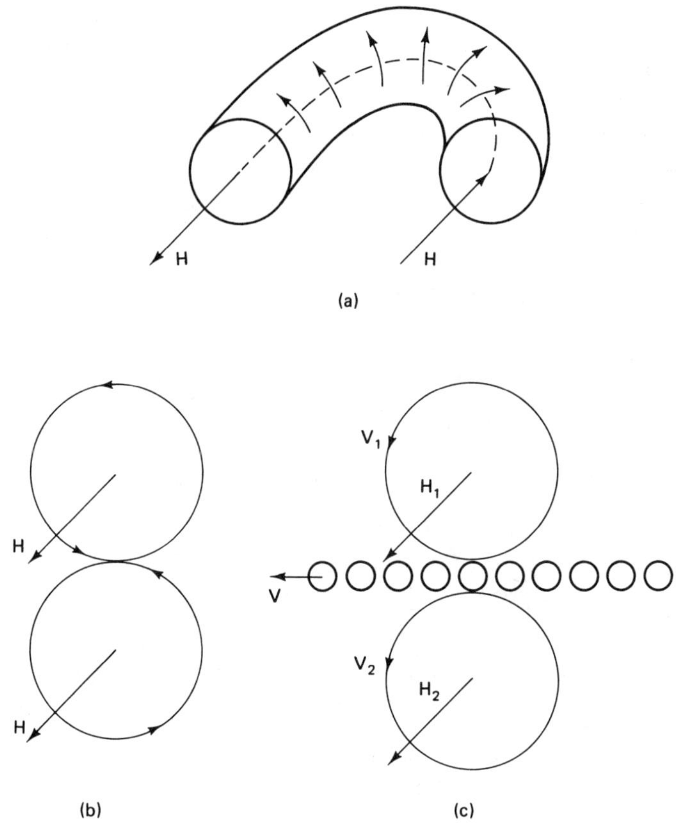

Figure 9–20 Maxwell's representation of a magnetic field as a spinning vortex.

charges, and their linear motion is a displacement. We see that a displacement current accompanies a change in magnetic field strength.

In Fig. 9–21 we show several neighboring vortices and assume that a disturbance, an increase in magnetic field strength, has occurred at the left side of the figure and is propagating to the right. An x,y,z coordinate system is indicated in the upper right of the figure. The x-coordinate is out of the plane of the figure, as are the magnetic field lines. The vortices in the first column have an increased field strength represented by an increased rim speed, v_1. The subscripts in the symbol for magnetic field strength, H_{x1}, indicate that the field strength is in the x-direction and is located at position y_1. Since the rim speeds v_1 are greater than the rim speeds of the vortices in the next column, where the field strength is less, there is a displacement velocity in the z-direction.

Eventually, the inertia or drag of the idler wheels causes the first column of vortices to lose speed, while the spin speed increases in the second column; thus the field strength decreases in the first column and increases in the second. In this manner a wave is propagated from left to right.

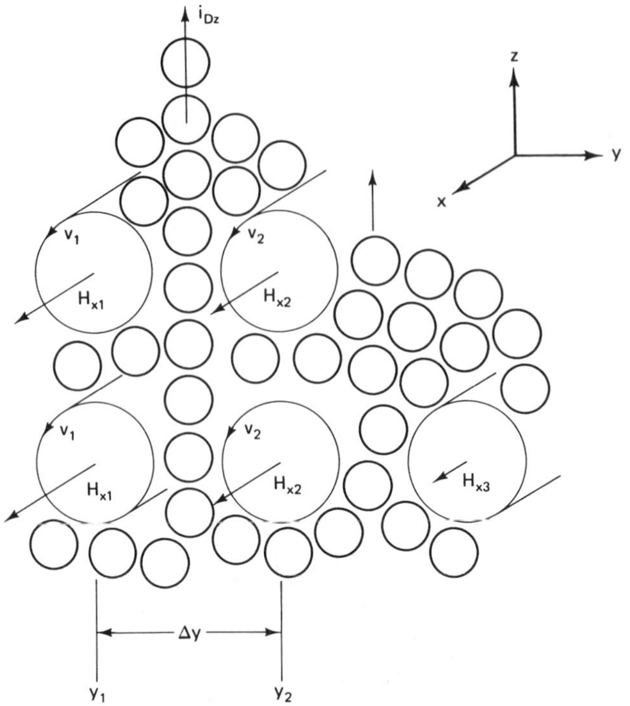

Figure 9–21 Maxwell's model of the ether.

We see that this model allows us to retain our original definition of a wave as a change of state propagating through a medium (Chapter 1). The ethereal medium does not propagate from left to right, but its state of rotation and magnitude of displacement current change as the wave passes through it.

Since the magnetic field strength is proportional to the spin speed of a vortex, let us choose a constant of proportionality such that we obtain the usual MKS units of amperes/meter. The right side of Eq. (9–42) becomes $H_{x2} - H_{x1}$. If we assume the idler wheels to carry an ethereal charge, the linear velocity on the left side of Eq. (9–42) becomes a displacement current density, i_{Dz}. We equate the change in field strength to the total current in the interval $y_2 - y_1$, and Eq. (9–42) becomes

$$(H_{x2} - H_{x1})\,(\text{amperes/meter}) = i_{Dz}\,(\text{amperes/meter}^2)(y_2 - y_1)\,(\text{meters})$$

or

$$\frac{-\Delta H_x}{\Delta y} = i_{Dz} \tag{9–43}$$

Equation (9–43) is valid in empty space and in dielectrics. If conductors are present, as near an antenna, a conduction current density can be added to the right side of the equation to give $-\Delta H_x/\Delta y = i_{Dz} + i_c$, and if we combine this result with Eq. (9–38), we have

$$\frac{-\Delta H_x}{\Delta y} = \frac{\Delta D_z}{\Delta t} + i_c \qquad (9\text{–}44)$$

Equation (9–44) is one component of one of the two electromagnetic equations that summarize and describe modern field theory. The corresponding component of the other equation is

$$\frac{\Delta E_z}{\Delta y} = \frac{-\Delta B_x}{\Delta t} \qquad (9\text{–}45)$$

Although Maxwell's only new contribution was the displacement current, the results usually are called *Maxwell's equations*.

When we apply Maxwell's equations to practical problems in antennas and propagation, we usually find that we are considering only sine waves at large distances from their sources. Under these conditions Eqs. (9–44) and (9–45) simplify to

$$H_x = \omega \bar{\lambda}_y \epsilon E_z + i_c \qquad (9\text{–}46)$$

and

$$E_z = \mu \omega \bar{\lambda}_y H_x \qquad (9\text{–}47)$$

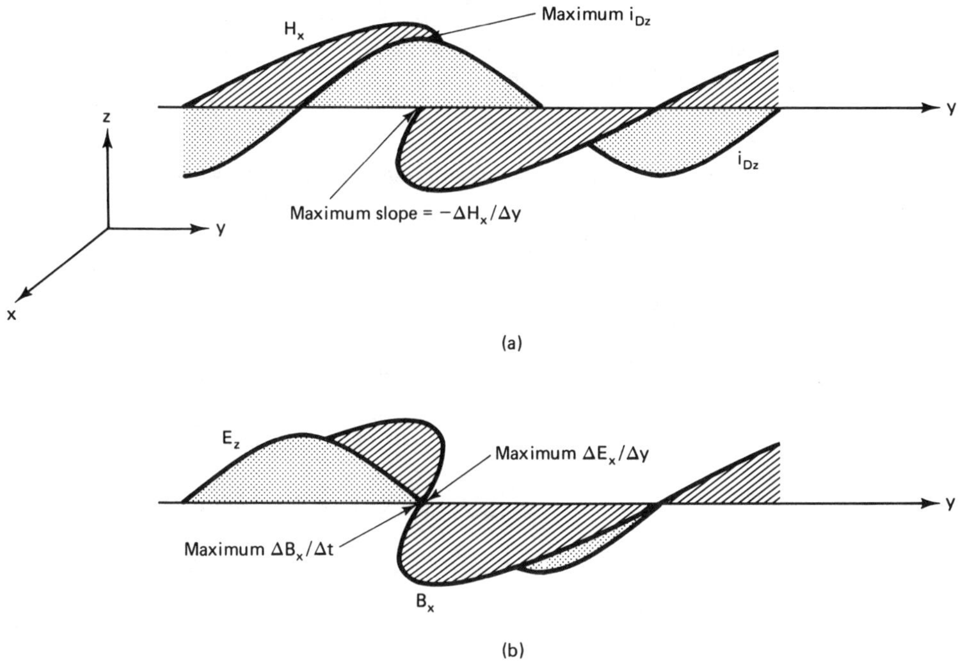

Figure 9–22 Schematic representation of the relationships between (a) the displacement current and the magnetic field and (b) the electric field and the magnetic displacement.

If we substitute $c = \omega\lambda_y$, $c = (\mu_0\epsilon_0)^{-1/2}$, and $Z_c = (\mu_0/\epsilon_0)^{1/2}$ into either Eq. (9–46) or (9–47) we obtain

$$E_z = Z_c H_x \tag{9–48}$$

a form of Ohm's law for field variables, with which we are already familiar.

In Fig. 9–22 we show sketches of sine waves and the relationships between displacement current and the electric and magnetic field variables.

EXAMPLE 9–9

A 2-GHz wave traveling in space has an electric field strength of 2.0 mV peak/m. Find the peak displacement current and the peak power delivered to a dipole antenna.

Solution The operating wavelength is $\lambda = c/f = (3 \times 10^8)/(2 \times 10^9) = 0.15$ m, and the effective quarter-wavelength is $ƛ = 0.15/2\pi = 2.39 \times 10^{-2}$ m. From Eq. (9–39) we find the displacement current

$$\begin{aligned} i_D &= -j2\pi \times (2 \times 10^9) \times (8.85 \times 10^{-12}) \times (2 \times 10^{-3}) \\ &= -j0.222 \text{ mA/m}^2 \end{aligned}$$

To find the power delivered to the dipole we first calculate the power density in the vicinity of the dipole: power per unit area $= p = i_D^2$ (amperes/m^2)$^2 ƛ^2$ (meters^2)Z (ohms), or

$$p = i_D^2 ƛ^2 Z \text{ (watts/meter}^2\text{)} \tag{9–49}$$

With our preceding numerical results and Eq. (9–49) we have

$$\begin{aligned} p &= (0.222 \times 10^{-3})^2(2.39 \times 10^{-2})^2 \times 377 \\ &= 1.06 \times 10^{-8} \text{ W/m}^2 \end{aligned}$$

Check: From

$$p = \frac{E^2}{Z} \text{ (watts/meter}^2\text{)} \tag{9–50}$$

we have

$$p = \frac{(2 \times 10^{-3})^2}{377} = 1.06 \times 10^{-8} \text{ W/m}^2$$

We complete the problem by finding the absorption area of the dipole and multiplying it by the power density. From Eq. (9–25) we have

$$\begin{aligned} A &= \frac{(0.15)^2(3\pi - 4)}{4\pi^2} \\ &= 3.092 \times 10^{-3} \text{ m}^2 \end{aligned}$$

and the peak power delivered to the dipole is

$$P = pA = (1.06 \times 10^{-8}) \times (3.092 \times 10^{-3})$$
$$= 32.8 \times 10^{-12}\ \text{W}$$

Check: Since the length of the antenna is approximately $\lambda/2 = 0.075$ m, the voltage impressed on it is $v \approx 2 \times 10^{-3}$ (volts/meter) $\times$ 0.075 (meter) $= 1.5 \times 10^{-4}$ V peak. The peak power delivered to the antenna is $Gv^2/R_2 \approx 1.63 \times (1.5 \times 10^{-4})^2/1099 = 33.4 \times 10^{-12}$ W.

Now that we have considered Maxwell's solution to the radiation problem, we shall examine the induction field, the field that represents the energy that remains on the dipole and does not travel into space.

Faraday's law of induction is

$$v = \frac{-\Delta\phi}{\Delta t} \tag{9–51}$$

where v is the voltage induced by a change of flux $\Delta\phi$ threading a circuit in time Δt. If we change the equation to field variables we find $E\lambda = -A\ \Delta B/\Delta t$, where A is the area of the circuit or loop, $A = \lambda^2$, and B is the flux density, ϕ/A. Continuing the change we have, for sinusoidal variations, $E_z = -\lambda\mu(-j\omega H_x) = j\mu cH = j(\mu/(\mu\epsilon)^{1/2})H_x$, or

$$E_z = jZ_cH_x \tag{9–52}$$

With Faraday's law in the form of Eq. (9–52) we see immediately the difference between the radiation field equation (9–48) and the induction field. The induction field is a standing wave, a wave that is trapped and can go nowhere. From Eq. (9–52) we see that its power factor is zero, $Ei_c = (Z_cH \cos 90°)i_c = 0$. In a standing wave the voltage and current are 90° out of phase; similarly, the associated field variables E and H are 90° out of phase. For a traveling wave, however, we see from Eq. (9–48) that E and H are in phase, the power factor is $\cos 0° = 1$, and power is delivered to a distant load. A sketch of the electric and magnetic field strengths and conduction and displacement currents for an induction field and a radiation field are shown in Fig. 9–23. The total field near an antenna is a composite of the fields shown in Fig. 9–22.

We are now able to describe the electromagnetic field in the vicinity of a dipole antenna, where the field components are more complicated than those of the far field. The displacement current, for example, has x, y, and z components. We shall not try to illustrate everything and shall omit the induction field. In Fig. 9–24a we show a line of magnetic vortices and the associated magnetic field lines and displacement currents. The maximum displacements are shown with broken lines. In sketching the conditions near a dipole we shall not try to show displacement current density but simply draw the maximum currents.

In Fig. 9–24b we show the maximum displacement current lines, in cross

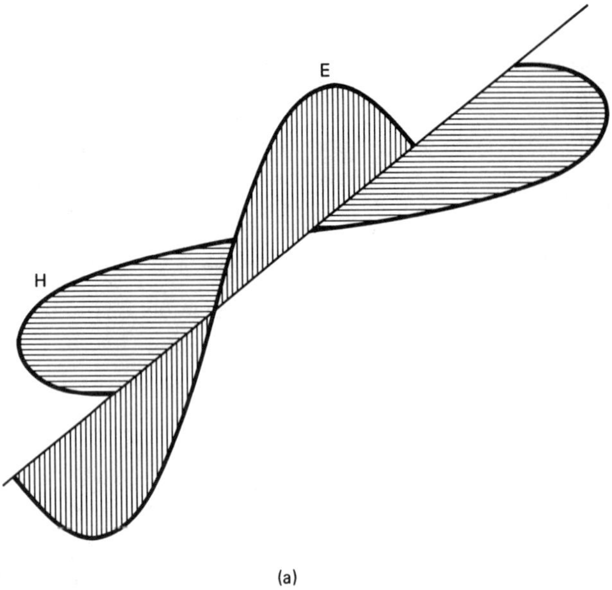

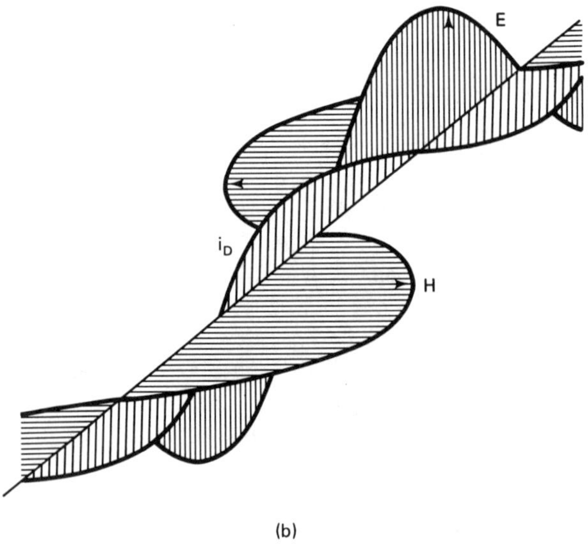

Figure 9–23 Sinusoidal variations in the field variables for (a) a standing wave or induction field and (b) a traveling-wave or radiation field.

section, in the proximity of the dipole. In cross section they are lines of maximum current; in three-dimensional space they are sheets of current. Like the skin of an apple, they have a dimple on the axis; for the dipole, in the end-fire direction. We have superimposed on this diagram the corresponding E- and H-fields in the direction of maximum radiation—the direction of the center of the radiated beam. Directions of maximum radiation are indicated by arrows marked "P." In Fig. 9–24c we

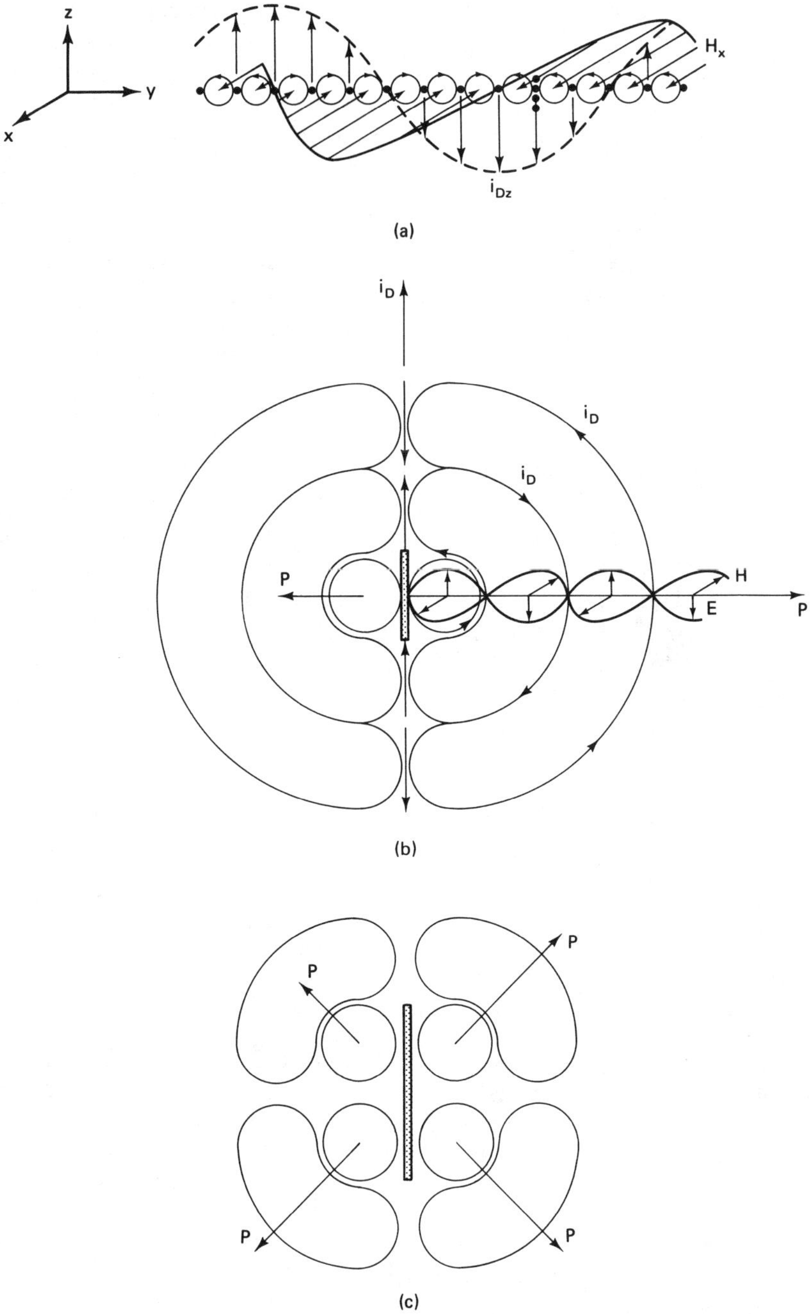

Figure 9–24 Maxwell's model of the ether and its relation to the field variables: (a) line of magnetic vortices; (b) displacement currents in the vicinity of a half-wave dipole; (c) displacement currents and principal directions of radiation for a full-wave (asymmetric current distribution) dipole.

show the displacement currents and principal directions of radiation for a full-wave dipole.

At the turn of the century an experiment was performed, the Michelson–Morley experiment, to determine the speed of the planet Earth with respect to the ether. At some time in its rotation around the sun the Earth ought to have a detectable velocity with respect to the ether. The results of the experiment were negative, and many people concluded that the ether does not exist. In the late twentieth century few textbooks describe the ether.

Whether it exists or not, the ether concept led Maxwell to the right answers; also, it gives us some insight into the launching of electromagnetic waves. Perhaps more important, Maxwell's mechanical model enables us to invent or create new types of antennas for practical applications.

9–11 THE DIPOLE AS AN OSCILLATOR

We have already obtained the input impedance of a half-wave, center-fed dipole by representing it as a series resonant circuit; however, it is the parallel resonant circuit that more closely resembles the equation for the atomic oscillator of classical optics and the parasitic elements used in antennas. By "classical" we mean the prequantum theory of the nineteenth century.

If we repeat the derivation of Section 9–6 for the equivalent parallel circuit, we obtain

$$\frac{Z_2}{Z_c} = \frac{Q}{1 + jQ(f - 1/f)} \tag{9–53}$$

This result for the normalized impedance of a parallel resonant circuit resembles the equation obtained in the nineteenth century for the field scattered by an atomic dipole,

$$\frac{a_s}{a_0} = \frac{j\sigma fQ}{1 + jQ(f - 1/f)} \tag{9–54}$$

where a_s and a_0 are the scattered and incident amplitudes and the scattering cross section σ is a factor that includes the radiation pattern, cos θ, of the dipole. This result for the atomic dipole was used in the theory of optical refraction and dispersion, which is discussed in Chapter 10. Since H. L. F. von Helmholtz made many contributions to the theory of refraction, we shall call the right side of Eq. (9–54) the Helmholtz function.

In comparing Eqs. (9–53) and (9–54) we see a difference of a factor jf. The missing j-operator is the 90° phase difference between a standing wave and a traveling wave. The circulating current in a tank circuit is part of a standing wave or induction field, whereas the displacement current scattered by a dipole is part of a traveling wave or radiation field. It was because of this difference that we added a 90° phase

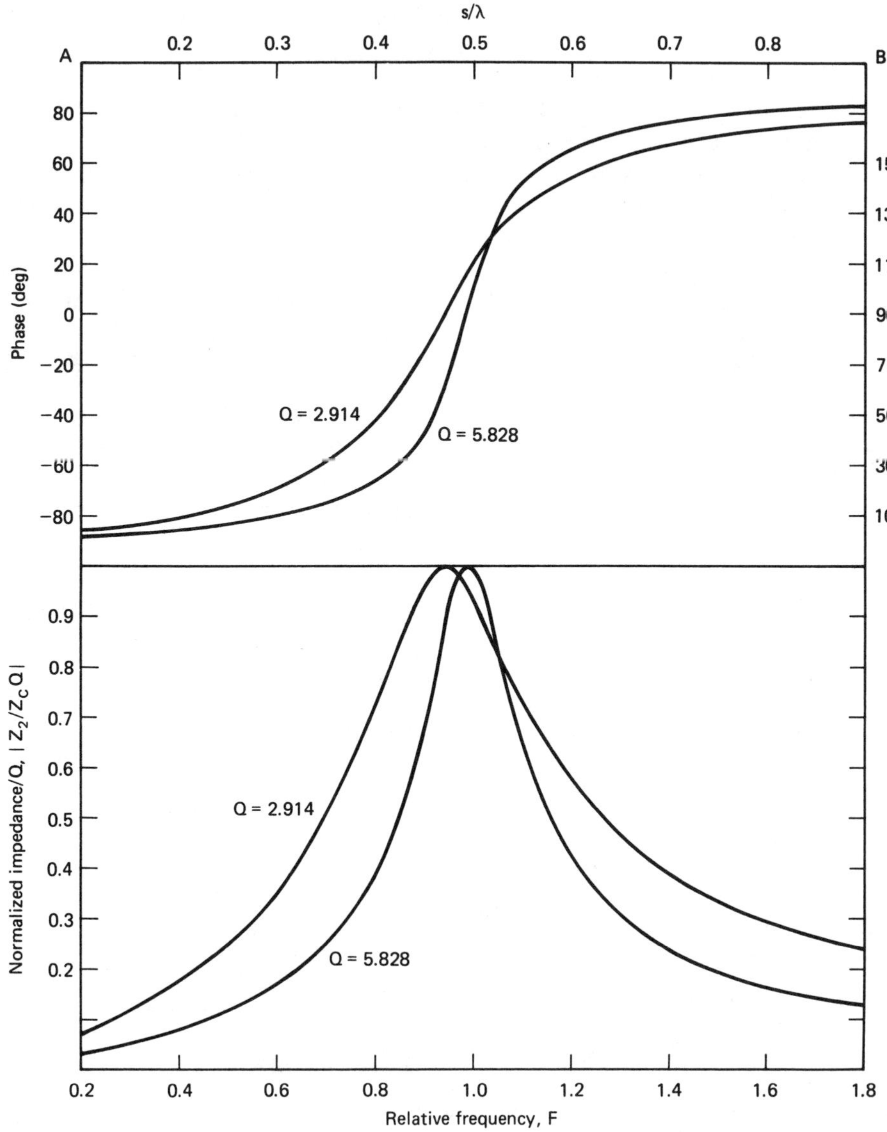

Figure 9–25 Magnitude and phase of (normalized impedance)/Q for a parallel resonant circuit. Adjusted phase for a wave scattered or radiated by a dipole is shown on scale B.

shift in the feedback network of an electronic oscillator in Section 8–8. The missing frequency factor, f, we have discussed in Section 9–8; it is part of the frequency dependence of radiation resistance.

We can express Eq. (9–53) in polar form as

$$\frac{Z_2}{Z_c Q} = \frac{1 \angle \tan^{-1}[-Q(f - 1/f)]}{[1 + Q^2(f - 1/f)^2]^{1/2}} \tag{9–55}$$

We have evaluated Eq. (9–55) for $Q = 2.914$ and 5.828 and the results, after adjustment for the end effect, are shown in Fig. 9–25. We can use the phase angle results for either resonant circuits or dipoles; we need only add 90° to account for the difference between standing and traveling waves. This has been done on scale B at the right side of the figure.

The amplitude of dipole radiation is somewhat different than the amplitude of a resonant circuit because of the frequency dependence of the radiation resistance of a dipole. Amplitudes adjusted for the frequency dependence are shown in Fig. 9–26. We shall find the results shown in these figures useful in designing dipole arrays and in the study of optical refraction.

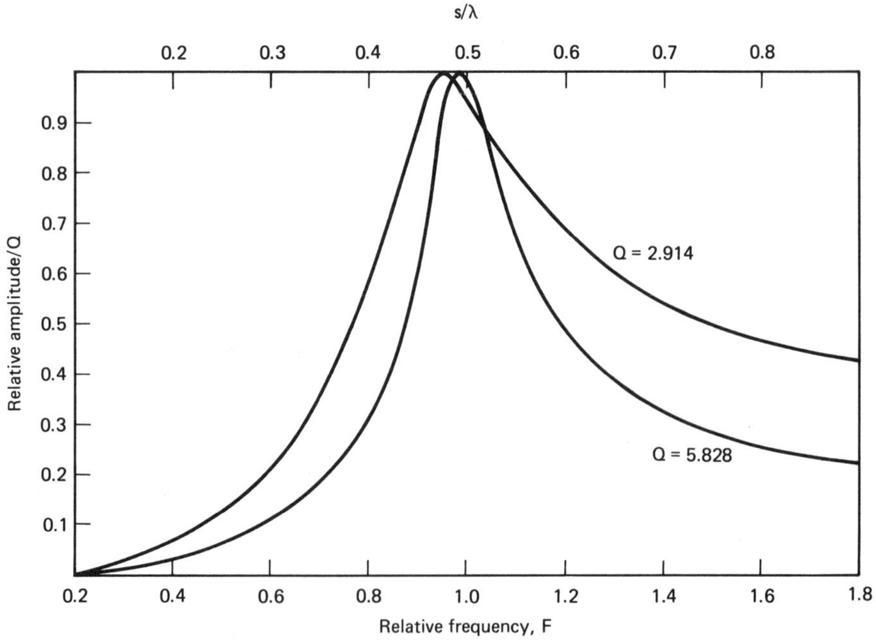

Figure 9–26 Relative amplitude/Q adjusted for the frequency dependence of the radiation resistance.

EXAMPLE 9–10

A passive optimum dipole is 0.6λ long and placed 0.15λ from a driven optimum half-wave dipole (Fig. 9–27). Estimate the power radiated in the broadside direction (x-direction) relative to the power radiated in the forward direction (y-direction).

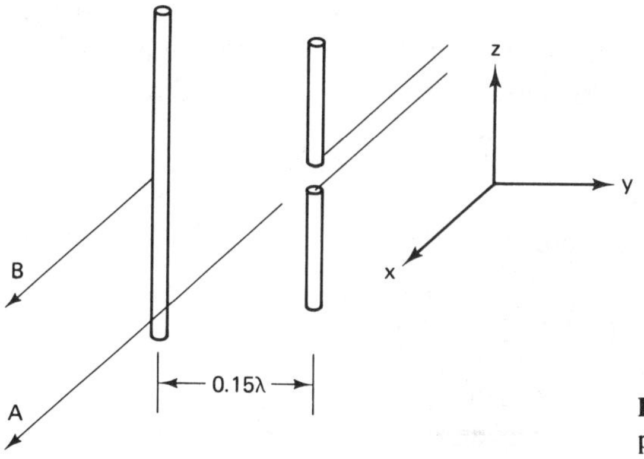

Figure 9–27 Half-wave dipole and parasitic reflector.

Solution At a distant point of observation in the broadside direction, radiation arrives directly from the driven dipole by path A, or by path B after being scattered by the passive dipole. In the latter case the radiation travels an extra distance 0.15λ, which is a phase lag of $0.15 \times 360° = 54°$. From the phase curve for $Q = 5.828$ in Fig. 9–25, we see that the wave scattered from the 0.6λ dipole has an additional phase lag of 155°; thus the total phase lag is $155° + 54° = 209°$.

The amplitude of the driven dipole is, from Fig. 9–26, $1 \times Q = 2.194$, and the amplitude of the passive dipole is $0.48 \times Q = 0.48 \times 5.828 = 2.80$. The resultant amplitude in the broadside direction is

$$\begin{aligned} a_x &= 2.914 + 2.80\underline{/209°} \\ &= 1.43\underline{/-71.1°} \end{aligned}$$

In the forward direction the scattered amplitude is twice delayed by the element separation, and the total phase delay is $2 \times 54° + 155° = 263°$. The amplitude in the forward direction is

$$\begin{aligned} a_y &= 2.194 + 2.80\underline{/263°} \\ &= 3.79\underline{/-47.2°} \end{aligned}$$

The relative power in the broadside direction is

$$P = \left(\frac{a_x}{a_y}\right)^2 = \left(\frac{1.43}{3.79}\right)^2 = 0.142 \quad \text{or} \quad -8.5 \text{ dB}$$

EXAMPLE 9–11

For the dipoles in Example 9–10, estimate the amplitude radiated in the backward direction ($-y$ direction) and calculate the front-to-back ratio—the ratio of the power radiated in the forward direction to that radiated backward.

Solution In the backward direction there is no path difference between the radiation from the driven dipole and the scattered radiation; thus the phase difference is 155°. The total amplitude in the backward direction is

$$a_{-y} = 2.914 + 2.80\underline{/155^\circ}$$
$$= 1.24\underline{/72.4^\circ}$$

With this and our result from Example 9–10 we find a front-to-back ratio of

$$\left(\frac{a_y}{a_{-y}}\right)^2 = \left(\frac{3.79}{1.24}\right)^2$$
$$= 9.34 \quad \text{or} \quad 9.7 \text{ dB}$$

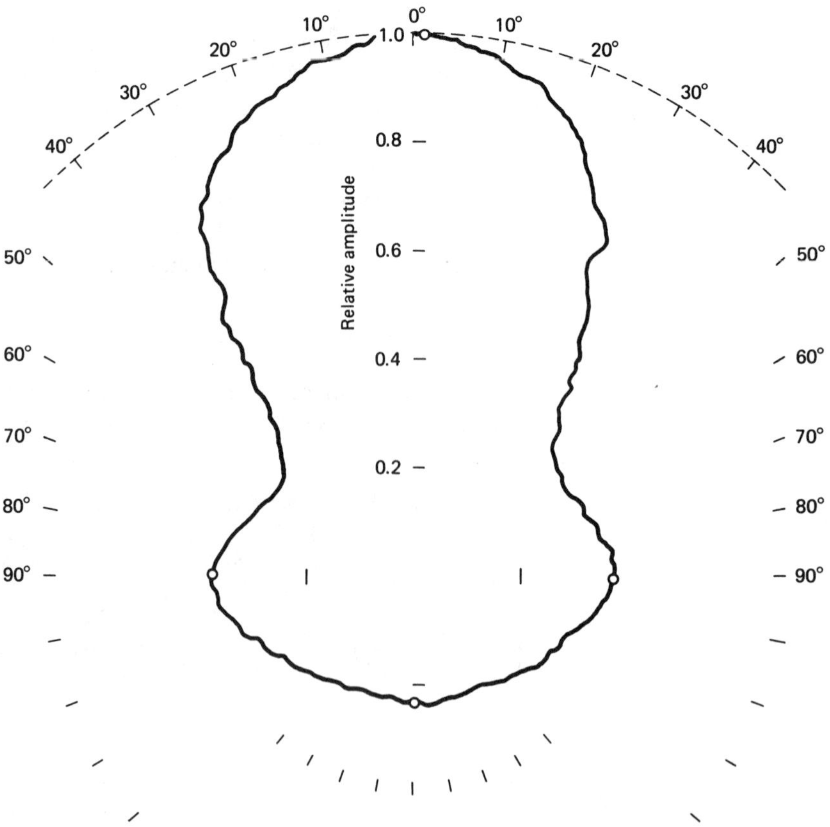

Figure 9–28 Quick sketch, based on only four points, of the radiation pattern of a dipole-reflector combination. Solid line, *H*-plane radiation pattern of driven dipole and reflector; dashed line, *H*-plane pattern of driven dipole only.

With the results of Examples 9–10 and 9–11 we can make a sketch of the radiation pattern of the driven and passive dipole combination. A more accurate pattern can be obtained by calculating additional points (Fig. 9–28).

We see that we have done a fair job of making a unidirectional antenna by combining a passive dipole with a driven dipole. In this case the passive dipole is called a reflector; in general, passive elements are called *parasitic elements* and usually are used as *reflectors* or *directors*. Our example is by no means optimum. In practice, the reflector spacing may be as small as 0.1λ, the reflector Q higher than what we have used, and the reflector length may be closer to a half-wavelength, say 0.55λ. If we attempt a design for minimum backward radiation, the broadside radiation may be increased; if we attempt minimum broadside radiation, backward radiation may be increased. A final design should be determined by the intended application.

We can design a director, a passive element cut shorter than the driven element and placed in front of it, using the same methods we used in Examples 9–10 and 9–11. A director is intended to increase forward radiation at the expense of backward and broadside radiation.

The above technique is inexact in that we have ignored multiple reflections and mutual impedances between the driven and parasitic elements; however, it serves as a useful guide for understanding parasitic arrays and making adjustments on experimental models.

9–12 YAGI ANTENNA

In Chapter 5 we see that a transmission line can be a sequence of lumped-parameter series inductances and shunt capacitances, and in Section 9–11 we see that a lumped-parameter resonant circuit has characteristics of a dipole antenna. Evidently, we can make a transmission line out of dipoles. In Fig. 9–29 we show a dipole transmission line excited by a driven dipole and reflector. At the receiving end the line is terminated in a receiving dipole and reflector. The transmission-line dipoles are cut shorter than the driven dipole. This means the line is operated below cutoff; it is a low-pass filter.

If we omit some of the dipoles at the center of the transmission line, a radiation field will jump across the gap, and what is left of the transmission line acts as transmitting and receiving antennas, as suggested in Fig. 9–29. These are called *Yagi antennas*.

If a Yagi antenna comprises a driver, reflector, and a few directors, it can be analyzed by the method we used in Section 9–11; however, if it is a long antenna with many directors, it is easier to analyze it as an end-fire, transmission-line antenna. The gain of a long Yagi is

$$G \approx 4\pi\left(\frac{S}{\lambda}\right)^{1/2} \tag{9–56}$$

where S is the length of the line of directors and λ is the operating wavelength.

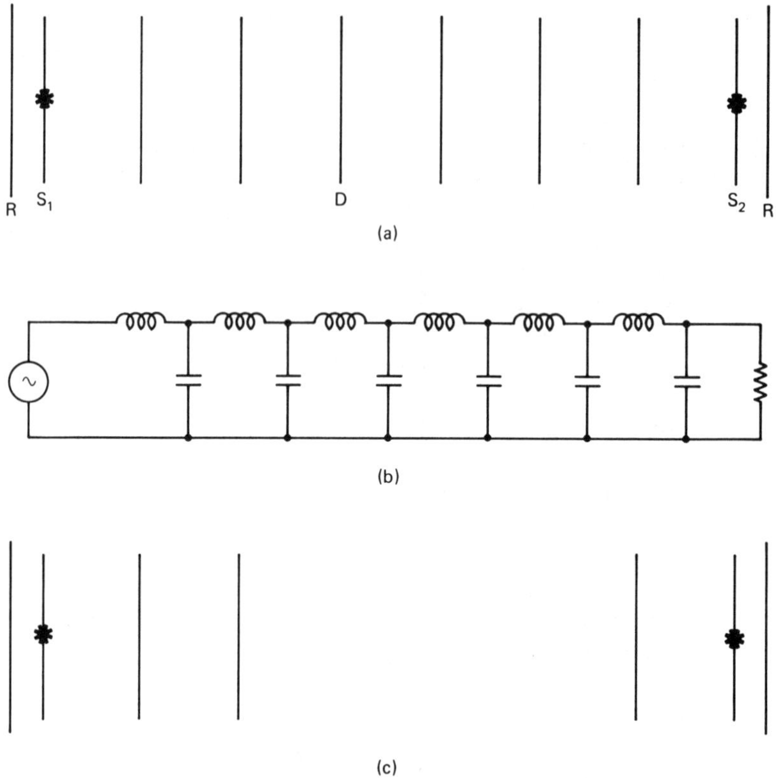

Figure 9–29 (a) Yagi transmission line; (b) lumped-parameter representation; (c) Yagi transmitting and receiving antennas.

Although Eq. (9–56) is a useful approximation, it gives us no information for analyzing or designing a Yagi. A further condition is

$$\eta - 1 \geq \left(\frac{\lambda}{S}\right)^{1/2} \tag{9–57}$$

where η is the refractive index of the dipole transmission line. From Eq. (9–56) we see that the gain is proportional to the square root of the length; if we double the length of the antenna we realize only a 40% increase in gain. Because of this diminishing return we prefer aperture-type antennas for very high gains; however, for moderate gains of 10 to 20 dB the Yagi is practical and inexpensive to build.

EXAMPLE 9–12

A 600-MHz Yagi has 30 directors spaced ⅓λ apart and each director is cut to a length of 0.4λ. Assume that the driver and directors are optimum dipoles;

also, assume that the reflector length and spacing have been optimized. Calculate the gain and evaluate the design.

Solution The length of the antenna is, including the driver, $31 \times (\lambda/3)$, or $S/\lambda = 10.33$. From Eq. (9–56) we find

$$G \approx 4\pi(10.33)^{1/2} = 40.4$$

To evaluate the design we must estimate the refractive index. We refer to Fig. 9–25 and find that the phase delay on scattering by an element 0.4λ long and with a Q of 2.914 is 43°. We use scale A when reading from the figure because we are concerned with an induction field, not a radiation field. Since there are three elements per wavelength, the extra phase delay is $3 \times 43 = 129°$ per wavelength. Since a free-space wavelength is 360°, the refractive index is

$$\eta = \frac{\lambda_0}{\lambda} = \frac{\lambda_0}{[360/(360 + 129)]\lambda_0} = 1.358$$

We now evaluate the inequality (9–27):

$$1.358 - 1 \geq \left(\frac{1}{10.33}\right)^{1/2}$$

or

$$0.358 \geq 0.316$$

The equal sign in the inequality gives the optimum condition, that is, maximum gain. We see, since the condition is close to equality, that we may expect the gain of 40.4 to be substantially realized; however, we probably can improve the performance somewhat by slightly increasing the director lengths.

We see, from Eq. (9–56) or (9–57), that we can create end-fire antennas by providing some kind of transmission-line structure with an appropriate refractive index; for example, a helical coil can be used to provide the appropriate phase delay. A sketch of a helical antenna is shown in Fig. 9–30.

We gain some insight into the Yagi antenna and end-fire antennas in general by considering the formation of an equivalent aperture, as shown in Fig. 9–31. Without the directors, the driver would radiate a diverging, circular wavefront $\widehat{AB}$. The presence of the directors causes a phase delay in that part of the field propagating near the line of directors. If the phase delay is of the right amount, the phase distribution becomes constant across the line $\overline{AB}$; that is, we have a plane wavefront across the aperture rather than a circular one. Since the rays are perpendicular to the wavefront, the emerging rays are parallel, as shown.

The input impedance of a Yagi is small because of the proximity of the reflector to the driving element. Input impedance can be increased by using a folded dipole

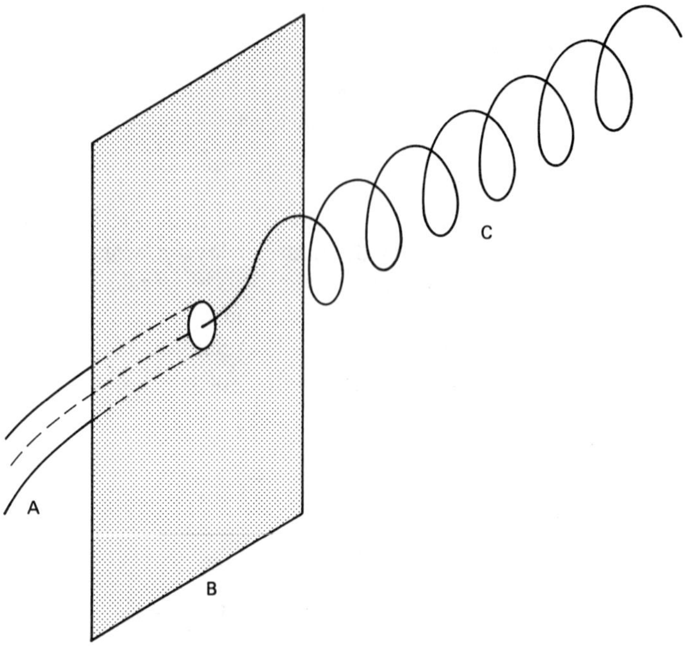

Figure 9–30 Helical antenna mounted on a reflecting screen.

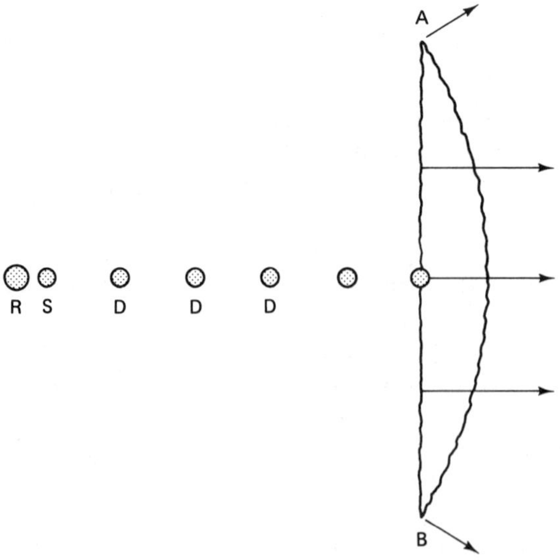

Figure 9–31 Formation of an equivalent constant-phase aperture by a Yagi antenna.

as the driver. It is often desirable to use a coaxial feed line to the antenna because of the low input impedance and to avoid transmission-line radiation loss. A problem arises from the grounded shield of the coaxial line, which cannot be allowed to ground an active part of the antenna. A transformer can be used to make a transition from "unbalanced" coaxial line to a "balanced" line or feed point (Fig. 9–32a). A transformer used for this purpose is called a *balun*. The lumped-parameter schematic in Fig. 9–32a is suitable for low-frequency applications, but at UHF and microwave frequencies there are more appropriate solutions to the balun problem.

In Fig. 9–32b we show a Yagi element driven on one side only from a coaxial line. The elements of the Yagi are held in place by being fastened at their centers to a conducting boom. Since there is no electric field at the center of a dipole, the boom is electrically neutral. The shield of the coax is grounded to the boom and the center conductor is connected to the feed point, as shown. The feed line can be connected to the driver with an adjustable clamp. The farther the feed point from the center, the greater the input impedance. By adjusting the clamp position we can match the input impedance to the characteristic impedance of the coaxial line.

Other types of baluns are shown in Fig. 9–32c and d. Here the idea is to put the feed point at the open end of a quarter-wavelength of short-circuited line. Since the impedance at the open end of a short-circuited, quarter-wave stub is infinite, the grounded shield of the coaxial line and the feed point are effectively isolated.

We can make a broadband balun by forming twin line from coax (Fig. 9–32c), but instead of short-circuiting the twin line, we load the Yagi elements directly onto it (Fig. 9–32e). A Yagi driven in this manner is superior to one driven by only one element. Gain is somewhat increased, bandwidth is increased, and sidelobes are reduced. As an example of this type of Yagi let us refer again to Example 9–12. We note that the elements are spaced $\lambda_0/3$, and the refractive index is 1.358. This means that the spacing in line wavelengths is $(1/3)(1.358)\ \lambda = 0.45\lambda$, or about a half-wavelength. A sketch of a portion of the electric field and charges on and around the directors is shown in Fig. 9–33. We see that the entire structure is resonant. Each element is driven by its neighbors as well as by the twin-conductor transmission line. For this reason the transmission line can be lightly coupled to the elements. It is neither necessary nor desirable to make metal-to-metal contact with the elements. If the twin line is formed from the coaxial lines, the outer insulation can be left intact to conveniently insulate the line from the elements.

Bandwidth of the Yagi antenna can be further increased by tapering the element lengths and adjusting their spacings in proportion (Fig. 9–34). Let us suppose that at an operating frequency f_1 the element marked D_1 is resonant and acts as a driver. The larger element behind it acts as a reflector, and the shorter elements in front of it act as directors. If a director length is much less than a half-wavelength, the scattering amplitude is much decreased (Figs. 9–25 and 9–26) and the element is ineffective; thus the active region of the antenna is limited as indicated by the distance marked A_1 in Fig. 9–34. If the operating frequency is decreased to f_2, such that the element D_2 is resonant and D_3 acts as a reflector, the active length of

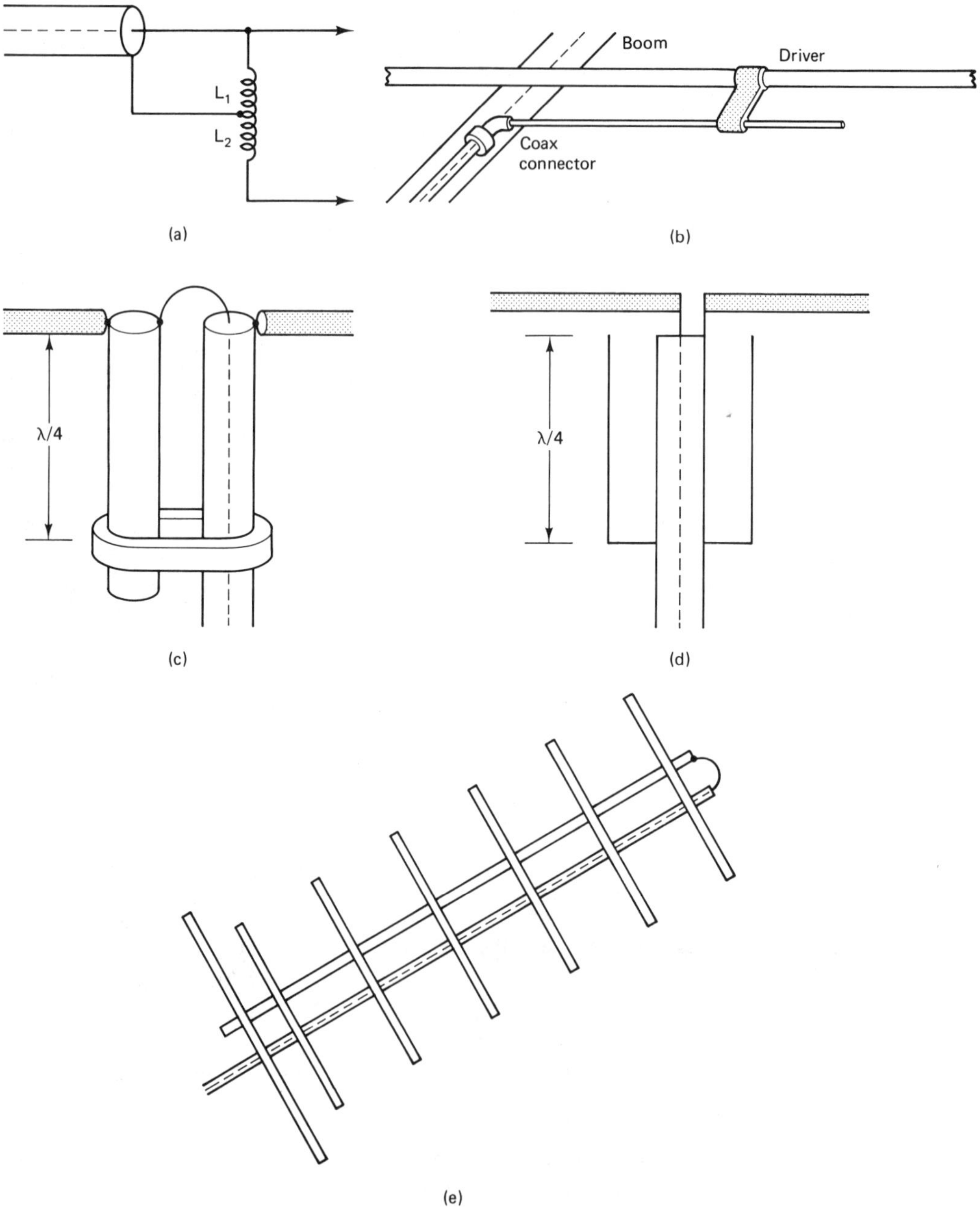

Figure 9–32 Baluns: (a) low-frequency, lumped-parameter; (b) "gamma" feed; (c) and (d) short-circuited quarter-wave stub baluns; (e) broadband balun formed from a coax to twin-line transition.

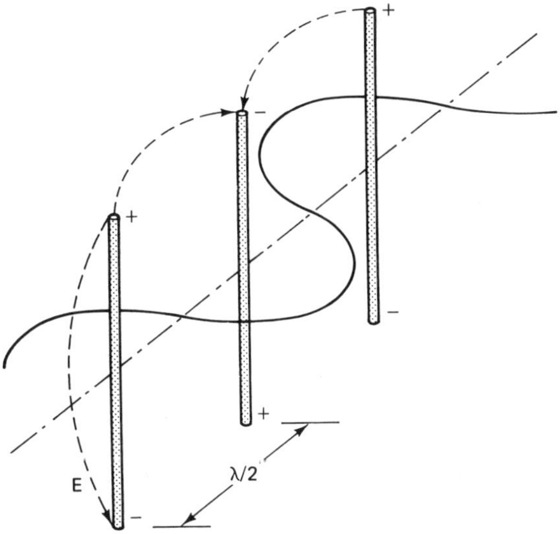

Figure 9–33 Electric field and charges near the directors of a well-designed Yagi antenna.

the antenna is proportionately increased to A_2, as indicated. Evidently the active length of the antenna in wavelengths is constant, $A_1/\lambda_1 = A_2/\lambda_2 = \cdots = S/\lambda =$ constant. From Eq. (9–56) we see that the gain remains constant as frequency is changed; consequently, radiation pattern and input impedance also are independent of frequency. The ultimate bandwidth is determined by the lengths of the first and last elements.

In the drawing shown in Fig. 9–34 each element and its spacing is 1.5 times larger than the preceding. The factor 1.5 is an example of a *logarithmic ratio*. Originally, various types of broadband end-fire arrays were given distinctive names, but today they are generally called *logarithmically periodics* or "LPs." The loga-

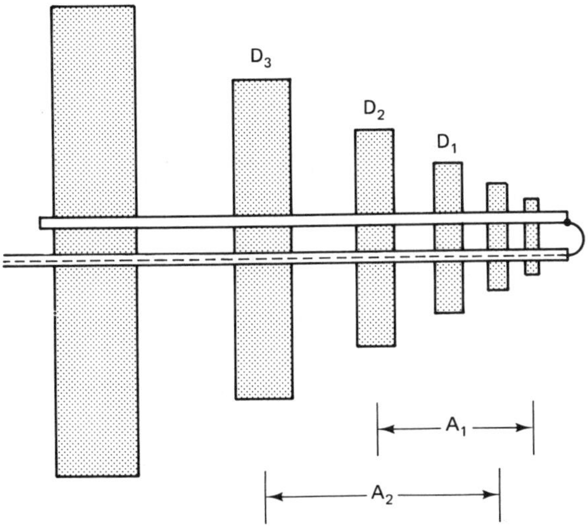

Figure 9–34 Broadband Yagi or logarithmically periodic dipole array.

rithmic ratio determines the length of the active region and the gain of the antenna: the smaller the ratio, the greater the gain.

EXAMPLE 9–13

The largest element of an LP antenna is 29.8 cm, the smallest 5.0 cm, and the logarithmic ratio is 1.25. The spacing between directors is $0.6d$, where d is the larger director length. Assume a loaded Q of unity and estimate the gain and bandwidth of the antenna.

Solution Our procedure is to assume a resonant frequency for one of the elements, calculate the amplitudes of the neighboring elements, decide which elements are contributing significantly, and so estimate the active length of the antenna. We can then use Eq. (9–56) to calculate the antenna gain. We can estimate the bandwidth by finding the lowest and highest frequencies that accommodate an active length.

Owing to the end effect we have, from Eq. (9–18),

$$f' = f\left(\frac{1 + Q^2}{Q^2}\right)^{1/2} = 1.414f$$

and Eq. (9–54) now yields, with $a = |a_s/ja_0\sigma|$,

$$a = \frac{1.414f}{1 + j(1.414f - 1/1.414f)}$$

We see that the resonant frequency of an element is $f = 0.707$. The elements in front of the resonant one are operating at relative frequencies of $f/1.25, f/1.25^2, f/1.25^3, \ldots = 0.707/1.25, 0.707/1.25^2, \ldots = 0.57, 0.45, 0.36, \ldots.$ We find the relative amplitude and power scattered by each element by evaluating Eq. (9–54) for the relative frequencies above:

f	a	a^2
0.707	1	1
0.57	0.92	0.85
0.45	0.67	0.45
0.36	0.39	0.13

and we see that only three directors are significantly active. We now can find the active length of the antenna and use Eq. (9–56) to estimate the gain.

Because of the end effect the length of a resonant element is $d = 0.707 \times \lambda/2 = 0.35\lambda$, and the length of the preceding element is found by dividing by the logarithmic ratio: $0.35\lambda/1.25 = 0.28\lambda$. Since the spacings are $0.6d$, we have spacings of $0.6 \times 0.35\lambda = 0.21\lambda$ and $0.6 \times 0.28\lambda = 0.17\lambda$, as shown in Fig. 9–35. We see that the length of the active region is $S/\lambda = 0.38$, and from Eq. (9–56) we have

$$G \approx 4\pi(0.38)^{1/2} = 7.75 \quad \text{or} \quad 8.9\text{ dB} \qquad \text{at most}$$

Since the reflector element may not be optimized, the actual gain may be 1 or 2 dB less.

The lowest operating frequency is determined from the length of the largest element. We have $0.35\lambda = 29.8$ cm or $\lambda = 85.1$ cm, which corresponds to an operating frequency of 0.35 GHz. Similarly, we find the highest operating frequency from the length of the third from the smallest element, $d = 1.25^2 \times 5$ cm $= 7.81$ cm. Since $0.35\lambda = 7.81$ cm, we have a smallest operating wavelength of $\lambda = 22.3$ cm, which corresponds to a maximum operating frequency of 1.35 GHz. The bandwidth is $1.35 - 0.35 = 1.0$ GHz or, expressed as a ratio, 3.9:1.

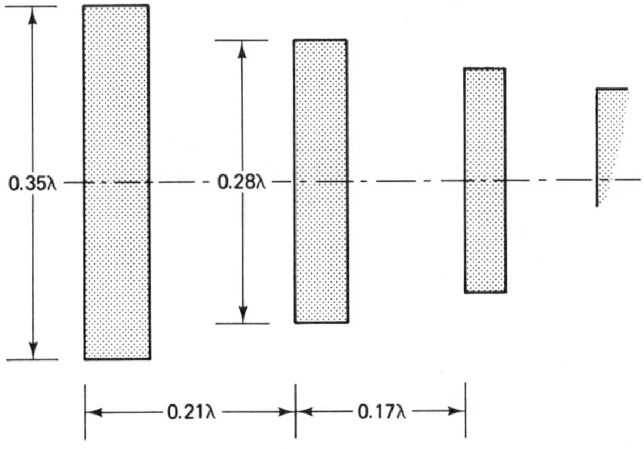

Figure 9–35 Plan view of LP elements, Example 9–13.

If the element spacing of an LP is about ⅓λ, the array can be driven by a loosely coupled twin line; however, if the elements are closely spaced, say 0.1 to 0.2λ, then in contrast to Fig. 9–33, each element is driven in opposition to its neighbors. To avoid this, the transmission line is crisscrossed (Fig. 9–36).

Do not hesitate to try an LP design. The mathematical theory is so complicated that even experts are reassured only when a model has been built and tested. With an intuitive understanding of how the antenna is supposed to work, the first measurement of a bad design will indicate what should be done to correct the error.

9–13 HUYGENS'S WAVELETS

If we reexamine the equivalent aperture of a Yagi antenna (Fig. 9–31), a puzzling question may occur to us. Why does a delay in phase at the center of an aperture cause a ray at the edge to change direction? The answer leads us to the wave theory of apertures.

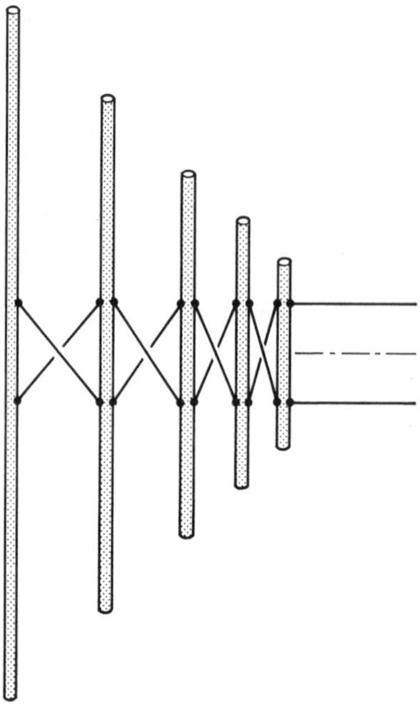

Figure 9–36 Crisscross feed arrangement for a close-coupled LP antenna.

In the seventeenth century Isaac Newton held to the corpuscular theory of light, but one of his contemporaries, Christian Huygens, was a proponent of the wave theory. *Huygens's principle* is that each point of a wavefront is the source of new wavelets, and it is the envelope of all the new wavelets that determines the next wavefront in the propagation of a wave. With Huygens's principle we can explain the bending of light around corners, the diffraction of light into a geometric shadow zone; in addition, we can understand why the direction of a ray is determined by the phase distribution across the whole wavefront, not just the conditions local to the ray. The diffraction pattern of light as it "bends" into the shadow regions is called a radiation pattern in antenna theory.

We have illustrated Huygens's principle in Fig. 9–37 by showing the spreading of wavelets and new wavefronts into a shadow zone. This effect is easily observed by watching water waves as they pass by an obstacle. In drawing the figure, we did not consider the possibility that the envelope of the circular wavelets could include a wavefront propagating in the backward direction. To avoid the backward wave, which is not observed in nature, we say that the amplitude of each wavelet is

$$a = \tfrac{1}{2}(1 + \cos\theta) \tag{9–58}$$

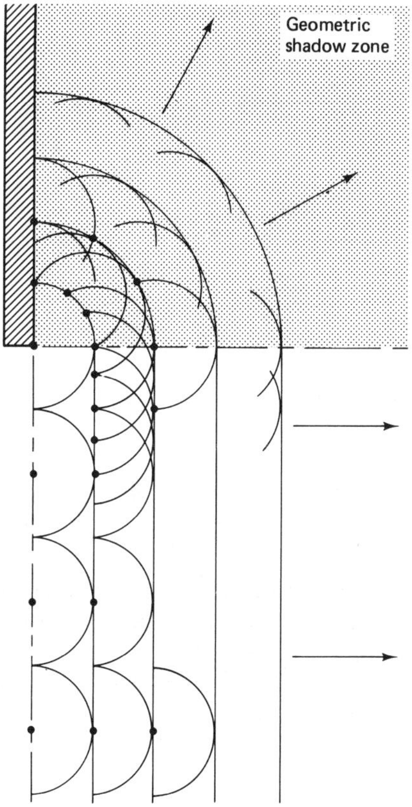

Figure 9–37 Huygens's wavelet construction showing the spreading of light into a shadow zone.

where θ is the angle of a point on the wavelet with respect to the forward ray (Fig. 9–38). The amplitude distribution, or radiation pattern, given in Eq. (9–58) is called the *obliquity factor*.

EXAMPLE 9–14

Calculate the amplitude of a Huygens's wavelet propagating in the backward direction.

Solution In the backward direction we have $\theta = 180°$. From Eq. (9–58) we have

$$a = \tfrac{1}{2}(1 + \cos 180°) = 0$$

With Huygens's wavelets we can answer the opening question. In Fig. 9–39 we show a circular wavefront generated by an omnidirectional antenna or *point source*. Let us suppose that some structure is present that reduces the propagation speed of the wavelets near the centerline of the figure. The structure may be a

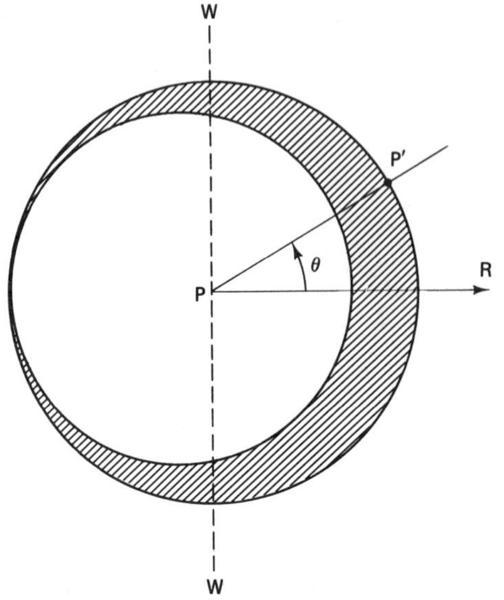

Figure 9–38 Huygens's wavelet, the wave analog of a photon. Amplitude is suggested by the thickness of the wavelet.

dipole transmission line, a dielectric rod, or anything that increases the refractive index. This means that the radius of a wavelet near the centerline is smaller than that of a wavelet at the edge, since it travels less distance in a given interval of time. Successive wavefronts start to flatten and the rays, which are perpendicular to the wavefronts, gradually rotate until they are parallel.

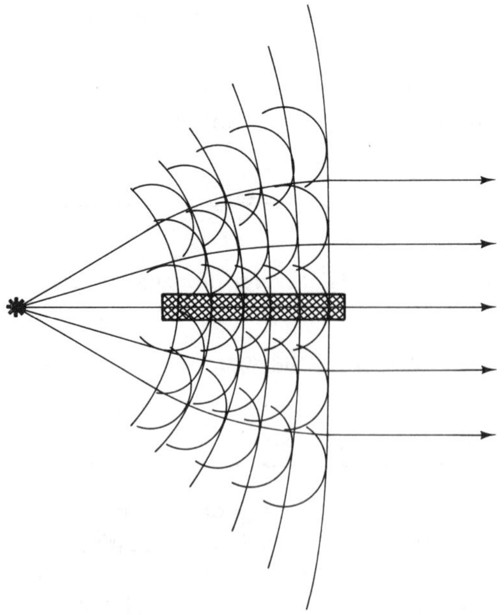

Figure 9–39 Use of Huygens's wavelets to show the development of a plane wavefront from an end-fire antenna.

With Huygens's wavelets we can explain why we can design an aperture antenna either by requiring all the rays across the aperture to be parallel, or by requiring that the phase across the aperture be constant. In Chapter 6 we designed a waveguide horn by requiring that all of the reflected rays in the aperture be parallel. More often, it is easier to complete a design by requiring that all points across an aperture be in phase.

As an example, consider a reflecting surface that is to produce a directed beam from a point source (Fig. 9–40). We could design the reflector by requiring that any ray from the source be reflected in the forward direction, a technique that would be rather tedious mathematically. A simpler approach is to require that all rays travel equal distances from the source to the aperture plane $\overline{AA}$; thus all points on the line $\overline{AA}$ are in phase. In particular, the wave reflected from point $P(x, y)$ must travel the same distance as a wave reflected from point $O(0, 0)$:

$$2f = a + b$$

We can find distance a with the rule of Pythagoras, and our equation becomes $2f = [y^2 + (f - x)^2]^{1/2} + f - x$, or $f + x = [y^2 + (f - x)^2]^{1/2}$. Squaring the equation and combining like terms, we have

$$y^2 = 4fx \tag{9–59}$$

which is the equation of a parabola.

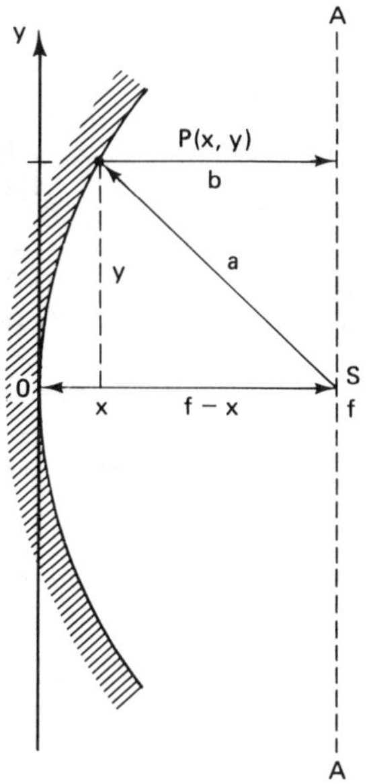

Figure 9–40 Geometry of a parabolic section: f, focal length; P, point on the reflector with coordinates x, y; $a + b$, ray path; $2f$, central ray path; $\overline{AA}$, aperture plane; S, point source.

EXAMPLE 9–15

Design a 2-m dish satellite antenna. By "two-meter dish" we mean a parabolic reflector 2 meters in diameter. By "satellite antenna" we mean an antenna for receiving signals from satellite relay stations, not an antenna borne by an Earth satellite.

Solution The first step in the design is to decide on a focal length-to-diameter ratio, f/D, where f is the focal length shown in Fig. 9–40 and D is the diameter of the aperture. The smaller f/D, the more shallow the dish; the greater f/D, the deeper the dish. A shallow dish is a more efficient reflector; a deep dish shields the feed antenna at the focal point from stray radiation and thus reduces noise. The f/D ratio also determines the beamwidth requirements of the feed antenna at the focal point. A too-wide beamwidth of the feed antenna will result in *spillover*—wasted energy—and high sidelobe levels. A too-narrow beamwidth will not utilize all of the aperture and result in reduced gain.

From the foregoing remarks we see that dish design is not routine. A particular design is determined by the application for which the dish is intended and by the experience of the designer in meeting compromises. We shall assume an f/D ratio of 0.55, a typical value.

If the f/D ratio is 0.55 and the dish is 2 m in diameter, the focal length is $f = 0.55 \times 2 = 1.10$ m. With Eq. (9–59) we now make a table of x-y values as follows:

$$y = 2(fx)^{1/2} = 2(1.10 \times x)^{1/2}$$

$x(m)$	$y(m)$
0	0
0.010	0.105
0.015	0.257
0.020	0.297
. . .	. . .
0.220	0.944
0.230	1.005

We can stop with $x = 0.230$ m since this gives slightly more than the required diameter of 2 m.

In Example 9–15 the design of the reflecting surface is complete, but the final design of the antenna is far from complete. Not only does a feed antenna have to be designed, but the tolerances for the mechanical construction of the dish have to be specified. The values of x and y in the table can be plotted full scale on sheet metal and a template cut to determine the shape of the reflector or to test the accuracy of the final product. Whether you build the dish yourself or are fortunate enough to turn the problem over to the mechanical section, you will have to determine the accuracy to which the dish should be constructed.

A rule of thumb is to require an accuracy of $\pm(\frac{1}{20})\lambda$. This means that at most the phase difference between waves reflected from different parts of the surface do not differ by more than $\lambda/10$, or 36° in phase. Two phasors of unit amplitude and in phase give an amplitude of 2. Two phasors of unit amplitude but differing in phase by 36° give an amplitude of $1 + 1\angle 36° = 1.902\angle 18°$. The reduction in power because of phase error is proportional to $(1.902/2)^2 = 0.90$, a loss of 10% in efficiency.

EXAMPLE 9–16

If the antenna in Example 9–15 is to be used for reception at 4 GHz, determine a tolerance for the reflecting surface and estimate the gain.

Solution At 4 GHz the operating wavelength is $(3 \times 10^8)/(4 \times 10^9) = 0.075$ m. A tolerance of $\pm(\frac{1}{20})\lambda$ corresponds to ± 0.375 cm.

We can find the directive gain from Eq. (9–24). The area of the aperture is πr^2, and the directive gain is

$$G = \frac{4\pi \times \pi \times 1^2}{(0.075)^2} = 7020 \quad \text{or} \quad 38.5 \text{ dB}$$

Since a tolerance of $\pm(\frac{1}{20})\lambda$ reduces the efficiency to 90%, the gain is reduced to $0.90 \times 7020 = 6320$. Further reductions due to aperture blocking by the feed antenna, spillover, sidelobes, or holes in the metal reflector or mesh to reduce wind load may result in an overall efficiency of 50 or 60%. The gain realized could be as low as 3510, or 35.5 dB.

9–14 RADIATION PATTERNS

The simplest aperture is one-dimensional: for example, a long slit in a screen (Fig. 9–41). We assume that the aperture is uniformly illuminated; that is, all points in the aperture are sources of Huygens's wavelets of equal amplitudes and all are in phase. To find the radiation pattern we sum the contributions of all the wavelets in a given direction.

Let Δx be the distance between neighboring wavelets. If there are n wavelets in the aperture, the width of the aperture is $s = n\,\Delta x$. In the forward direction, $\theta = 0$, all wavelets arrive in phase at a distant point of observation and the total amplitude is $a = nk$, where k is the amplitude contributed by one wavelet. At an angle θ, each wavelet travels a greater distance, $\Delta x \sin\theta$, than its neighbor, as shown in Fig. 9–41. This means that successive contributions at a distant point are delayed in phase by an amount $\Delta\phi = 2\pi(\Delta x/\lambda)\sin\theta$. We can find the total amplitude by summing all the phasors $k\angle\Delta\phi$ (Fig. 9–42). We can simplify the summation

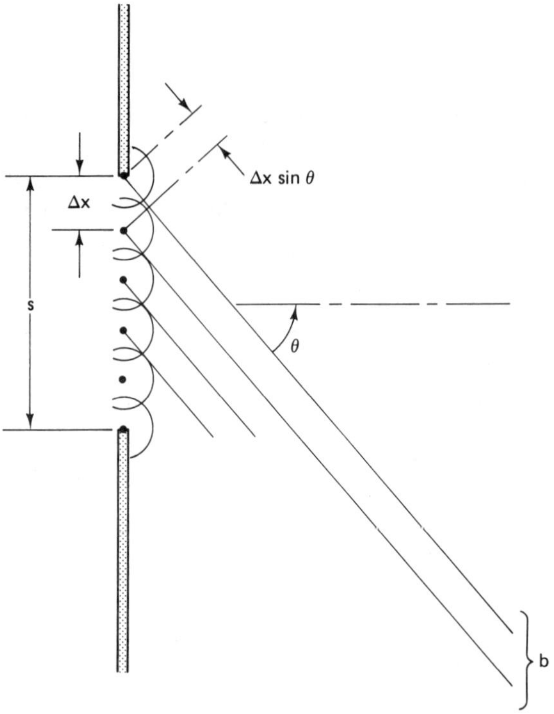

Figure 9–41 Radiation geometry for a long slit, shown in section.

problem by noticing that if the aperture contains a large number of small wavelets, the phasor diagram is almost an arc of a circle, and the problem of finding the total amplitude, b, is reduced to the problem of finding the length of the chord of an arc.

In the right triangle formed by the radius, chord, and centerline, we have $b/2 = r \sin \phi/2$. Since $\phi = a/r$, we can eliminate r to obtain

$$b = 2\left(\frac{a}{\phi}\right)\frac{\sin \phi}{2} \qquad (9\text{–}60)$$

Let the relative amplitude be $A = b/a$; then Eq. (9–60) becomes

$$A = \frac{\sin \phi/2}{\phi/2} \qquad (9\text{–}61)$$

From Fig. 9–41 we have $\phi = n\,\Delta\phi = 2\pi n(\Delta x/\lambda) \sin \theta = 2\pi(s/\lambda) \sin \theta$, and we find that Eq. (9–61) can be written in the form

$$A = \frac{\sin u}{u} \qquad (9\text{–}62)$$

where $u = \pi(s/\lambda) \sin \theta$. A graph of Eq. (9–62) is shown in Fig. 9–43. The right side of Eq. (9–62) should be multiplied by the obliquity factor [Eq. (9–58)], but for high-gain, narrow-beam applications this is usually ignored.

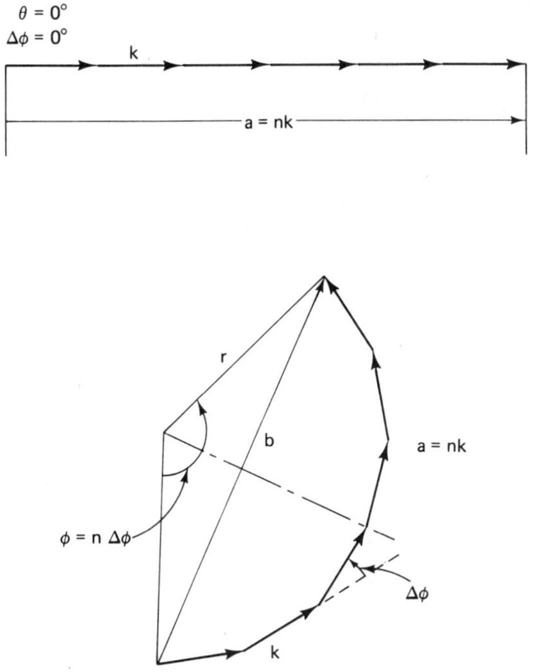

Figure 9–42 Phasor construction to determine the radiation pattern of a uniformly illuminated aperture.

EXAMPLE 9–17

Estimate the half-power beamwidth of the satellite antenna of Examples 9–15 and 9–16.

Solution From Fig. 9–43 we see that half-power occurs at $u = 1.39 = \pi(s/\lambda)\sin\theta_{1/2}$. Solving for $\theta_{1/2}$, we have

$$\theta_{1/2} = \sin^{-1}\frac{1.39\lambda}{\pi s} \tag{9–63}$$

$$= \sin^{-1}\frac{1.39 \times 0.075}{\pi \times 2}$$

$$= 0.0166 \text{ rad}$$

This is the angle to one side of the centerline. The full beamwidth between the half-power levels is

$$2 \times 0.0166 = 0.0332 \text{ rad} \quad \text{or} \quad 1.90°$$

For many applications the beamwidth is so small that we can use the approximation $\sin\theta \approx \theta$. With this approximation we can reduce Eq. (9–63) to

$$\theta_{HP} \approx \frac{0.882\lambda}{s \text{ rad}} \tag{9–64}$$

EXAMPLE 9–18

A waveguide pyramidal horn has an aperture of 4λ by 5.5λ. Estimate the half-power beamwidths in the *E*- and *H*-planes.

Solution With Eq. (9–64) we have, for the *E*-plane,

$$\theta_{HP} \approx \frac{0.882\lambda}{s} = \frac{0.882\lambda}{4\lambda}$$

$$= 0.221 \text{ rad} \quad \text{or} \quad 12.6^\circ$$

In the *H*-plane we have

$$\theta_{HP} \approx \frac{0.882\lambda}{5.5\lambda}$$

$$= 0.160 \text{ rad} \quad \text{or} \quad 9.2^\circ$$

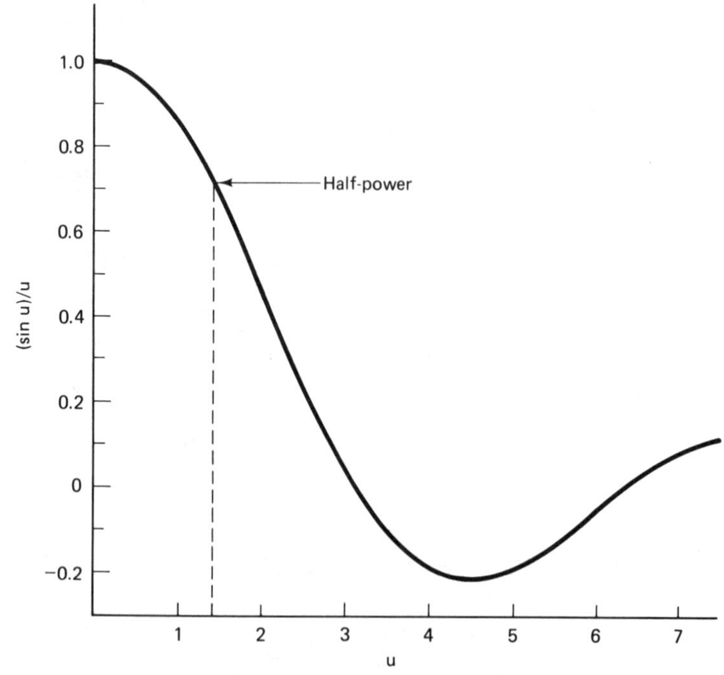

Figure 9–43 Graph of (sin *u*)/*u*, the Fourier transform of a rectangular, equiphase amplitude distribution.

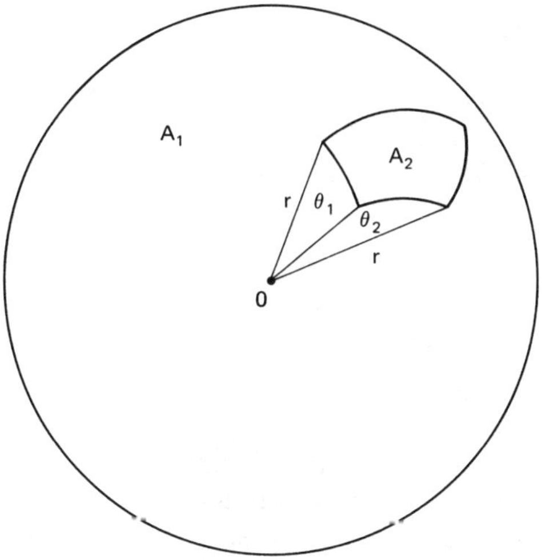

Figure 9–44 A directive antenna delivers power to a restricted area A_2, where A_1 is the area of a sphere of radius r.

We have defined gain as the ratio of the received signal intensity with and without use of the antenna. If we apply this definition to a transmitting antenna, the gain is the ratio of the maximum signal intensity delivered to a distant point in space to the signal intensity that would be received if the same transmitting station power were radiated uniformly in all directions. In Fig. 9–44 we show the area A_2 in which power is delivered by an antenna with E- and H-plane beamwidths θ_1 and θ_2. If the antenna were not used, the power would be spread over a sphere of radius r, theoretically. Since signal intensity, or power density, is inversely proportional to the area over which the power is spread, the gain of the antenna is approximately $G = A_1/A_2 = 4\pi r^2/\theta_1 r\theta_2 r$, or

$$G \approx \frac{4\pi}{\theta_1\theta_2} \tag{9–65}$$

A source that radiates uniformly in all directions is called a *point source* or *isotropic radiator*. Gain defined with reference to an isotropic radiator is called *absolute gain*.

EXAMPLE 9–19

Estimate the absolute directive gain of the horn described in Example 9–18.

Solution From Eq. (9–65) we have

$$G \approx \frac{4\pi}{0.160 \times 0.221}$$

$$= 355 \quad \text{or} \quad 25.5 \text{ dB}$$

Check: From Eq. (9–24) we have

$$G = \frac{4\pi A}{\lambda^2} = \frac{4\pi}{\lambda^2}(4\lambda \times 5.5\lambda)$$

$$= 276 \quad \text{or} \quad 24.4 \text{ dB}$$

EXAMPLE 9–20

If the horn antenna of Example 9–19 is used to transmit a power of 500 W at 3 GHz, find the power received by a dipole at a distance of 5 km.

Solution The area covered by the transmitting horn is $A_2 = r^2\theta_1\theta_2 = (5 \times 10^3)^2 \times 0.160 \times 0.221 = 8.84 \times 10^5$ m². The power density in that area is $p = P/A_2 = 500/(8.84 \times 10^5) = 5.66 \times 10^{-4}$ W/m².

From Eq. (9–25) we now find the absorption area of the dipole. Our plan is to multiply the power density by the absorption area of the dipole to find the power delivered by the dipole to its receiver. The absorption area of the dipole is $0.137\lambda^2 = 0.137 \times 0.1^2 = 1.37 \times 10^{-3}$ m². The power delivered to the dipole is $pA_d = (5.66 \times 10^{-4}) \times (1.37 \times 10^{-3}) = 7.75 \times 10^{-7}$ W. Half of this will be delivered to a matched load.

Check: The power density delivered by a transmitting antenna is the power density of an isotropic source multiplied by the gain of the transmitting antenna,

$$p_t = \frac{P_t G_t}{4\pi r^2}$$

The gain of a receiving antenna is, from Eq. (9–24), $G_r = 4\pi A_r/\lambda^2$, from which we find $A_r = G_r\lambda^2/4\pi$. Combining equations we find that the received power is

$$P_r = p_t A_r = \frac{P_t G_t}{4\pi r^2}\frac{G_r \lambda^2}{4\pi} \quad \text{or} \quad P_r = \frac{P_t G_t G_r}{(4\pi r/\lambda)^2} \tag{9–66}$$

With the values appropriate to our problem, we have

$$P_r = \frac{500 \times 355 \times 1.73}{[4\pi \times (5 \times 10^3)/0.1]^2} = 7.78 \times 10^{-7} \text{ W}$$

We can reverse our analysis of the radiation from a uniformly illuminated aperture and calculate the radiation received when the aperture is uniformly illuminated by a distant source. By the principle of reversibility, we should find that the far-field angular radiation distribution shown in Fig. 9–43 becomes the angular distribution near the focal point, the image of a distant point source. The half-amplitude width of the image is the diameter of a photon.

9–15 INTERFERENCE PATTERNS

The radiation pattern from two slits can be regarded as the diffraction pattern of an aperture that has a central portion blocked, or as a pattern of interference from two separate but coherent (phase-related) slits or sources. We shall consider the simple case in which the width of each slit is small compared to their separation; in other words, each slit is considered to be a line source of a Huygen wavelet.

In Fig. 9–45 we show the distance of separation of the slits as s, the angle of observation as θ, and the phase difference between the amplitude contributed by the wavelets as $2\pi(s/\lambda)\sin\theta$. If k is the amplitude contributed by each source and b the total amplitude at the point of observation, the total normalized amplitude at the point of observation is

$$A = \frac{b}{2k}\frac{1+\cos\theta}{2}$$

$$= \frac{k + k\angle\phi}{2k}\frac{1+\cos\theta}{2}$$

or

$$A = \tfrac{1}{2}\left(\frac{1+\cos\theta}{2}\right)\left(1 + 1\underline{/2\pi\frac{s}{\lambda}\sin\theta}\right) \qquad (9\text{–}67)$$

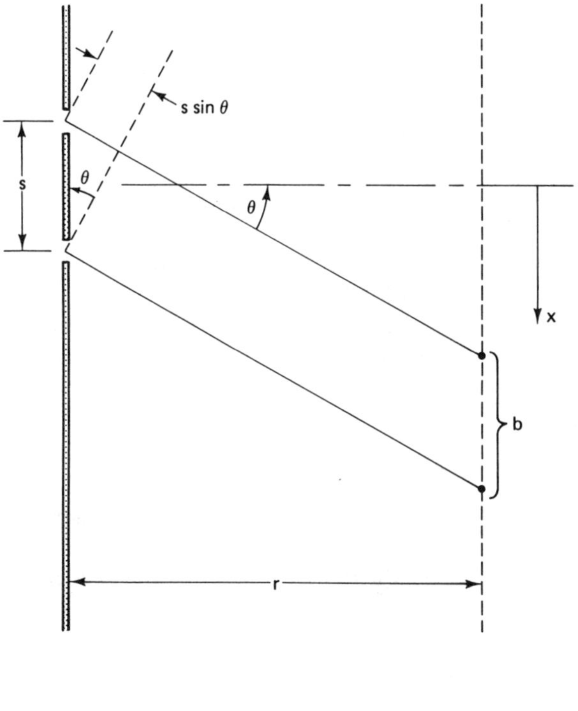

Figure 9–45 Geometry for calculating the radiation pattern of a double slit.

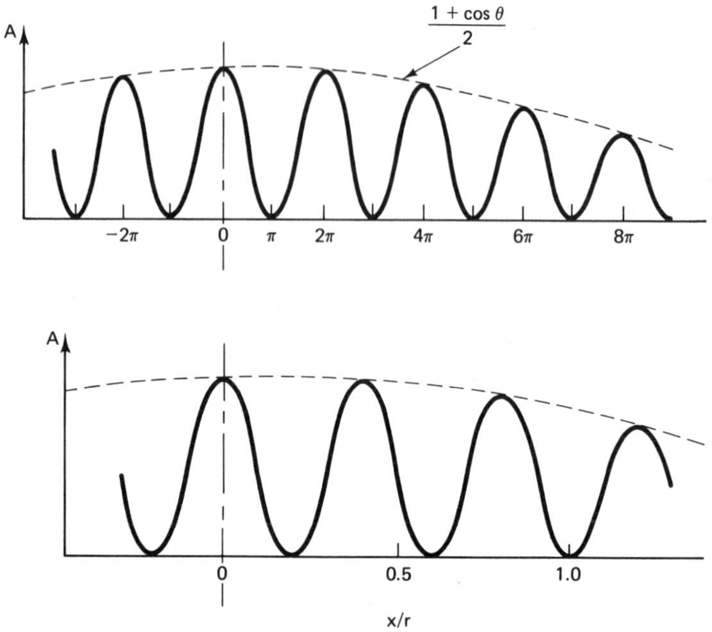

Figure 9–46 Interference patterns from two slits or two coherent sources.

where $(1 + \cos \theta)/2$ is the obliquity factor. We see that the amplitude received at the point of observation is zero whenever $\phi = 2\pi(s/\lambda) \sin \theta$ is $n\pi$, and a maximum if ϕ is 2π or $n2\pi$. We show a sketch of the interference pattern in Fig. 9–46.

The patterns shown in Fig. 9–46 are what would be observed if a photographic plate were placed parallel to the screen at a distance r; the positions of the maxima are measured in the x-direction from the centerline, where $x << r$. An observer traveling in the x-direction would see a series of maxima, and with increasing time the maxima would appear less frequently.

A more familiar example occurs when the observer is stationary and one of the sources is moving (Fig. 9–47). For example, the lower source is a TV broadcasting station, the upper source is a reflection from an aircraft flying away from the station, and the observer is viewing a television receiver at b. The result is a signal flutter at the receiving site, the signal going through a maximum whenever the direct path and the path by reflection differ by $n\lambda$,

$$s + (r^2 + s^2)^{1/2} - r = n\lambda$$

where $s = vt$ (v is the speed of the aircraft and t is time). We can determine the fluctuation pattern in time by simplifying and solving the equation for s.

We rearrange the equation so that only the radical is on the left side, then remove it by squaring the equation:

$$r^2 + s^2 = (r - s + n\lambda)^2$$

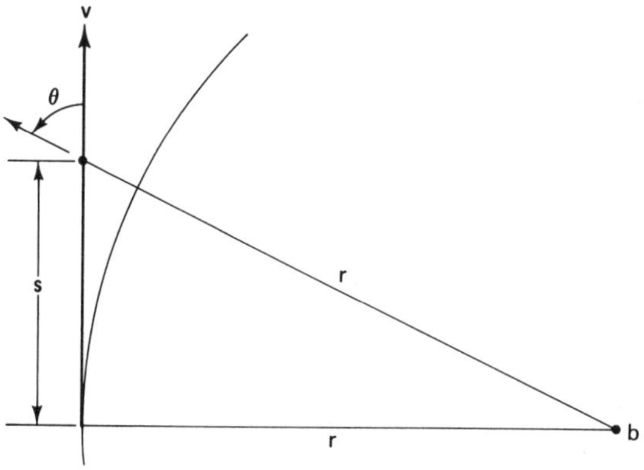

Figure 9–47 The geometry of signal flutter caused by reflections, an example of the Doppler effect.

or

$$(r + s)(r - s) = (r - s)^2 + 2n\lambda(r - s) + n^2\lambda^2$$

$$r + s = r - s + 2n\lambda + \frac{n^2\lambda^2}{r - s}$$

or

$$s - n\lambda - \frac{n^2\lambda^2}{2(r - s)} = 0$$

Clearing the fraction, we have

$$-s^2 + s(r + n\lambda) - n\lambda r - \frac{n^2\lambda^2}{2} = 0$$

If the range is much greater than the distance between source and reflector, $r >> s$, the first and last terms can be ignored and we have $s = n\lambda r/(r + n\lambda)$, or

$$t = \frac{nr}{v(r/\lambda + n)} \tag{9–68}$$

EXAMPLE 9–21

For the geometry shown in Fig. 9–47, make a sketch of the flutter pattern when the operating wavelength is 10 m, the range is 100 m, and the speed of the aircraft is 25 m/s.

Solution From Eq. (9–68) we have

$$t = \frac{100n}{25(100/10 + n)}$$

$$= \frac{4n}{10 + n}$$

For integral values of n we have

n	t (s)
0	0
1	0.36
2	0.66
3	0.92
4	1.14
5	1.34

from which we can make the sketch shown in Fig. 9–48.

The flutter pattern shown in Fig. 9–48 is an example of a *signature*, a characteristic pattern in space or time by which we attempt to recognize an event. Another example of pattern recognition is that used by an intrusion alarm. A room or building is illuminated with microwave radiation and monitored by microwave receivers. A moving intruder will cause a flutter pattern in a receiver, recognized as an "event," and an alarm is activated.

The flutter pattern is an example of the *Doppler effect*, and the frequency of the flutter pattern is called the Doppler shift. A somewhat simpler example is a moving source rather than a moving reflector. If v is the velocity of the source, we see from Fig. 9–47 that $\pm v \cos\theta$ is the velocity of approach or separation of source and observer. Let us assume that the observer mixes or heterodynes the received signal with a local oscillator tuned to the same frequency, f. If the result is a maximum, the next maximum will occur when the separation of source and observer is changed by one wavelength, and the difference frequency will be

$$\Delta f = \frac{\pm(v \cos\theta)}{\lambda}$$

or, since $\lambda = c/f$,

$$\Delta f = \frac{\pm fv(\cos\theta)}{c} \tag{9–69a}$$

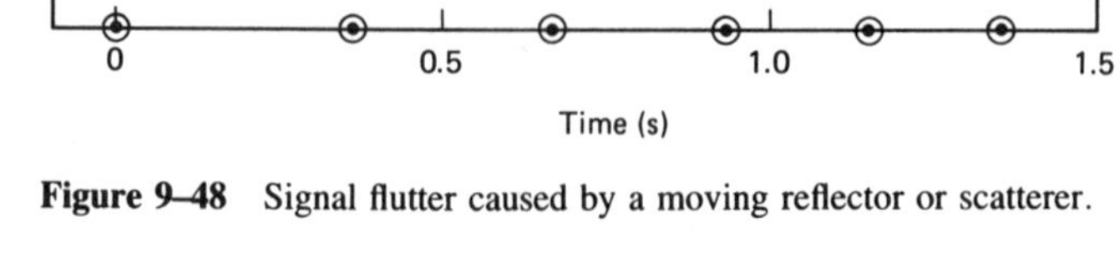

Figure 9–48 Signal flutter caused by a moving reflector or scatterer.

The difference frequency can be regarded as the difference between the frequency when the source is at rest and when it is moving, $\Delta f = f - f'$; accordingly, we can write Eq. (9–69a) as

$$f' = f\left(1 \mp \frac{v}{c}\cos\theta\right) \tag{9–69b}$$

The Doppler effect occurs in acoustic as well as electromagnetic waves. Consider, for example, a horn sounding at a frequency of 880 Hz on a train approaching at 26.8 m/s (60 mph). From Eq. (9–69a) the Doppler shift is $\pm 880 \times 26.8/340 = \pm 69.4$ Hz, where we have assumed the speed of sound to be 340 m/s. The apparent pitch of the horn is $880 + 69.4 = 949$ Hz on approach, and 811 Hz after it has passed by. The change in pitch is easily heard, and it was in this way that the Doppler effect was first experimentally verified.

If we compare Figs. 9–45 and 9–47, we see that in the first case we assumed the range to be so great that the rays were parallel at the point of observation—not so in the second case. The first case is called the *far-field* condition. For historical reasons this is also called *Fraunhofer diffraction* in optics, and the second case is an example of *Fresnel diffraction*.

9–16 SURFACE WAVES

If we look again at Fig. 9–33 we notice that the electromagnetic field has a longitudinal electric field component. This is an example of a TM surface wave mode, a mode that is guided along a surface instead of being radiated into space as a TEM mode. In the case of the Yagi antenna, both a surface wave mode and a TEM mode are present, and the Yagi can be considered to be a leaky wave guide.

A simpler example is a dielectric-clad cylindrical conductor placed near the end of a dipole antenna (Fig. 9–49a). Part of the displacement current is trapped in the dielectric and a TM mode is launched on the cylinder. This is an inefficient surface wave launcher since most of the energy is radiated broadside in a TEM or "dipole" mode.

A more efficient surface wave launcher can be made by placing the dipole near a dielectric-coated ground plane (Fig. 9–49b). The proximity of an image dipole in the ground plane greatly decreases broadside radiation in the TEM mode, and increasing the width and length of the dipole enhances radiation in the end-fire direction.

Terrestrial surface waves are more likely observed at VHF to microwave frequencies than at lower frequencies. At the higher frequencies the earth acts more like a lossy dielectric and less like a conducting ground plane. At one time the author buried a VHF dipole in a trench 1.3 m deep, connected it by a shielded coaxial line to a communications receiver, and was surprised to find that several local commercial stations were broadcasting as much signal below ground as above.

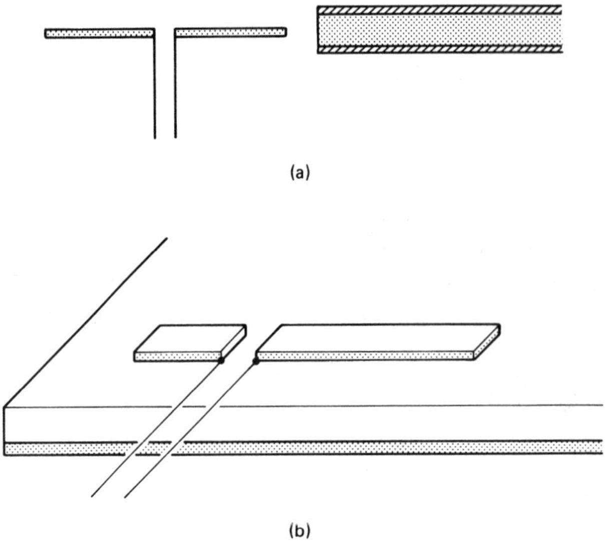

Figure 9–49 Surface-wave launchers: (a) the dipole as an end-fire launcher; (b) a launcher for surface waves guided on a dielectric-clad plane conductor.

The launcher need not be buried to detect a surface wave; it can simply be laid on the ground. Detection of a surface wave can be verified by observing maximum signal strength when the axis of the dipole is pointed at the transmitting station. We can obtain the launcher radiation pattern by changing its position on the ground. If a launcher is used for transmitting, surface-wave propagation can be verified by observing an exponential decrease in electric field strength with increase in height above the ground plane.

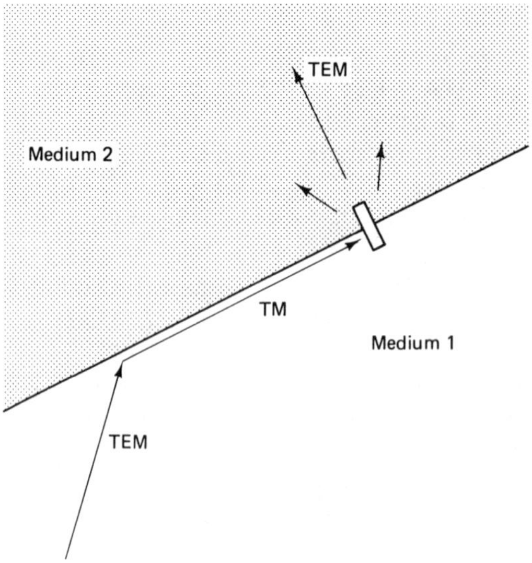

Figure 9–50 Generation of a "lateral wave" by displacement in a surface-wave mode.

Investigators working at lower frequencies usually have not recognized the surface wave as a distinct mode but instead call it a ground wave, often considering it to be a diffracted sky wave. Not recognizing it as a separate, orthogonal mode, some workers have attempted to detect surface waves by using vertical dipoles or monopoles.

The roughened surface of a conductor—water waves, for example—can also serve to guide radio surface waves. It is possible for a TEM wave to generate a surface wave at an interface. If such a surface wave propagates along the interface until it encounters an obstacle and is then reradiated, the net effect is that of a lateral displacement (Fig. 9–50). Because of this effect some investigators call surface waves *lateral waves*.

Sound waves can also be trapped or guided by a surface, and these are called *surface acoustic waves* or SAWs. The interaction of SAWs and radio surface waves has applications in integrated optics, which are discussed in Chapter 10.

9–17 UNCERTAINTY PRINCIPLE

In our discussion of the tunnel diode we referred to the "uncertainty principle," and in this chapter we have used the same principle, although we have not always said so. We shall close with a discussion of the origins of the uncertainty principle in Fourier transform theory.

In our study of transmission lines (Chapter 5) we saw that a series of rectangular pulses has a frequency spectrum of $(2/\pi)(\cos \omega t - \frac{1}{3}\cos 3\omega t + \frac{1}{5}\cos 5\omega t - \cdots)$. This is called a line or discrete spectrum and is shown by the arrows in Fig. 9–51a.

The frequency spectrum of a single pulse is much greater than that of a train of pulses because many more frequency components are required to cancel all amplitudes to zero at times greater than the pulse width; in fact, continuous distributions of bands of frequencies are required. The distributions of frequencies fill the shaded areas under the curve shown in Fig. 9–51a. The curve is the $(\sin u)/u$ function that we derived from aperture theory [Eq. (9–62)] and which also appeared in the Baden-Powell velocity formula, Eq. (5–6).

The corresponding amplitude distributions, pulse width in one case and illuminated aperture width in the other, are shown in Fig. 9–51b. The mathematical operation by which a spectrum is obtained from an amplitude distribution is called a *Fourier transform*, and the amplitude and spectrum variables are called *conjugate variables*. A consequence of transform theory is that the product of the expected values of conjugate variables is a constant. For example, the expected values of the conjugate variables of an aperture antenna are, approximately, the aperture width in wavelengths and the beamwidth, and their product is

$$\frac{a}{\lambda}\theta \approx \text{constant} \tag{9–70}$$

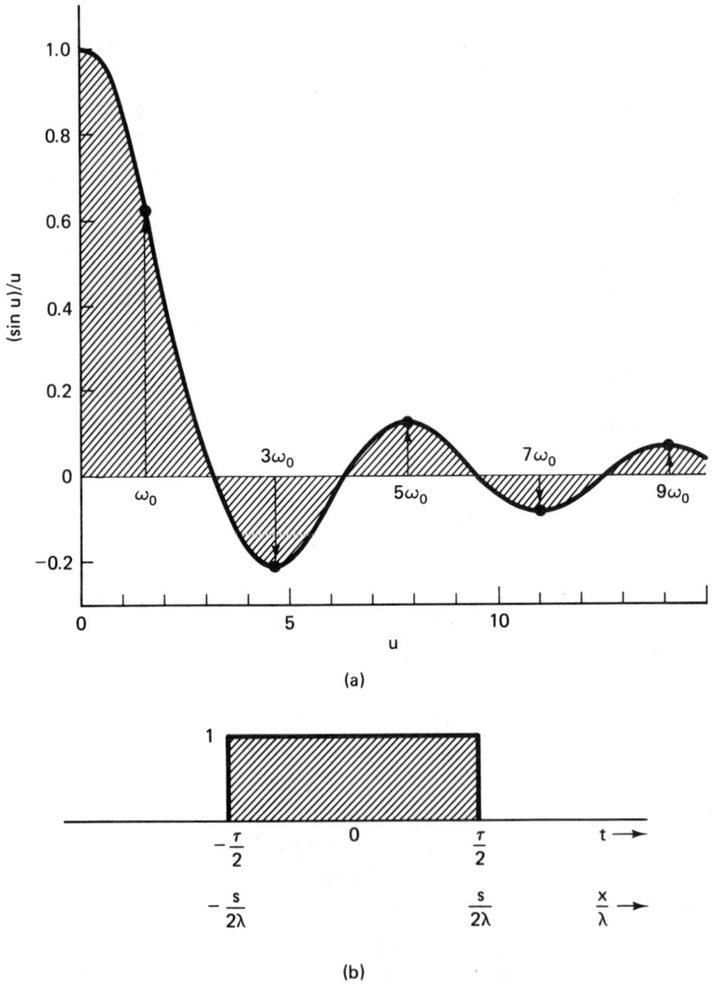

Figure 9–51 Line spectrum of a train of rectangular pulses and its relation to the continuous spectrum of a single pulse.

From Eq. (9–64) we find the expected value of θ, the half-power beamwidth, is $0.8\lambda/a$, and we have

$$\frac{a}{\lambda}\left(\frac{0.8\lambda}{a}\right) = 0.8$$

a constant. Similarly, we can show that the product of pulse length and the half-power width of the corresponding spectrum is a constant. In general, the smaller the aperture, the greater the beamwidth; the shorter the pulse, the wider the corresponding frequency spectrum.

In the second quarter of the twentieth century the mathematics of Fourier transforms was applied to atomic physics and the result is called the "new quantum theory," to distinguish it from what went before. One of the results of the new theory is the *Heisenberg uncertainty principle*,

$$\Delta p \, \Delta x \geq h \tag{9–71}$$

where Δp and Δx are the expected values of conjugate variables and h is Planck's constant (Section 8–14). In statistics an "expectation value" is what we call in electronics an rms value. The mathematical proof of the uncertainty principle follows from Fourier transform theory.

We can apply the uncertainty principle to a photon; for example, consider a photon that is changed in momentum by an amount Δp when passing through an aperture of width $\Delta x = s$. In Fig. 9–52 we show that the expected total beam width is $\theta \approx \Delta p/p$. With this result and the momentum of a photon, h/λ, we find from Eq. (9–71), $p\theta s \geq h$, or

$$\frac{\theta s}{\lambda} \geq 1$$

which is Eq. (9–64), previously obtained by wave theory. For a rectangular aperture we can multiply two equations like the above and obtain Eqs. (9–65) and (9–24).

As a second example, let us consider a photon traveling along a Yagi antenna of length s_0. The change in momentum is

$$\Delta p = \frac{h}{\lambda} - \frac{h}{\lambda_0} = h\frac{\lambda_0/\lambda - 1}{\lambda_0} = \frac{h}{\lambda_0}(\eta - 1)$$

The change in length is the electrical length minus the physical length, $s - s_0 = \eta s_0 - s_0 = s_0(\eta - 1)$. From the uncertainty principle we have $(h/\lambda_0)(\eta - 1)s_0(\eta - 1) \geq h$, or $\eta - 1 \geq (\lambda_0/s_0)^{1/2}$, which is Eq. (9–57).

For a third example we use the uncertainty principle in the form

$$\Delta E \, \Delta t \geq h \tag{9–72}$$

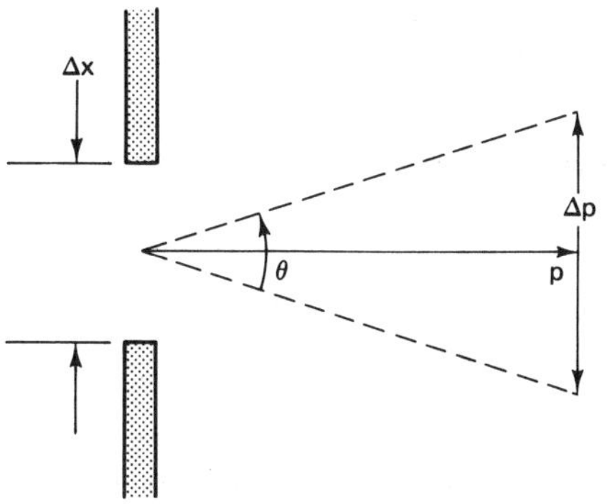

Figure 9–52 Deflection of a photon passing through a narrow slit.

where ΔE is change in energy and Δt is change in time. If E_1 is energy input, the energy change is $\Delta E = E_1 + GE_1 = E_1(G + 1)$, where G is the gain of a device or system. We relate change in time to the bandwidth of the device as $\Delta t = 1/f_1 - 1/f_2 = (f_2 - f_1)/f_1 f_2 = -\Delta f/f_1 f_2$. With Eq. (9–72) we now find $-E_1(G + 1)(\Delta f/f_1 f_2) \geq h$,

$$\frac{G+1}{Q} \leq 1$$

or

$$\frac{G}{Q} \leq 1 - \frac{1}{Q} \tag{9–73}$$

This agrees numerically with Eq. (9–33a) for an optimum dipole with a loaded Q of 2.914.

If the gain is high or Q is high, we can write Eq. (9–73) as $G/Q \leq 1$, or

$$G \times \text{BW} \sim \text{constant} \tag{9–74}$$

where BW is bandwidth. This general result is exemplified by the bandwidth in the cutoff region of a low-pass filter or by the gain–bandwidth product of a transistor.

APPLICATION 9–1: Gain Measurement with Identical Antennas

In this application we shall see that it is possible to measure the absolute gain of either of two identical antennas without having a reference or standard gain antenna.

Equipment and Materials

- copper wire, bare, heavy gauge, No. 8 or No. 10 AWG
- S-band signal generator, modulated, with precision attenuator
- receiver, or spectrum analyzer, or VSWR meter/amplifier, or a large bandwidth oscilloscope
- pyramidal horn, as in Chapter 6
- hand tools
- polystyrene foam insulation board

Procedure

Build two identical Yagi antennas for operation at 3 GHz. Design the antennas to have four or five directors. The elements can be formed from heavy, bare, copper wire, AWG No. 8 or No. 10, or from thin brass tubing. Solder the elements to a central boom made of the same wire or tubing. Mount the antennas on polyfoam blocks or construct a wooden or plastic vertical support. The driven elements may be folded dipoles. Connect miniature coaxial feedlines to the driven elements as shown in Fig. 9–32b, c, or d. Fit BNC connectors to the other ends of the feed lines. The coaxial lines should be positioned perpendicular to the elements to avoid unnecessary interference.

Use one Yagi for transmitting and the other for receiving, at a range of 3 to 5 m, and determine the gain of the antennas in the following manner. Use Eq. (9–66) in the form

$$G_r G_t = \frac{P_r}{P_t}\left(\frac{4\pi r}{\lambda}\right)^2$$

or, since $G_r = G_t$,

$$G_t = \left(\frac{P_r}{P_t}\right)^{1/2} \frac{4\pi r}{\lambda}$$

The receiver can be any device that gives a signal-strength indication: a receiver with a meter, a spectrum analyzer, or if the signal is modulated, a VSWR meter or an oscilloscope.

If reflections are a problem, move the receiving antenna toward or away from the transmitter and observe the maxima and minima, and when measuring the distance r, measure to a position halfway between a maximum and a minimum.

EXAMPLE 9–22

Find the gain of two identical antennas from measurements made at 3 GHz. The received signal indication was 3.5 at a distance of 4 m.

Solution The operating wavelength is 10 cm and the range is 400 cm; the only other information needed is the power ratio P_r/P_t. To find the ratio we connect the receiver directly to the transmitting generator and increase the attenuator until we again have a signal indication of 3.5. If the signal generator is not equipped with an attenuator, connect the receiver to the generator through a precision attenuator.

Let us suppose that the attenuator was set at −5 dB (power) when the antenna measurement was made, and was increased to −45 dB when the receiver was connected directly to the generator. This means that the attenuation in the antenna measurement was $P_r/P_t = -45 + 5 = -40$ dB. We now have

$$G_t = \left(\log^{-1}\frac{-40}{10}\right)^{1/2}\left(4\pi \times \frac{400}{10}\right)$$

$$= 5.03 \quad \text{or} \quad 7.01\text{ dB}$$

Repeat the measurements two or three times and average the results. Compare the average gain with what would be expected from Eq. (9–56).

If the expected and measured results are greatly different, and if time and equipment are available, measure the VSWR of the antennas, using the technique described in Chapter 6. Either correct the gain measurements to allow for mismatch, or attempt to achieve a better match by adjusting the reflector spacing. Show all measurements in tabular form in your report and always include sample calculations.

APPLICATION 9–2: Gain Measurement with Respect to a Dipole

In this type of measurement the gain of an antenna is measured with respect to a dipole. At the receiving site, for example, the received signal strength obtained with the test antenna is recorded, the antenna under test is replaced with a dipole and the signal strength is again recorded. The gain of the dipole is assumed to be known, and the gain of the antenna under test is determined from the ratio of the signal strengths.

Equipment and Materials

The equipment required is the same as in Application 9–1.

Procedure

Build a Yagi antenna as described in Application 9–1 or, if you wish, build a similar Yagi with 8 to 12 directors. Also construct two optimum dipole antennas. All antennas should be designed for the same operating frequency, say 3 GHz, or whatever frequency is convenient with equipment that is available.

Use one dipole antenna as a transmitter, the other as a receiver, and determine their gain using the procedure given in Application 9–1. A mismatch of antenna impedance of 64 Ω into a 52-Ω or 75-Ω coaxial line is acceptable if this is a laboratory class and time is limited.

After the dipole gains have been determined, again set up the equipment as in Application 9–1 and measure the signal strength received with the dipole. Now replace either dipole with the Yagi antenna and record the increased received power. The ratio of the two signal strengths is the gain of the Yagi with respect to a dipole. For example, if the increase in the received power is 9.8 dB and the dipole gain is 1.6, the absolute gain of the Yagi is

$$\begin{aligned} G_y &= G_1 G_d \\ &= 9.8 \text{ dB} + 10 \log 1.6 \\ &= 11.8 \text{ dB} \end{aligned}$$

Record and report all data; include sample calculations in your report.

APPLICATION 9–3: Radiation Pattern Measurement

Equipment and Materials

The same equipment is required as in the previous applications; in addition, we need string, measuring tape, and protractor, or some other means of measuring angles.

Procedure

Build an optimum dipole for the operating frequency of the optimum horn antenna described in Chapter 6, or for whatever antenna is available. Mount the dipole on a long stick of polystyrene foam so that it can be supported with minimal interference from reflections from nearby objects.

Use the dipole as a transmitting antenna and the horn as a receiving antenna. Determine the radiation pattern of the horn by rotating it about a vertical axis through the throat of the aperture. Record power received for various angles of rotation, normalize the results, and draw a radiation pattern on rectangular or polar graph paper. Determine both E- and H-plane radiation patterns. The long axis of the dipole must always be in the E-plane of the horn; otherwise, the antennas are *cross-polarized* and little or no signal is received.

If the patterns are lopsided because of reflections, repeat the measurements at different locations and at different ranges, and average the results.

From the radiation patterns determine the E- and H-plane half-power beamwidths. Use these data and Eq. (9–65) to determine the directive gain of the horn. Measure the aperture dimensions of the horn and use Eq. (9–24) to calculate again the gain of the horn.

In your report compare results from Eqs. (9–65) and (9–24) and discuss why you believe one result or the other to be the more reliable. Points to consider are: aperture efficiency, mismatch, and reflections.

APPLICATION 9–4: Radiation Safety

Match an S-band dipole to a calibrated spectrum analyzer. Place the dipole near the closed door of an operating microwave oven and determine whether or not the safety standard of 0.1 mW/cm^2 is satisfied.

SUMMARY

Because of its long history the dipole antenna appears in many forms under many different names. We review these so that the reader may not be confused by references in current technical literature to very different antennas that are called dipoles. The "dipole" may be:

A conductor, a half-wavelength long

A conductor less than a half-wavelength long

A full-wavelength conductor with either a symmetric or an antisymmetric current distribution

A quarter-wavelength conductor excited above a ground plane—a "unipole" or "whip" antenna

A conducting element of any length

A transition between atomic energy levels that produces "dipole radiation"

A scattering element such as a dust particle suspended in the atmosphere

A probe, as in a coax-to-waveguide transition

The radiation resistance of a dipole, or any antenna, is a resistance which, when multiplied by the square of the current at the feed point, produces the power lost by radiation,

$$P = i^2 R_1$$

If a dipole is end fed, rather than center fed, the radiation resistance equation becomes

$$P = \frac{v^2}{R_2}$$

If we regard the half-wave dipole as a transmission line, it is a quarter-wave transformation from endpoint to center point, and the radiation resistances R_1 and R_2 are related by a quarter-wave rotation on a Smith chart.

We define an optimum dipole as one which, in accordance with the maximum power transfer theorem, absorbs or radiates half of the incident power. An optimum dipole has an unloaded Q of 5.828 and a loaded Q of 2.914. The characteristic impedance of an arm of a dipole is

$$Z_c = 120\left(\ln\frac{2s}{a} - 1\right)$$

and is 377 Ω for an optimum dipole. The radiation resistances of a half-wave, resonant, optimum dipole are

$$R_1 = 64.68\ \Omega\text{, series fed}$$
$$R_2 = 2197\ \Omega\text{, end fed}$$

An optimum full-wave dipole can be regarded as two shunt, end-fed, half-wave dipoles and has a radiation resistance of

$$\frac{R_2}{2} = 1099\ \Omega$$

The dipole antenna also can be regarded as a resonant LCR circuit. The resonant frequency of such a circuit is decreased by loading,

$$\omega_0' = \omega_0 \left(\frac{Q^2}{Q^2 + 1} \right)^{1/2}$$

In antenna theory this is called an "end effect," and it is common practice to shorten the length of the dipole to compensate for the end effect,

$$2s' = 2s \frac{\omega_0'}{\omega_0} = 2se$$

where e, the end-effect factor, is 0.946 for an optimum dipole. By modifying the LCR circuit equation to account for the frequency dependence of the radiation resistance, we obtain the approximate input impedance of a half-wave dipole as the frequency is varied,

$$Z_1 = R_1 \left(\frac{f}{e} \right)^4 \left[1 + jQ \left(\frac{f}{e} - \frac{e}{f} \right) \right]$$

The directive gain of an antenna can be defined as the power received at a great distance from the antenna, divided by the power that would be received if the power had been radiated uniformly in all directions. From the photon theory of radiation we find an equation for gain,

$$G = \frac{4\pi A}{\lambda^2}$$

that is valid for all antennas, transmitting and receiving; in particular, for an optimum, resonant, half-wave dipole antenna we have

$$G = 1.630$$

Gain and radiation resistance are not independent. We can find a relation between the two by introducing a field concept, the displacement current, and for a dipole we find

$$G = \frac{Z_c}{\pi R_1}$$

With the introduction of the displacement current, James Clerk Maxwell was able to formulate electromagnetic theory in terms of field variables. Two components of the celebrated field equations are

$$\frac{-\Delta H_x}{\Delta y} = \frac{\Delta D_z}{\Delta t} + i_c$$

$$\frac{\Delta E_z}{\Delta y} = \frac{-\Delta B}{\Delta t}$$

A review of the manner in which Maxwell described the theory helps us to visualize the radiation of energy from an antenna.

The near field of a dipole antenna comprises radiation and induction components. Electric field and displacement current are 90° out of phase in the induction field, a characteristic of a standing wave.

If we regard the dipole as a resonant *LCR* circuit we find that its impedance is

$$\frac{Z_2}{Z_c} = \frac{Q}{1 + jQ(f - 1/f)}$$

which resembles the result obtained in the nineteenth century for the amplitude of radiation scattered by an atomic dipole,

$$\frac{a_s}{a_0} = \frac{jfQ}{1 + jQ(f - 1/f)}$$

This result was used to formulate the classical theory of optical refraction and dispersion. In this chapter we use these results as a guide in designing dipole arrays, Yagi antennas, and logarithmically periodic (LP) antennas. The gain of a long Yagi antenna is

$$G \approx 4\pi \left(\frac{S}{\lambda}\right)^{1/2}$$

subject to the condition

$$\eta - 1 \geq \left(\frac{\lambda}{S}\right)^{1/2}$$

A fundamental principal of wave theory was formulated by Huygens: each point of a wavefront is the source of new wavelets, and the envelope of all such wavelets is the next wavefront of a propagating wave. Huygens's principle provides a basis for diffraction theory in optics and the aperture theory of antennas. The amplitude of a Huygens wavelet is

$$a = \tfrac{1}{2}(1 + \cos\theta)$$

which is called the obliquity factor. An example of an aperture antenna is the familiar paraboloidal "dish." The equation of a section through the surface of a dish is the parabola

$$y^2 = 4fx$$

Paraboloidal antennas frequently are used in radar systems. An example of a radar system is assigned in problems at the end of this chapter. The radiation pattern of an aperture antenna is

$$A = \frac{\sin u}{u}$$

which is of the same form as the Baden-Powell equation.

A far-field pattern is called Fraunhofer diffraction; near-field radiation or diffraction patterns are called Fresnel diffraction. When a radiation field is produced from two or more coherent sources, the diffraction or radiation fields produce an interference pattern. If one or more of the sources is moving, the pattern will change in time. The frequency change caused by a changing interference pattern is called a Doppler shift. The simplest case is the radiation frequency from a single moving source, the Doppler frequency being

$$f' = f\left(1 \mp \frac{v}{c}\cos\theta\right)$$

The propagation of waves near surfaces is complicated by the generation of surface waves. Unlike TEM waves, surface waves have a longitudinal field component and propagate in TE or TM modes. An antenna designed to produce surface waves is called a surface-wave launcher. Early investigators, working at low frequencies and near the earth's surface, referred to the combined diffracted and surface waves as a "ground wave" to distinguish it from the TEM "sky wave." Surface acoustic waves, SAWs, are widely used in electronic devices.

One of the results of Fourier transform theory is called, in the new quantum mechanics, the Heisenberg uncertainty principle. It can be expressed as

$$\Delta p \, \Delta x \geq h$$

or

$$\Delta E \, \Delta t \geq h$$

As examples of the principle, we use it in this chapter to estimate the half-power beamwidth of an aperture antenna,

$$\frac{\theta s}{\lambda} \geq 1$$

and to show that the gain–bandwidth product of a device is constant,

$$G \times \text{BW} \sim \text{constant}$$

PROBLEMS

9–1. Two identical, optimum, half-wave dipoles are used for transmitting and receiving. The operating frequency is 300 MHz, transmitted power is 40 W, and the range is 16.1 km. What power is delivered by the dipole at the receiving site? If the dipole is matched to a preamplifier input of 50 Ω, what voltage appears at the preamplifier input?

9–2. If, in Problem 9–1, the receiving dipole is replaced by a Yagi with a gain of 12 dB absolute, by what ratio will the received power be increased?

9–3. A 2-W 4-GHz signal generator is connected to a matched, half-wave dipole. How far from the dipole should you be in order not to expose yourself to radiation greater than the 0.1 mW/cm^2 safety standard?

9–4. What is the input impedance of an optimum, half-wave dipole if it is operated at a frequency 10% lower than its resonant frequency?

9–5. A half-wave dipole antenna is designed for operation at 500 MHz. If the Q of the dipole is to be 12, what conductor diameter is required?

9–6. An S-band microwave system is used as an intrusion detector. Suppose that an intruder is anyone or anything moving at a speed of 0.3 m/s or faster. Can the receiver output be connected to 555 timers to sound an alarm? Explain why or why not.

9–7. Two vertical dipoles are used for transmitting and receiving. How much is the received power decreased if the receiving dipole is rotated until it is inclined 45° to the vertical?

9–8. A satellite dish is 2.14 m in diameter. Calculate the gain and beamwidth at 3.5 GHz. Assume 60% aperture efficiency.

9–9. A ''billboard'' antenna is a broadside array of dipoles mounted above and parallel to a reflecting screen. Assume that there are two rows of eight dipoles each mounted over a reflecting screen as shown in Fig. 9–53. The dipoles are one-quarter wavelength above the screen and spaced about one-half wavelength from one another. Assume that the dipoles are optimum, the operating wavelength is 2 m, and the dipoles are all fed in phase from transmission lines of appropriate lengths.

Find the array power gain by adding the amplitude from each dipole and squaring the sum. Your answer should agree with

$$G = n^2 G_1 \tag{9–75}$$

where n is the number of dipoles and G_1 is the gain of one dipole-reflector combination. Check your answer by calculating the area of the billboard and using Eq. (9–24).

9–10. Two Yagi antennas are stacked, one above the other, and matched to a common feedline. If each Yagi has 12 directors spaced ⅓λ apart, estimate the gain of the entire array.

9–11. A radar (*ra*dio *d*irection *a*nd *r*ange) is a system in which a short pulse is transmitted from a high-gain antenna. Usually, the same antenna is used to detect the pulse after it has been scattered or reflected from a target. The receiver is protected by an isolator during transmission. The directivity of the antenna determines the direction of the target, and the time delay between transmission and reception of the pulse determines the range to the target. If the time delay is 66.7 μs, how far away is the target?

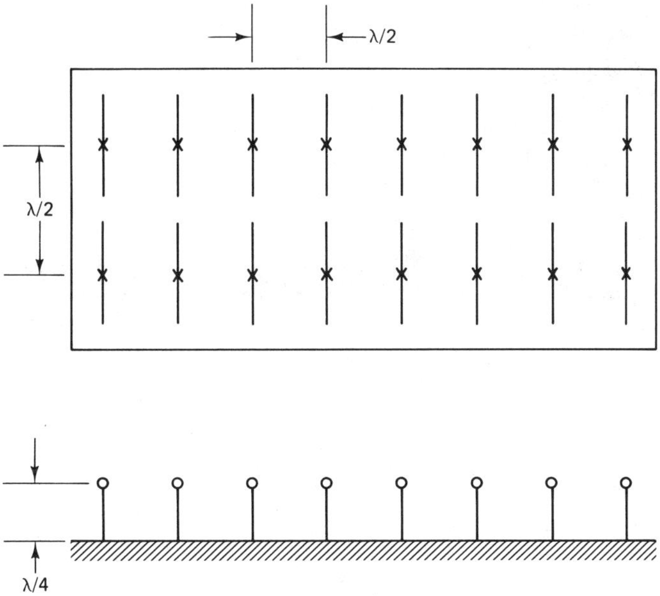

Figure 9–53 Plan and profile views of a billboard antenna.

9–12. If the radar in Problem 9–11 is operating at 3 GHz and the received pulse has a difference frequency of 4.92 kHz, what is the velocity of approach of the target? Express the answer in mph. Can you identify the probable target? If you use Eq. (9–69a), you must divide the answer by 2 because a reflecting target shortens the optical path twice as fast as a moving source.

10

Optical Frequencies

10–1 INTRODUCTION

We can use a few pounds of glass to replace several tons of copper and provide thousands of communications channels. The economic factor alone makes fiber optic technology a replacement or competitor for metal transmission lines, transoceanic cables, and satellite communications systems; in addition, fiber optic systems provide greater security from intrusion and certain types of interference, for example, large electromagnetic pulses. Optical frequency systems are also being used for signal processing. Small, compact, and relatively inexpensive devices are becoming available in the form of integrated optics.

The techniques for generating, transmitting, and detecting signals at optical frequencies are sometimes quite different from those used at microwave and lower frequencies; nevertheless, we shall find that the same fundamentals of wave theory are applicable.

10–2 REFRACTION AND DISPERSION

At optical frequencies refraction is a useful tool for measuring propagation speed in a medium. We introduced refraction in Chapter 2 as the bending of a ray across an interface separating two media, the velocity of propagation being different in the two media. This change in direction of propagation can be explained in terms of Huygens's wavelets. As radiation enters a medium in which the wave speed is slower, the diameters of the wavelets become smaller, and the envelopes of the

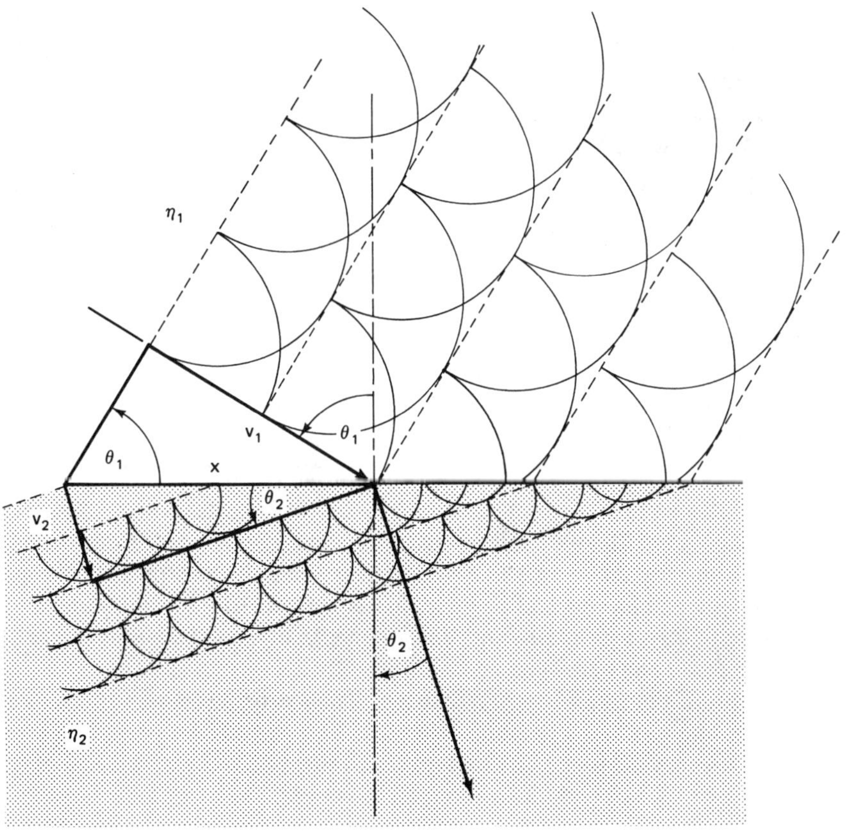

Figure 10–1 Huygens's wavelet construction of refraction at an interface.

wavelets, the wavefronts, change direction as do the rays (Fig. 10–1). From the velocity triangles shown in the figure we have $\sin \theta_1 = v_1/x$ and $\sin \theta_2 = v_2/x$, and after dividing these equations we have

$$\frac{\sin \theta_1}{\sin \theta_2} = \frac{v_1/c}{v_2/c} = \frac{\eta_2}{\eta_1}$$

which is Snell's law [Eq. (2–2)].

We see from Fig. 10–1 that as a ray enters the medium in which the wave speed is slower, it is bent toward the normal. By the principle of reversibility, a ray entering a less refractive medium is bent away from the normal. In the reversed case there evidently is an angle of incidence, θ_2, at which the angle of refraction is grazing the interface, $\theta_1 = 90°$, and from Snell's law we have $\sin \theta_2 = (\eta_1/\eta_2)$ $\sin 90°$, or

$$\theta_t = \sin^{-1} \frac{\eta_1}{\eta_2} \tag{10–1}$$

We call the angle θ_t the *critical angle* or the *angle of total reflection*, since for greater angles of incidence there is no refracted ray; all the radiation is internally reflected.

EXAMPLE 10–1

Calculate the critical angle at an interface between glass and air. Assume that the refractive indices are 1.55 and 1.00.

Solution From Eq. (10–1) we have

$$\theta_t = \sin^{-1}\frac{1.00}{1.55} = 40.2°$$

At all internal angles of incidence greater than 40.2° the radiation is totally reflected.

Fiber optics communications are possible because of internal total reflections in dielectrics. Metallic reflections would be impractical. Consider, for example, the number of reflections that must occur for light to propagate through 100 m of fiber optic cable. Typically, the cable is 100×10^{-6} m in diameter, the radiation wavelength is 1.3×10^{-6} m, and the angle of incidence may be as small as 70°. These parameters are illustrated in Fig. 10–2.

From the figure we see that the distance between reflections is $s = d \tan \theta$; that is, there is one reflection when the light travels a distance s. In a distance of 1m there are $N = 1/s = 3.64 \times 10^3$ reflections. As long as the reflections are total there is no problem. Suppose, however, that the reflections occur inside a hollow, silver-coated tube of the same diameter. In Example 5–18 we see that for yellow light the reflecting power of silver is 0.95 at normal incidence. The reflecting power at an angle of 70° is not much different. After 3.64×10^3 reflections the initial intensity is reduced to 0.95^{3640} or $10 \times 3640 \log 0.95 = -813$ dB. Evidently, communications in small metallic waveguides is not practical at optic frequencies.

A *step-index* fiber is a cylindrical glass core surrounded by a layer of glass of lower refractive index, called *cladding*; in addition, there is an outer layer of plastic to provide mechanical protection and an outer jacket that contains aramid strength members for cable pulling. At typical operating frequencies the core diameter

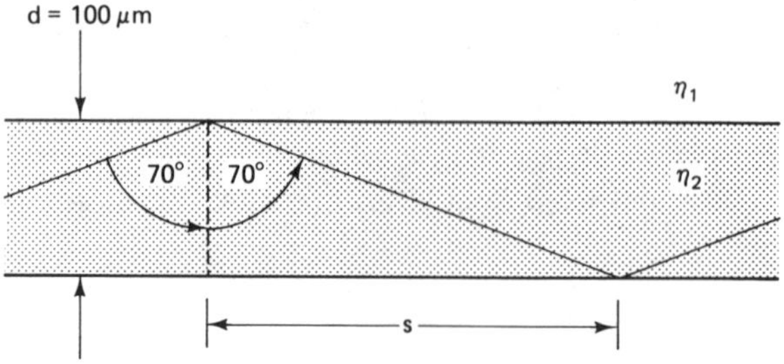

Figure 10–2 Internal reflections in a fiber optic core.

TABLE 10–1 WAVELENGTH AND FREQUENCY FOR VARIOUS RADIATIONS[a]

Radiation	Wavelength		Frequency	
X-rays				
Ultraviolet				
Upper	4.00	mm	7.5×10^{16}	Hz
Lower	4000	Å		
Visible light				
Violet	4000	Å	7.5×10^{14}	Hz
Yellow	5890	Å		
(Sodium D-lines)	5893	Å		
Red	0.72	μm	4.2×10^{14}	Hz
Infrared				
Upper	1	μm	3×10^{14}	Hz
Minimal	1.3	μm		
Low loss	1.6	μm		
Lower limit	1	mm	300	GHz
Microwave				
Upper limit	1	mm	300	GHz
Lower limit	1	m	0.3	GHz
UHF				

[a] Units: Å, angstrom, 10^{-10} m; nm, nanometer, 10^{-9} m; μ or μm, micron or micrometer, 10^{-6} mm; millimeter, 10^{-3}m; GHz, gigahertz, 10^9 Hz.

is about 70 wavelengths. To orient ourselves with respect to dimensions at optic frequencies, we show wavelengths and frequencies for microwaves and light in Table 10–1.

The angles at which light can enter the end of a fiber, and subsequently propagate in trapped modes, are determined by the refractive indices of the core and cladding. Let η_2 and η_1 be the indices of refraction of the core and cladding, as shown in Fig. 10–3a. If the angle of incidence at the end of the core is θ, then from Eq. (2–2) we have $\sin\theta/\sin(90° - \theta_t) = \sin\theta/\cos\theta_t = \eta_2/1$. By combining this with Eq. (10–1) we obtain $\sin\theta/\cos\theta_t = \sin\theta/\cos\sin^{-1}(\eta_1/\eta_2) = \eta_2$. From Fig. 10–3b we see that $\cos\sin^{-1}(\eta_1/\eta_2) = (\eta_2^2 - \eta_1^2)^{1/2}/\eta_2$, and we have

$$\sin\theta = (\eta_2^2 - \eta_1^2)^{1/2} \tag{10–2}$$

All rays entering the end of the core at angles less than θ are totally reflected internally; therefore, 2θ is the *acceptance angle*, and $\sin\theta$ is called the *numerical aperture*, NA.

EXAMPLE 10–2

If the refractive indices of the core and cladding are 1.55 and 1.50, find the acceptance angle and numerical aperture of the fiber.

Solution From Eq. (10–2) we have

$$\theta = \sin^{-1}(1.55^2 - 1.50^2)^{1/2}$$
$$= 23°$$

The acceptance angle is $2\theta = 46.0°$, and NA $= \sin 23° = 0.391$.

By the principle of reversibility the acceptance angle is also the maximum angle of emergence of rays from the end of the core. At large distances from the end of the core, $R >> d$, where d is the core diameter, the angle 2θ defines the far-field radiation pattern and is analogous to the beamwidth of an aperture-type antenna. Unlike the radiation pattern of an antenna, the beamwidth of a fiber core is determined geometrically by reflections, not by diffraction. The term "directive gain" is not applicable; *optical gain* is defined as the ratio of radiated power to current density in the source and will be discussed later.

Because waves in different modes travel different distances between reflections, as indicated by the rays in Fig. 10–3a, there is considerable multimodal or modal dispersion in a step-index fiber. Dispersion is reduced and usable bandwidth increased in a *graded-index* fiber, in which the refractive index decreases continuously from the center of the core to the outer edge. We see in Fig. 10–4 that the rays are refracted continuously. The longer ray paths travel faster through regions of lower refractivity. Because of grading, faster propagation partly compensates for a longer path, and all rays have nearly equal optical path lengths. If all of the modal components of a pulse or group of waves have nearly equal travel times, "spreading out" or dispersion is reduced, in practice, to about half of that in step-index fibers.

More bandwidth—less dispersion—is obtained with *monomode* fibers. The core of a monomode fiber is much smaller than that of a step-index fiber (10 to 20 μm is typical), and the difference between the refractive indices of the core and

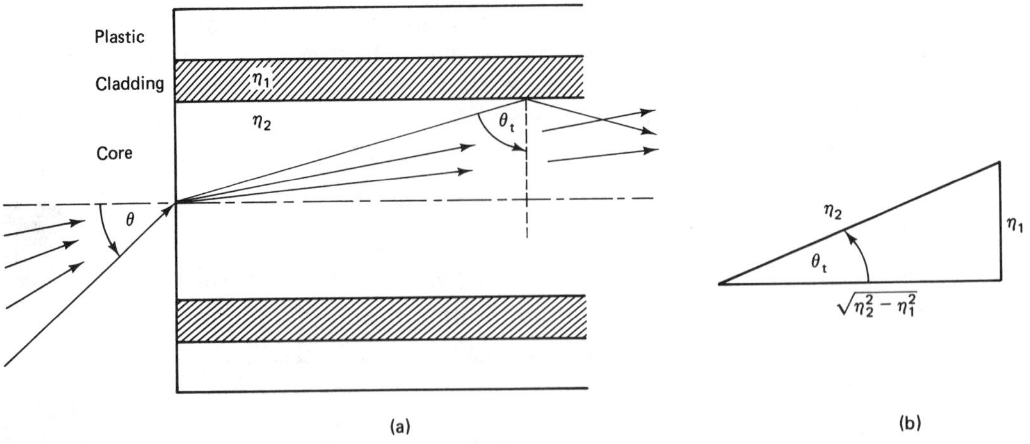

Figure 10–3 Radiation incident at the end of a fiber optic cable.

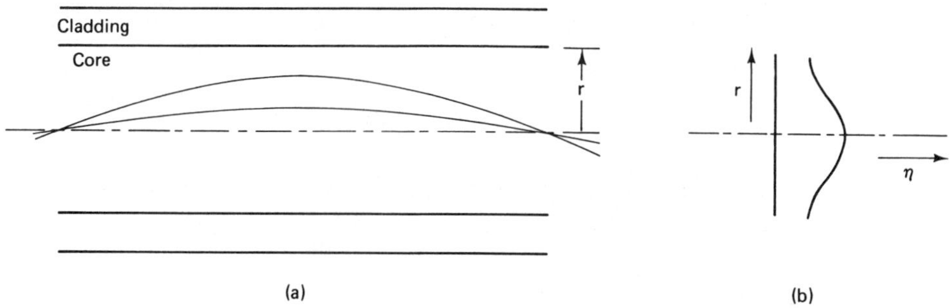

Figure 10–4 (a) Refracted rays in a graded-index core; (b) refractive index profile in the core.

cladding is reduced. Because of these reductions, only one mode can propagate, somewhat like the fundamental mode in waveguide.

We can estimate the modal dispersion in a step-index fiber by calculating the path difference between a ray reflected at the critical angle and a ray propagating along the axis of the core. We see from Fig. 10–5 that the path difference is $\Delta s = s_1 - s = (s^2 + d^2)^{1/2} - s = s[(1 + d^2/s^2)^{1/2} - 1]$. For $d << s$ we have, approximately, $\Delta s \approx s[(1 + \beta^2)^{1/2} - 1] \approx s(1 + \frac{1}{2}\beta^2 - 1)$, or

$$\Delta s \approx \tfrac{1}{2}\beta^2 s \tag{10–3}$$

From Snell's law we have $\sin\theta/\sin\beta = \eta$, or $\sin\theta \approx \eta\beta$, and

$$\beta \approx \frac{\text{NA}}{\eta} \tag{10–4}$$

We combine Eqs. (10–3) and (10–4) to obtain the path difference in terms of the numerical aperture and refractivity,

$$\Delta s \approx \frac{1}{2}\frac{(\text{NA})^2 s}{\eta^2} \tag{10–5}$$

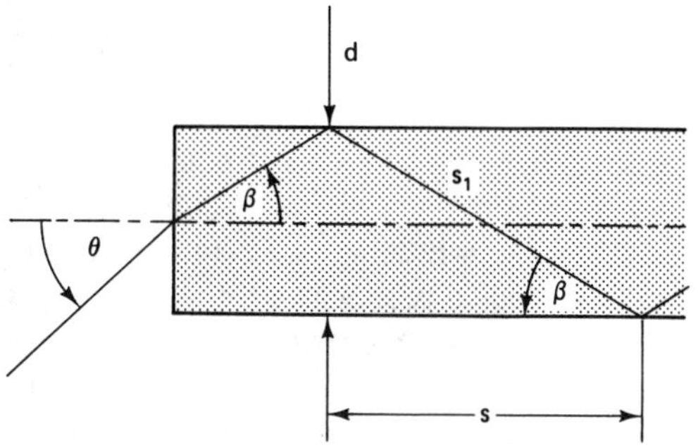

Figure 10–5 Refraction and reflections at the end of a fiber optic core.

Modal dispersion is approximately the difference in propagation times for the two extremal rays, $\Delta t = \Delta s/v \approx \frac{1}{2}(\text{NA})^2 s/v\eta^2$, per unit length, or

$$\frac{\Delta t}{s} \approx \frac{\frac{1}{2}(\text{NA})^2}{c\eta} \tag{10–6}$$

Although we derived Eq. (10–6) for step-index fibers, it is also approximately valid for graded-index fibers.

EXAMPLE 10–3

Estimate the dispersion and bandwidth in a fiber optic cable with an acceptance angle of 20° and a refractive index 1.5.

Solution From Eq. (10–6) we have

$$\frac{\Delta t}{s} = \frac{\frac{1}{2}(\sin 10°)^2}{3 \times 10^5\ (\text{km/s}) \times 1.5}$$

$$= 33.5\ \text{ns/km}$$

From the uncertainty principle we have $\Delta U\ \Delta t \approx h$, $h\ \Delta f\ \Delta t \approx h$, or

$$\Delta f \approx \frac{1}{s}\frac{s}{\Delta t}$$

$$= \frac{1}{s} \times (33.5 \times 10^{-9})$$

$$= 29.9\ \text{MHz}$$

If we use rms or "expectation" values, the bandwidth is $\Delta f = 29.9/(2)^{1/2} =$ 21.1 MHz.

In Example 10–3 our concept of bandwidth is slightly different from what we have had before. In Chapters 5 and 6 we regarded transmission lines as "windows" that passed some frequencies and rejected others; the passband of frequencies was the bandwidth. In some cases bandwidth was limited not by rejection but by the occurrence of an undesirable event, for example, the generation of a higher-order mode. In Chapter 9 we defined bandwidth as a frequency band in which a minimum power level is maintained; the Q^{-1} of a dipole, for example.

We define dispersion-limited bandwidth as the frequency range over which the power level of a pulse can be maintained above half of its initial value. The pulse has a total energy of $U = p_0\ \Delta t_0$, where p_0 is the initial power level and Δt_0 is its time length or duration. As the pulse propagates through a dispersive medium its time length increases, and conservation of energy requires that the power level decrease, $U = p_0\ \Delta t_0 = p_1\ \Delta t_1$, where $\Delta t_1 > \Delta t_0$ and $p_1 < p_0$. This is illustrated in Fig. 10–6. Since a short pulse comprises a wide band of Fourier frequencies, it

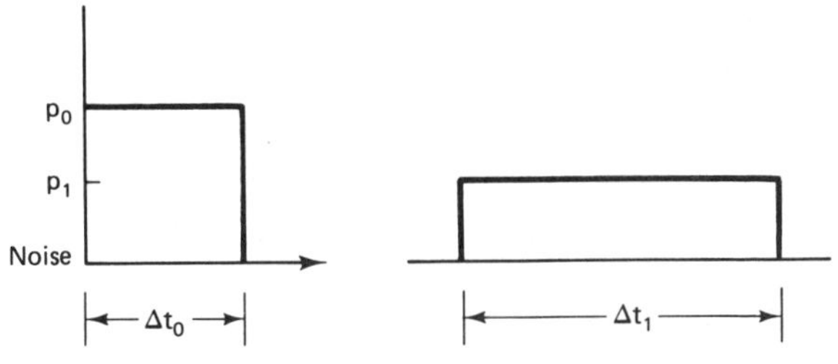

Figure 10–6 Pulse dispersion.

will disperse in a short distance. A long pulse, with a corresponding narrow frequency spectrum, will disperse more slowly and maintain its power level over a greater distance. We have an optimum trade-off between power drop and bandwidth if the pulse length is equal to the dispersion it suffers in traveling a given length of transmission line. Under this condition the pulse length will have approximately doubled and the power level will have dropped to half-value (-6 dB electrical or -3 dB optical power). We define bandwidth as the reciprocal of the duration of an initial optimum pulse:

$$\text{BW} \approx \frac{1}{\Delta t_0} \tag{10–7}$$

which, as shown in Example 10–3, also follows from the uncertainty principle.

Dispersion-limited bandwidth is defined with the implicit assumption that as a pulse is dispersed its power level drops and eventually it disappears in the noise level of the system. This assumption is not always realized. Consider, for example, a medium or transmission line in which a pulse is normally dispersed; that is, the higher-frequency components travel more slowly than the lower. The power level drops to the noise level before the signal reaches the receiver. At the receiving station the signal is processed through a complementary medium in which anomalous dispersion occurs: the higher-frequency components travel more rapidly than the lower. The random noise fluctuations are unaffected, but the pulse is reverse-dispersed or compressed until its power level rises above noise and is detectable. This process is sometimes called ''chirp'' transmission and can be used for covert communications.

Thus far we have been discussing modal dispersion, except for the ''chirp'' example. Frequency-dependent dispersion is called *chromatic dispersion*, different colors being the optical equivalent of different frequencies. Chromatic dispersion can be the result of *material dispersion*, which is the change of refractive index with frequency in the bulk material of the medium; or it can be caused by *waveguide dispersion*, which is discussed in Chapter 6.

A familiar example of material dispersion is the separation of white light into its rainbow spectrum of colors by a glass prism. From Snell's law [Eq. (2–2)] we see that the angle of refraction is proportional to the refractive index. If the refractive index increases with frequency, the higher frequencies are refracted more than the

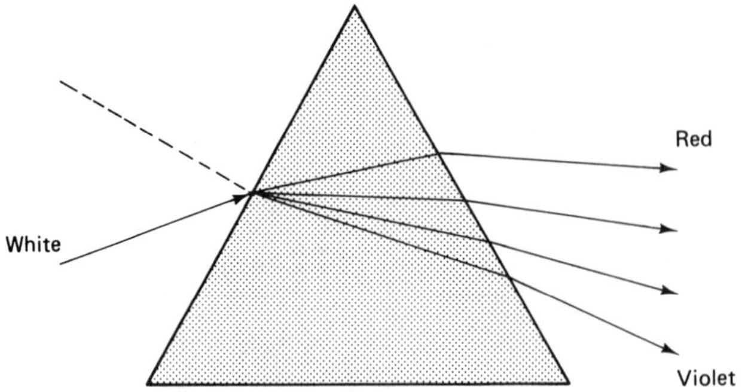

Figure 10–7 A prism disperses white light into its component frequencies.

lower; thus white light is dispersed into its component frequencies or colors (Fig. 10–7). We see that the prism is a form of spectrum analyzer.

A prism can be used to determine refractive index accurately at a single frequency. By "single frequency" we mean that we use a narrow-band source, which in optics is called a *monochromatic source*, literally a "one-color" source. A discharge

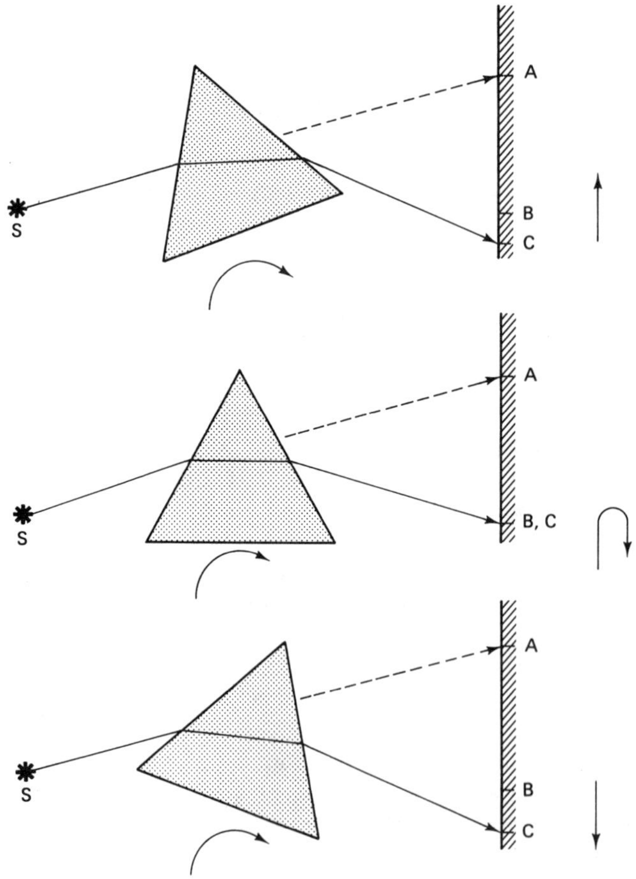

Figure 10–8 Determination of the angle of minimum deviation.

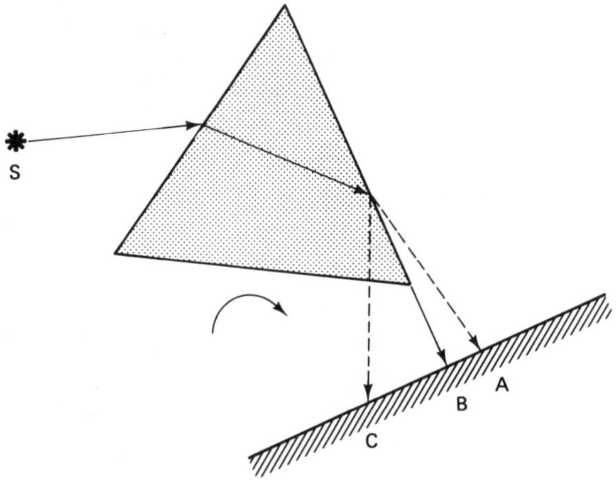

Figure 10–9 Determination of the critical angle in a prism.

tube or laser can be used as a monochromatic source. One method of using a prism to determine the index of refraction is to rotate the prism to find the *angle of minimum deviation*. As shown in Fig. 10–8, the deviation is the apparent angular deflection of a beam as it is refracted by the prism. The emerging beam can be observed by placing a screen to the right of the prism, as shown. A laser can be used as a source and the undeflected position of the beam, point *A* on the screen, can be determined before the prism is in place. As the prism is rotated the deviation decreases, appears to stop as point *C* moves to point *B*, and then appears to turn around and start increasing again. At the position of minimum deviation the angles of incidence and emergence from the prism are equal, and a measurement of the angular deviation can be used to determine the refractive index. The required geometric analysis is given in Application 10–2.

A second method for measuring refractive index is the determination of the *critical angle*. As shown in Fig. 10–9, we rotate the prism and observe the refracted beam on a screen. The angle of total reflection first occurs at grazing emergence of the refracted ray, as indicated at point *B* in the figure. By measuring the corresponding angle of incidence we can determine the refractive index. The geometric analysis is given in Application 10–1.

10–3 ATTENUATION

Material dispersion is associated with loss or attenuation. We assume a bulk medium to contain a myriad of small atomic or molecular dipoles. And we assume, as we did in analyzing the Yagi antenna, that each dipole absorbs and re-broadcasts or "scatters" part of the incident wave. The scattered wavelets are delayed in phase and so is the resultant wave, which is the sum of the incident and scattered wavelets. The phase delay is measured as a relative decrease in wave speed—the refractive index.

We can obtain an expression for the refractive index by assuming that the presence of the dipoles causes a change ΔZ in the characteristic impedance of the medium, $Z = Z_0 + \Delta Z$, or $z = 1 + \Delta Z/Z_0$. By comparing Eqs. (2–7) and (2–10), we see that the refractive index is the reciprocal of the normalized impedance, $\eta = 1/z = 1/(1 + \Delta Z/Z_0) \approx 1 - \Delta Z/Z_0$. We take the increase in normalized impedance to be the expression for dipole impedance [Eq. (9–53)], and we have

$$\eta \approx 1 - \frac{jQ}{1 + jQ(f - 1/f)} \tag{10–8}$$

where we have multiplied $\Delta Z/Z_0$ by the j-operator to account for the difference in phase between the oscillator induction field and the radiation field.

If we separate Eq. (10–8) into its real and imaginary parts, we have

$$\begin{aligned} \eta &\approx 1 - \frac{Q^2(f - 1/f)}{1 + Q^2(f - 1/f)^2} - \frac{jQ}{1 + Q^2(f - 1/f)^2} \\ &\approx \eta' \quad jK \end{aligned} \tag{10–9}$$

where $K = \alpha\lambda/2$ is the extinction coefficient and α is the absorption coefficient, discussed in Chapter 5. A graph of the magnitude of the refractive index and the absorption index ηK for $Q = 1.75$ is shown in Fig. 10–10. We see that at low frequencies the refractivity increases with frequency, which is normal dispersion. In the absorption region the refractivity has a negative slope, a region of anomalous dispersion, and beyond resonance the dispersion is again normal. This is why we say that material dispersion is associated with loss or attenuation. In a bulk medium there may be several atomic or molecular resonances, and the dispersion may resemble the sketch shown in Fig. 10–11.

In a region well below resonance, $f << 1$, the expression for the extinction coefficient becomes $K = -Qf^2/[f^2 + Q^2(f^2 - 1)] \approx f^2Q$. We see that the relative

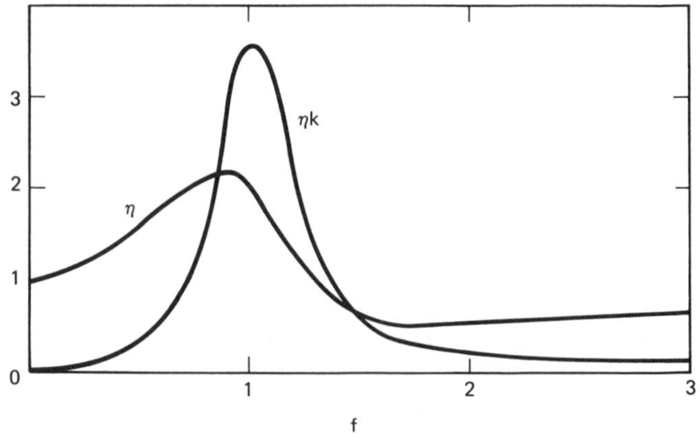

Figure 10–10 Refraction and absorption in a bulk material.

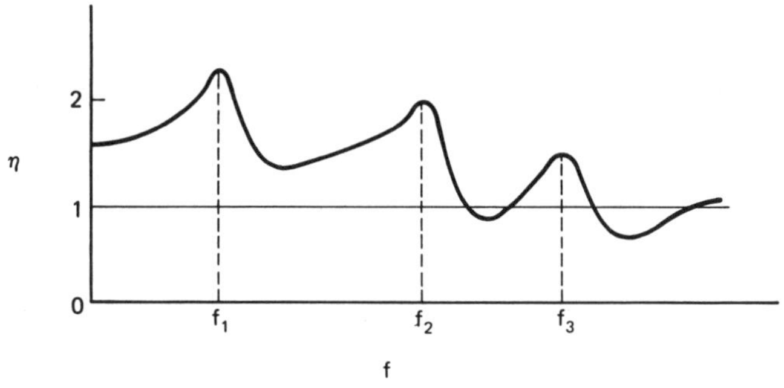

Figure 10–11 Refraction in a bulk medium.

amplitude of a wave decreases with the square of the frequency and conclude that the relative intensity or power decreases as

$$\frac{p}{p_0} \approx \frac{f^4}{Q^2} \tag{10–10}$$

That is, the intensity decrease due to scattering is proportional to the fourth power of the frequency. This is known as Rayleigh's law and accounts for the blue color of the sky. Since the high frequencies are the most scattered, the sky appears blue, except when looking toward the sun. In looking at a sunset, we see orange or red because the blue has been scattered out of the line of sight.

In glass there is a resonance in the infrared region due to the $Si0_2$ molecule. This absorption band is a low-frequency limitation, and Rayleigh scattering is a high-frequency limitation, as indicated in Fig. 10–12. For this reason many fiber optic communications channels are located in the "window" between wavelengths of 1.3 and 1.7 μm. It is customary to take a kilometer as the unit of distance, and in the IR window the attenuation coefficient varies between 0.2 and 1.0 per kilometer, depending on the exact wavelength and the fiber. The expressions for attenuation are

$$\frac{p}{p_0} = \exp(-\alpha x) \tag{10–11}$$

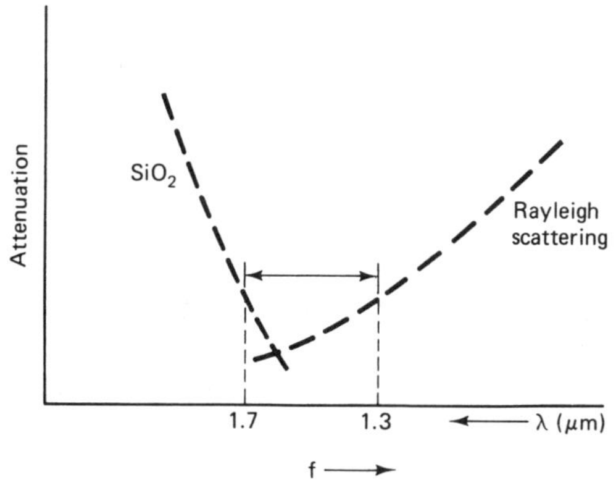

Figure 10–12 Attenuation in glass fiber.

or

$$\frac{p}{p_0} = 10^{-0.434\alpha x} \tag{10–12}$$

EXAMPLE 10–4

Find the attenuation due to material dispersion in 1.5 km of fiber cable when the operating wavelength is 1.5 μm. Assume that the attenuation coefficient is 0.6/km.

Solution From Eq. (10–12) we have

$$\frac{p}{p_0} = 10^{-(0.434\times 0.6\times 1.5)}$$

$$= 0.407 \quad \text{or} \quad -3.91 \text{ dB}$$

On a scientific calculator enter $-(0.434 \times 0.6 \times 1.5) = -0.391$; then use the INV and LOG keys to obtain 0.407.

10–4 SOURCES

An atom or molecule, unlike an antenna, is not continuously supplied with energy from a generator. When radiation is emitted, the energy is supplied by a change in the internal energy of the atom. The energy levels of electrons in atoms are discrete or "quantized," much as the standing waves on a rope line are quantized into a fundamental mode and its various harmonics. Radiation is emitted when an electron drops from a higher energy level to a lower one. The energy of the emitted photon is determined by the energy difference of the two levels,

$$hf = U_2 - U_1 \tag{10–13}$$

where h is Planck's constant, $h = 6.623 \times 10^{-34}$ *J-s*. In one respect an atom is somewhat like the front end of a superheterodyne receiver. As shown in Fig. 10–13, the local oscillator—the higher-energy level—is mixed with the amplified radio-frequency signal, and the difference or "intermediate frequency" wave propagates to the IF amplifier. The analogy is better than might be expected. In its rest state an atom does not radiate. The electrons, rotating or oscillating in stable energy levels, are analogous to standing waves or induction fields. A traveling wave is generated when a transition to a lower energy level occurs.

Light sources are created when atoms or molecules are excited into unstable or transient energy levels, and radiation is emitted as the electrons drop back to stable states. The excitation can be caused by electrical heating, as in an incandescent lamp, electrical discharge through a gas (fluorescent lamps, lightning) or by current flow through a semiconductor [light-emitting diodes (LEDs)]. In some cases it takes

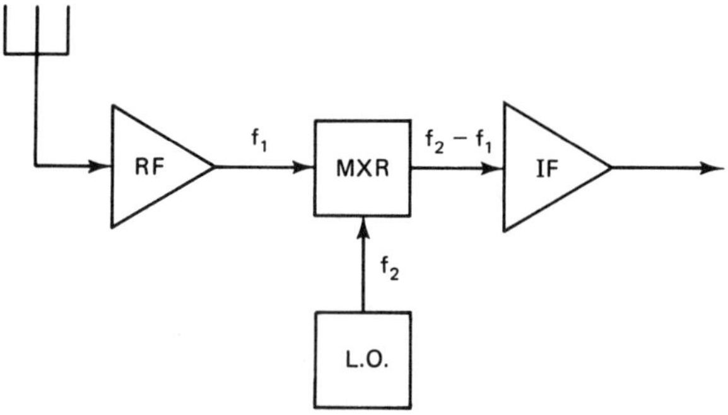

Figure 10–13 Front end of a superheterodyne receiver.

so long for the electrons to drop back into stable energy levels that the radiation can be seen by the human eye as an afterglow or fluorescence. The persistence of the green glow on the CRT of an oscilloscope is a familiar example.

In semiconductors, radiation and heat are generated when a free or "conduction" electron drops from the conduction energy band down to an energy level in the valence band. By "valence band" we mean the energy levels of the outermost electrons of an atom, which are called "valence" electrons. In solid-state electronics this action is described as the recombination of a negative carrier and a positive hole, a sort of pair annihilation. The difference in energy between the conduction and valence bands is called *bandgap energy*. Bandgap energies for a few materials are listed in Table 10–2.

TABLE 10–2 BANDGAP ENERGY FOR VARIOUS MATERIALS

Substance	Energy gap (eV)[a]
Germanium, Ge	0.67
Silicon, Si	1.1
Gallium arsenide, GaAs	1.4
Gallium phosphide, GaP	2.0
Silicon dioxide, SiO_2	8

[a] One electron volt (1 eV) is 1.60×10^{-19} J.

When a diode is forward-biased and current flows, some recombination occurs, and if the emitted radiation is in the visible range it is called a light-emitting diode (LED). In fiber optics systems the radiation may be in the infrared region and invisible, but the diode may still be called an LED.

EXAMPLE 10–5

Calculate the wavelength of radiation from a GaP diode. Is the radiation visible?

Solution From Eq. (10–13) and Table 10–2 we have

$$f = \frac{2 \times (1.60 \times 10^{-19})}{6.623 \times 10^{-34}}$$
$$= 4.83 \times 10^{14} \text{ Hz}$$

and the wavelength is $\lambda = c/f = (3 \times 10^8)/(4.83 \times 10^{14}) = 0.621\ \mu\text{m}$. Referring to Table 10–1 we see that this wavelength is in the visible region, between red and yellow.

Another way to create a light source is to expose a medium to high-intensity light. The atoms in the medium absorb the light and eventually, as the electrons in excited energy levels decay to stable levels, light is reemitted. This process is called *optical pumping*.

Optical pumping may seem pointless since we already have a light source to begin with; however, it is possible to cause all the reemission to occur in-phase, a phenomenon called "laser action" or "lasing." Ordinarily, an atom in an excited state will drop back to a stable state and emit radiation at a random time; however, if a wave or oscillation is present at the same frequency as the radiation to be emitted, that radiation will be emitted in phase with the wave, a process called *stimulated emission*. To appreciate the significance of this phenomenon, recall the billboard antenna of Problem 9–9. For n dipoles in phase, the total gain is n^2G_1, where G_1 is the gain of one dipole. This is an appreciable increase in gain for even a small number of dipoles, but for atomic or molecular dipoles the number per unit volume may be billions. The increase in gain is incredible. For example, if 1 million atoms reemit radiation at random times, the energy density is only 10^6 times the energy emitted by one atom, but if the atoms reemit in phase, there is an increase in energy density of 10^{12}. The acronym *laser* comes from "*l*ight *a*mplification and *s*timulated *e*mission of *r*adiation."

The driven oscillator is a rough analogy of stimulated emission. Consider an oscillator that does not quite satisfy the Barkhausen conditions and does not, by itself, oscillate. If we have another oscillator at the same frequency and couple it to the nonperforming oscillator, the extra input may make up for the insufficient feedback and enable the circuit to oscillate in phase with the driver.

We need to provide a wave that will act as a driver in the optically pumped medium. One way to do this is to confine the medium between two parallel mirrors. Although radiation may be lost from the sides, waves are generated that reflect

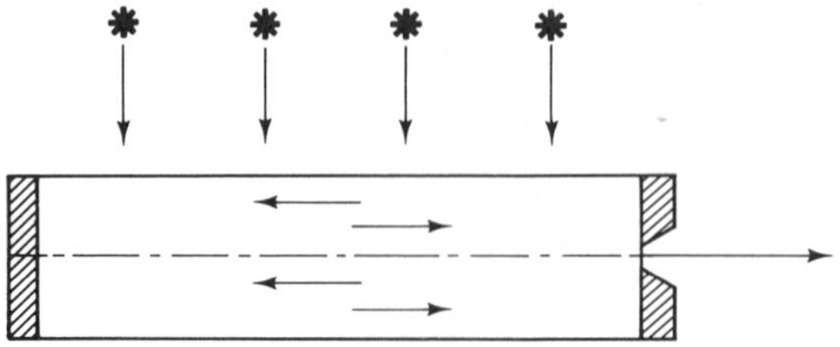

Figure 10–14 Optical pumping of a laser.

back and forth between the mirrors. If there is sufficient optical pumping, the reflected wave amplitudes become large enough to cause stimulated emission or lasing. The amplified, in-phase radiation can be allowed to escape by making a portion of the mirror at one end a partial reflector, as indicated in Fig. 10–14.

EXAMPLE 10–6

Estimate the far-field half-power beamwidth and gain of a laser if the aperture diameter is 2 mm and the operating wavelength is 1.5 μm.

Solution From Eq. (9–64) we have

$$\theta_{HP} \approx \frac{0.882 \times (1.5 \times 10^{-6})}{2 \times 10^{-3}}$$

$$= 6.62 \times 10^{-4} \text{ rad} \quad \text{or} \quad 0.0379°$$

and from Eq. (9–24)

$$G = \frac{4\pi(\pi/4)(2 \times 10^{-3})^2}{(1.5 \times 10^{-6})^2}$$

$$= 1.75 \times 10^7 \quad \text{or} \quad 72.4 \text{ dB directive gain}$$

Laser radiation is said to be *coherent*, which means that any phase relationship is maintained over an extended period of time or distance, called the *coherence time* or *length*. The ordinary random-phase radiation from excited atoms is called *incoherent*. The double-slit experiment shown in Figs. 9–45 and 9–46 could not be done if ordinary light sources were used at each of the slits. The random phases of the light would "wash out" the interference pattern. The interference pattern can

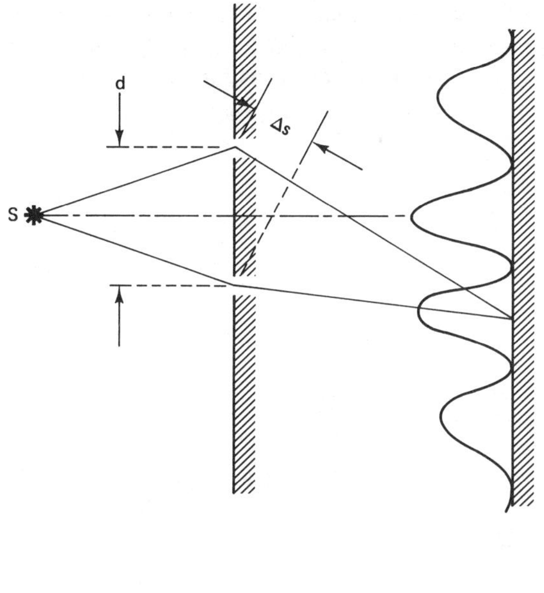

Figure 10–15 Two-slit interference pattern.

be observed using an ordinary monochromatic light source provided that the same source is used to illuminate both slits (Fig. 10–15). Although the phase of the wavelets generated by the source is random in time, at any instant related phases arrive at each slit and an interference pattern can be observed.

The double-slit experiment again fails if the distance between the slits or the angle of observation is increased until the path difference between two rays exceeds the coherence length of atomic emissions. This type of experiment can be used to measure coherence length.

EXAMPLE 10–7

A laser beam is split by a half-silvered mirror and after a second reflection the two beams are brought together to interfere at the point of observation, as shown in Fig. 10–16. Maxima and minima can be observed as the second mirror is rotated slightly. As the angle θ is increased, the interference pattern "washes out" when $\theta = 2.87°$; that is, maxima and minima are no longer observable when the second mirror is rotated. Find the coherence length of the radiation. Neglect the extra optical path length of one of the rays that passes through the half-silvered mirror.

Solution We have to calculate the path lengths traveled by the two rays when $\theta = 2.87°$; the coherence length is the difference in path lengths at "washout." The path length traveled by the upper ray is $R \tan \theta + R/\cos \theta = 1 \times \tan 2.87° + 1/\cos 2.87° = 1.050$ m, and the path length of the lower ray is approximately 1.000 m. The difference is the coherence length, 5.0 cm.

Laser action can be achieved by optically pumping various gaseous, liquid, or solid media; it can also occur in semiconductors, not by optical pumping but by maintaining sufficiently high current densities and partially confining the emitted radiation. If the current density and confinement cause the radiation to build up to amplitudes sufficient for stimulated emission, the device changes from an ordinary

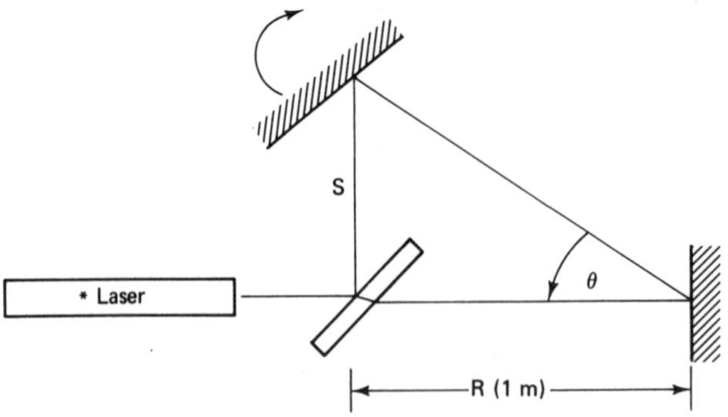

Figure 10–16 Measurement of coherence length.

incoherent LED to a coherent light source—a laser diode or LD. As we indicate in Fig. 10–17, current density is confined and increased by shaping the electrode. An active region confined in this way is said to be *gain guided*. In addition, radiation can be partially confined by using layers of semiconductor materials that have refractive indices less than the refractive index of the active layer; thus the radiation is confined by internal reflections. Radiation confined in this manner is said to be *index guided*. Additional index guiding can be achieved by providing lateral variations in refractive index. Finally, confinement is increased by covering the end faces with reflecting and partially reflecting mirrors.

Laser diodes have greater bandwidth and coupling efficiency than LEDs. The greater efficiency in coupling to optical fibers is a result of the narrower beamwidth

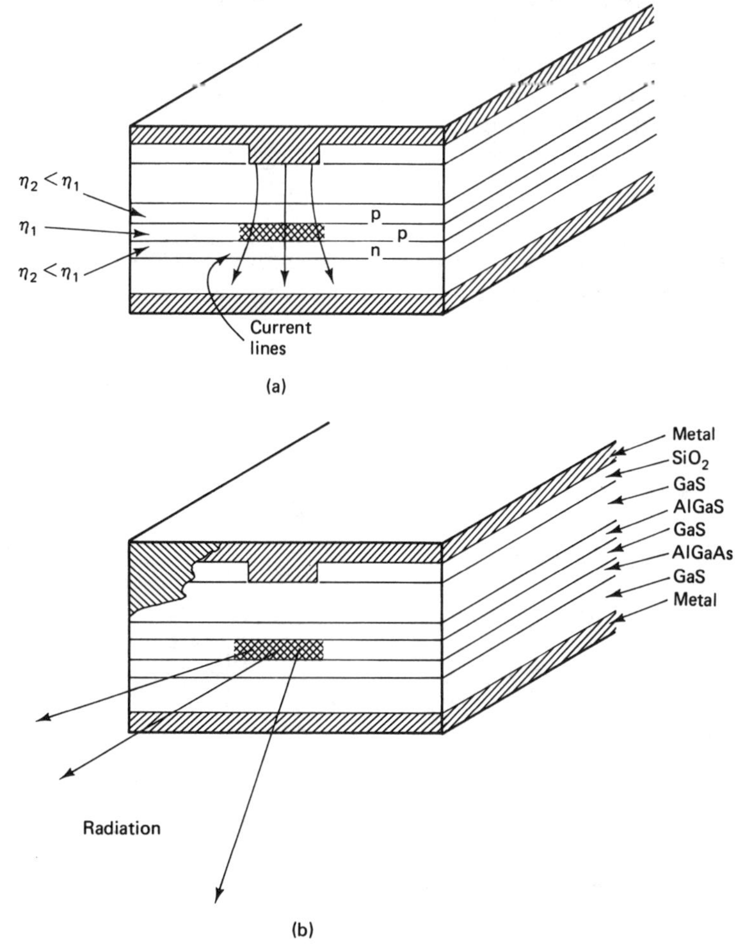

Figure 10–17 Solid-state laser diode: a, showing current lines; b, showing radiation directions.

TABLE 10–3 LED AND LD CHARACTERISTICS

Characteristic	LED	LD
Gain	50 W/mA	5 W/mA
Bandwidth	200 MHz	2 GHz
Coupling efficiency	10%	50%
Polarization	None	90%
Lifetime	10^7 hr	10^6 hr

of an LD. The gain of an LED is greater than that of an LD, where by "gain" we mean here the ratio of the optical power output to the input or drive current; this ratio also is called the optical efficiency. Characteristics of typical LEDs and LDs are summarized in Table 10–3. The characteristic curve of an LD is similar to that of an ordinary diode, and a representative curve is shown in Fig. 10–18. Here the voltage is not the potential impressed across the diode but the ratio of the power output to the drive current,

$$v = \frac{p}{i} \tag{10–14}$$

Disadvantages of LDs are their temperature sensitivity, which requires operation under temperature-controlled conditions, and their shorter lifetime, which is about a tenth that of an LED.

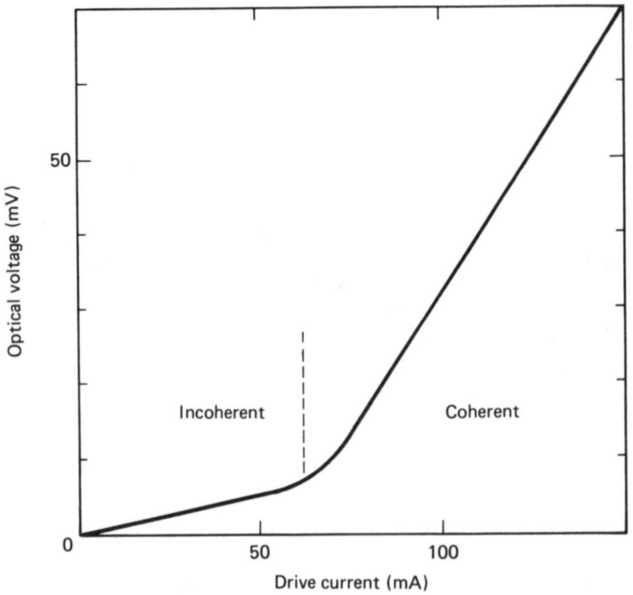

Figure 10–18 Characteristic curve of an LD.

EXAMPLE 10–8

Calculate the optical power output for a typical LD when the drive current is 120 mA.

Solution From Fig. 10–18 we find that the corresponding optical voltage is 47 mW. From Eq. (10–14) we calculate the optical power as

$$\begin{aligned} p &= (47 \times 10^{-3}) \times (120 \times 10^{-3}) \\ &= 5.64 \text{ mW} \end{aligned}$$

10–5 DETECTORS

The detection of optic signals depends on electrical or chemical effects caused by the interaction of photons and matter; for example, the removal of an electron from its parent atom by the absorption of the energy of a photon. The remaining atom has a net positive charge and is an ion; the atom has been ionized. The photon has been annihilated, and its energy has been used to pull the electron away from the positive nucleus of the atom. The energy or work required to free the electron is called the *ionization potential*. It is usually measured in electron volts (1 eV = 1.6×10^{-19} J). Some workers express ionization potentials in volts, it being understood that the charge involved is that of an electron.

The removal of an electron from the surface of a bulk metal requires less energy than the ionization of an atom, partly because the electron may already be in a conduction band and partly because of the image effect. As the electron approaches

TABLE 10–4 TYPICAL ENERGY VALUES

Energy	Magnitude (V)
Ionization potential	1–25
Work functions	
Ag	4.6
W	4.5
Cu	4.3
Ba0	2.0
Ce	1.8
Gap energies	
GaS	1.4
Si	1.1
Ge	0.67
Barrier potential	
Si	0.6
Ge	0.3

the surface, the oppositely charged image helps it to cross the boundary. The energy required to remove an electron from a surface is called the *work function* of the material. To orient ourselves, we list typical energy values in Table 10–4.

Excitation potential is the energy per unit charge required to raise an electron from one atomic energy level to another. In the reverse process radiation is emitted with a wavelength characteristic of the difference in energy levels.

EXAMPLE 10–9

Calculate the excitation potential of the sodium atom for the emission of yellow light.

Solution From Table 10–1 we select 5890 Å as one of the spectral lines that is characteristic of the yellow light emitted by sodium. The corresponding frequency is

$$f = \frac{c}{\lambda} = \frac{3 \times 10^8}{5890 \times 10^{-10}}$$

$$= 5.09 \times 10^{14} \text{ Hz}$$

and from the energy of a photon at this frequency we find

$$U = hf = (6.62 \times 10^{-34}) \times (5.09 \times 10^{14})$$
$$= 3.37 \times 10^{-19} \text{ J}$$

or

$$hf = \frac{3.37 \times 10^{-19}}{1.60 \times 10^{-19}}$$

$$= 2.11 \text{ V}$$

A familiar application of a light detector is the photoelectric cell, used to open a shop door or sound an intrusion alarm when a light beam is interrupted. A more sensitive device is the electron amplifier. The light to be detected is incident on a surface with a low work function, for example, a surface covered with cesium or one of its compounds, and the ejected electrons are attracted to a series of plates, each plate at a higher voltage than the preceding plate (Fig. 10–19). The electrons are accelerated between the plates and their increased energy ejects more electrons when they strike the next plate. The resultant avalanche produces an amplified current output from the last plate. One application of this device is the astronomical measurement of weak radiation levels from distant stars.

The avalanche photodiode (APD) is a solid-state version of the electron multiplier and is often used in communications receivers. It is basically a solid-state diode reverse-biased to a relatively high voltage. As indicated in Fig. 10–20, photons strike the *n*-type material and produce electron–hole pairs. The high voltage accelerates

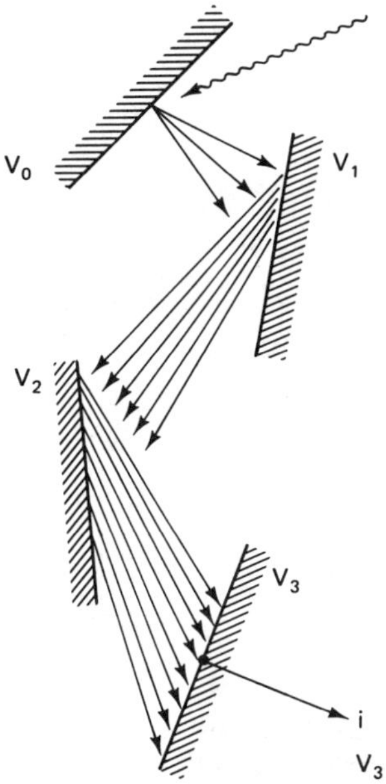

Figure 10–19 Electron multiplier.

the electrons into the *p*-type material, where subsequent collisions produce an avalanche of additional electrons and a multiplied or amplified output current.

A PIN diode differs in construction from an APD in that a low-conductivity layer of silicon is sandwiched between the two highly doped *p*- and *n*-type materials that form the diode. The low-conduction layer is called the intrinsic zone, whence the name PIN. As indicated in Fig. 10–21, incident photons create electron–hole pairs in the intrinsic zone, which subsequently are attracted to the reverse-biased *p*- and *n*-type layers. The PIN diode can be considered to be a photo-controlled or current-controlled resistor. As a photo-controlled resistor it can be used as a light detector in communications receivers. As a current-controlled resistor its resistance

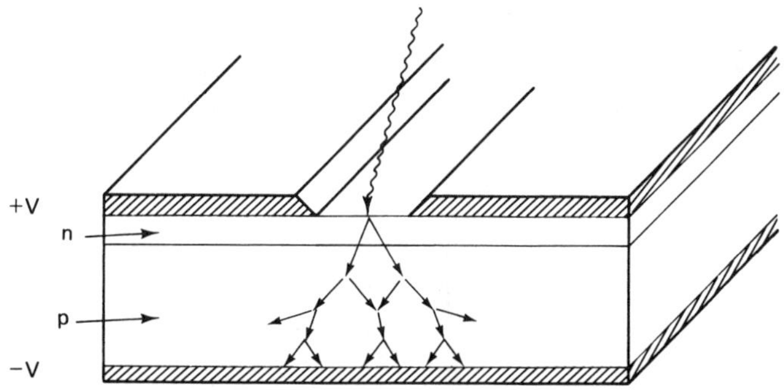

Figure 10–20 Avalanche photodiode, APD.

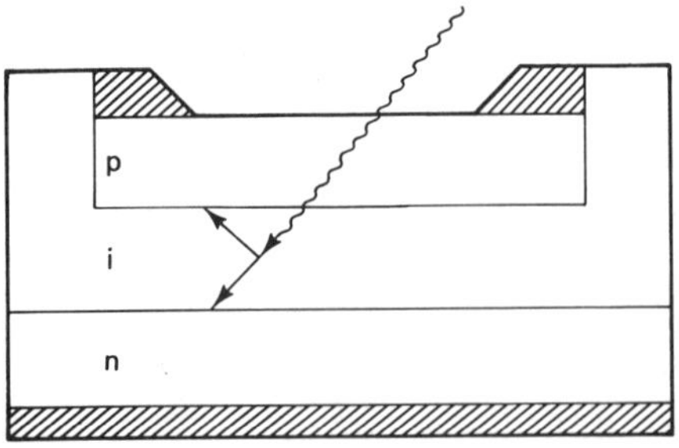

Figure 10–21 PIN diode.

can be varied from thousands of ohms to a few ohms by changing the bias current, and it can be used as a modulating or switching device at VHF to microwave frequencies.

10–6 SPECTRUM ANALYSIS

A prism can be used as a spectrum analyzer, as we can see in Fig. 10–7. Light waves of different frequencies are refracted at different angles, and by inspecting the resultant radiation pattern we can determine the amplitudes of the various frequency components of the input signal. A grating can be used as a spectrum analyzer more versatile than a prism, and in this section we show the grating as an example of an integrated optics (I-O) system.

An optical grating can be a multitude of slits and, as indicated in Fig. 10–22, can be analyzed in much the same manner as in our analysis of a double-slit diffraction pattern. A lens can be used to focus all the rays at a convenient, nearby point of observation, thus eliminating the necessity for moving the observation point to a great distance to observe the far-field radiation pattern. Since the phase difference between neighboring rays is $(2\pi d \sin \theta)/\lambda$, there is a maximum intensity at the point of observation whenever $(2\pi d \sin \theta)/\lambda = 2\pi n$, or

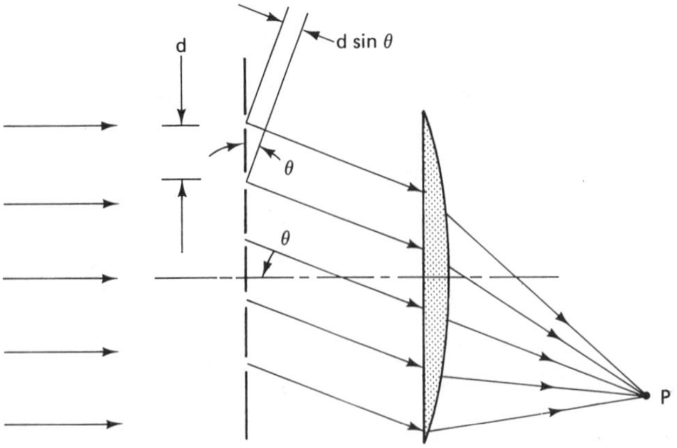

Figure 10–22 Observation of a spectrum formed by a grating.

$$d \sin \theta = n\lambda \qquad n = 0, 1, 2, \ldots \tag{10–15a}$$

where d is the slit spacing, called the *grating constant*, the wavelength is λ, and θ is the angle of observation. We see that there are maxima whenever n is an integer, and the nth maximum is said to be in the nth order of the spectrum.

EXAMPLE 10–10

A grating is illuminated with yellow light from sodium, 5890 Å. The grating constant is 3.5 μm. Find the angle of observation of the maximum in the first- and second-order spectra.

Solution From Eq. (10–15a) we have

$$\theta_1 = \sin^{-1}\frac{n\lambda}{d} = \sin^{-1}\frac{1 \times (5890 \times 10^{-10})}{3.5 \times 10^{-6}}$$

$$= 9.69°$$

and

$$\theta_2 = \sin^{-1}\frac{2 \times (5890 \times 10^{-10})}{3.5 \times 10^{-6}}$$

$$= 19.67°$$

One of the advantages of a grating over a prism or double-slit implementation is the increase in gain. If s is the illuminated width of a grating and d the grating constant, there are $s/d = N$ in-phase slits contributing to the radiation pattern. This is a gain N^2 times greater than the gain of one slit, and the narrower line widths in the spectrum are approximately, from Eq. (9–64),

$$\theta_n \approx \frac{0.882\lambda}{Nd} \tag{10–16}$$

If only two slits were used, the overlapping and large beamwidths would make spectral analysis difficult; however, two slits or two coherent sources are sometimes used at radio frequencies for direction finding. Whether used at radio or optical frequencies, the two-source device is called an *interferometer*.

EXAMPLE 10–11

A microwave grating is illuminated by a 10-GHz PRF square wave. The illuminated area is 50 cm wide, the grating constant is 5 cm, and the slit widths are 2.5 cm. Make a sketch of the first- and second-order spectra, showing only amplitudes greater than 0.2 maximum.

Solution From Eq. (5–9) we know that the leading terms in the Fourier spectrum are cos ωt, ⅓ cos $3\omega t$, and ⅕ cos $5\omega t$, which correspond to the fundamental and third and fifth harmonics at 10, 30, and 50 GHz ($\lambda = 0.03$, 0.01, and 0.006 m). From Eq. (10–15a) we have

$$\theta_1 = \sin^{-1}\frac{0.03, 0.01, 0.006}{0.05}$$

$$= 36.9°, 11.5°, 6.9°$$

$$\theta_2 = \text{missing}, 23.6°, 13.9°$$

We can find the line widths from Eq. (10–16):

$$\theta_{HP} \approx \frac{0.882\lambda}{Nd}$$

$$= \frac{0.882(0.03, 0.01, 0.006}{10 \times 0.025}$$

$$= 6.1°, 2.0°, 1.2°$$

A sketch of these results is shown in Fig. 10–23. We see that the spectral orders overlap.

In the preceding discussion we assumed that the grating was illuminated at normal incidence. If the grating is illuminated at an angle of incidence α, the grating equation is

$$d(\sin\alpha + \sin\theta) = n\lambda \tag{10–15b}$$

Evidently, at some angles of incidence the phase difference between slits will be appropriate for end fire, and the grating will resemble a Yagi antenna. End fire from an optical grating is called *Wood's anomaly*.

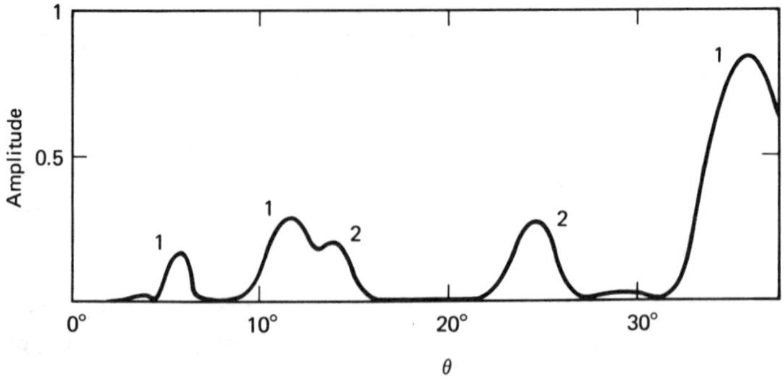

Figure 10–23 Grating spectrum, Example 10–11.

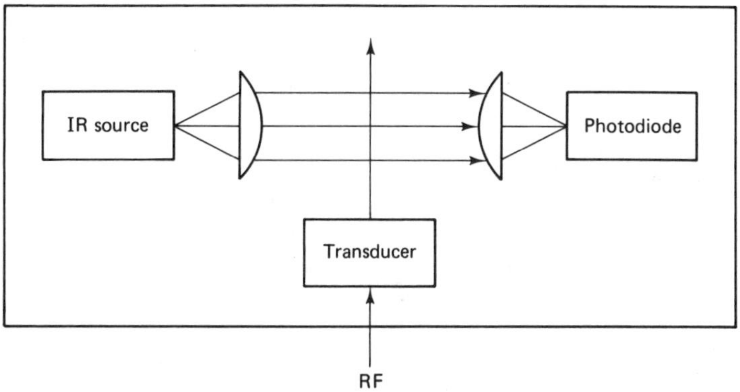

Figure 10–24 Integrated optics device comprising an LD infrared source, two-dimensional lenses, an EM-acoustic transducer and a photodiode, or an array of photodiodes, deposited on a chip.

Optical, microwave, and radio-frequency techniques are brought together with the introduction of surface acoustic waves (SAWs). Since the ratio of the speed of light to the speed of sound is about 1 million to 1, the ratio of the wavelengths at a given frequency is also 1 million to 1. This means, for example, that a 500-kHz radio signal with a wavelength of 600 m can be reduced by a transducer to an acoustic signal with a wavelength of about 6×10^{-4} m, which corresponds to the wavelength of an electromagnetic signal in the infrared region. This provides the possibility of wave–wave interaction between the IR and acoustic signals.

Consider, for example, an IR laser and photodiode deposited on a silicon chip, together with lenses or reflectors to provide a collimated beam, as suggested in Fig. 10–24. A transducer at one edge of the chip converts an RF signal into surface acoustic waves that propagate through the IR beam. The relatively slow-moving acoustic waves produce condensations—sequential variations in density—that act as a grating for the light waves. The device can be used as a modulator or a spectrum analyzer to analyze either the incident IR light or the RF input. These are possible examples of integrated optics.

APPLICATION 10–1: Connectoring Fiber Optic Cable

Fiber optic connecting should not be attempted without first reading the manufacturer's instructions. Allow at least 2 hours for a first attempt. A cutaway sketch is shown in Fig. 10–25 in order that we may identify the various parts of a cable.

As you follow the step-by-step procedure given in the manufacturer's instructions, record in your notebook the part numbers and procedure. Include sketches showing dimensions required for proper connecting.

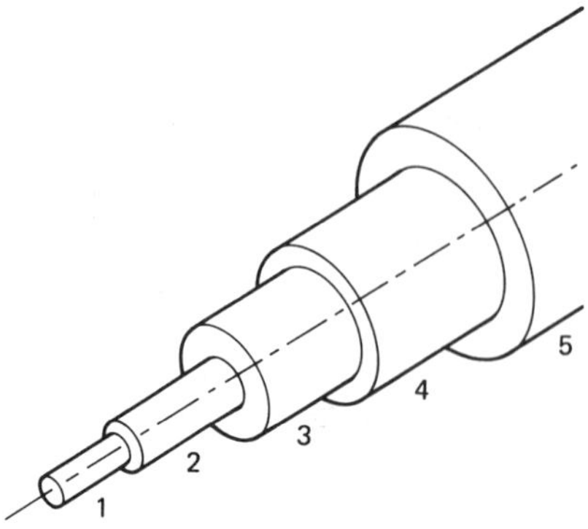

Figure 10–25 Cutaway view of a fiber optic cable: 1, glass fiber; 2, silicone buffer; 3, inner jacket; 4, aramid strength members; 5, polyurethane jacket.

The procedure below is not complete, cannot be substituted for the manufacturer's instructions, but indicates the variety of tasks and skills required.

Equipment and Materials

fiber optic connector assembly kit; for example, Hewlett-Packard HFBR-0100 and consummables kit HFBR-0101

two- or 3-meter length of single, glass fiber optic cable, HFBR-3200

incandescent light source

optional: an optic link, for example, a Hewlett-Packard HFBR-1502 transmitter and an HFBR-2502 receiver; signal generator; oscilloscope

Caution. Fiber optic cable has a minimum bending radius of 5 cm; greater bending damages the cable mechanically. Connecting requires the use of Loctite or "super glue," which sticks instantly. Avoid contact. Do not panic if stuck; eventually, it can be removed. Read the manufacturer's instructions *before* starting.

Never examine a fiber optic cable that may be connected to a source. Examination under magnification can damage the eye. Continuity can be checked using an ordinary incandescent light bulb.

If the glass fiber is broken in step 6, it must be discarded and the procedure started again from the beginning.

Procedure

As a continuity check, look at an incandescent lamp through the cable. Slide the body nut, washer, and crimp sleeve onto the cable; if the other end of the cable is connected or unavailable, you will not be able to do this later.

Expose the buffer jacket, peel back the aramid strength members to the dimensions specified, and slide the crimp support sleeve into place. Use the crimping tool to crimp the sleeve, and use scissors to trim the aramid strength members.

Use the special stripping tool to peel off the buffer jacket and clean the silicone from the exposed glass fiber with the solvent provided in the kit.

Mount the loaded syringe (hypodermic needle) in the bench vise and inject epoxy into the body of the connector, insert the cable, and assemble with the body nut, washer, and a drop of Loctite. *Caution*: If the fiber is broken in the attempt to insert it into the connector, we have to start again from the beginning. Cure the epoxy in the oven supplied in the kit.

Cleave off any excess length of glass fiber with the sapphire ribbon and polish the fiber until examination under the low-power microscope is satisfactory. If the polishing has removed the chamfer on the ferrule, discard the connector and start over.

Check for light transmission using an incandescent lamp as a source. As a final test, replace a commercially connected cable in a fiber optic link with one that you have connected and compare the performance.

APPLICATION 10–2: Refractive Index from the Angle of Minimum Deviation

The index of refraction of a prism can be measured accurately by the method of minimum deviation, described in the text. The prism is rotated until the deviation is minimal, as indicated in Fig. 10–8. Our analysis is for a 90° plastic prism, a type readily available since it is used in binoculars. If a 60° glass prism is available, the analysis is the same with the substitution of 60° in place of 45°.

At the angle of minimum deviation the geometry is symmetrical and the angle of incidence, α, is equal to the angle of emergence, as shown in Fig. 10–26. The incident ray is deflected through an angle $\alpha - \beta$, where β is the angle of refraction. By symmetry, the ray is again deflected $\alpha - \beta$ on emerging from the prism, and the total deviation is

$$\theta_m = 2(\alpha - \beta) \qquad (10\text{–}17)$$

which is the angle to be measured.

The angle of refraction, β, can be determined from the geometry. The interior angles of the triangle formed by the internal ray are $45° + 67.5° + 67.5° = 180°$. Since β is the complement of the 67.5° angle, we have $\beta = 90 - 67.5 = 22.5°$.

From Snell's law and Eq. (10–17) we now have

$$\eta = \frac{\sin\alpha}{\sin\beta} = \frac{\sin(\tfrac{1}{2}\theta_m + \beta)}{\sin\beta} = \frac{\sin(\tfrac{1}{2}\theta_m + 22.5°)}{\sin 22.5°} \qquad (10\text{–}18)$$

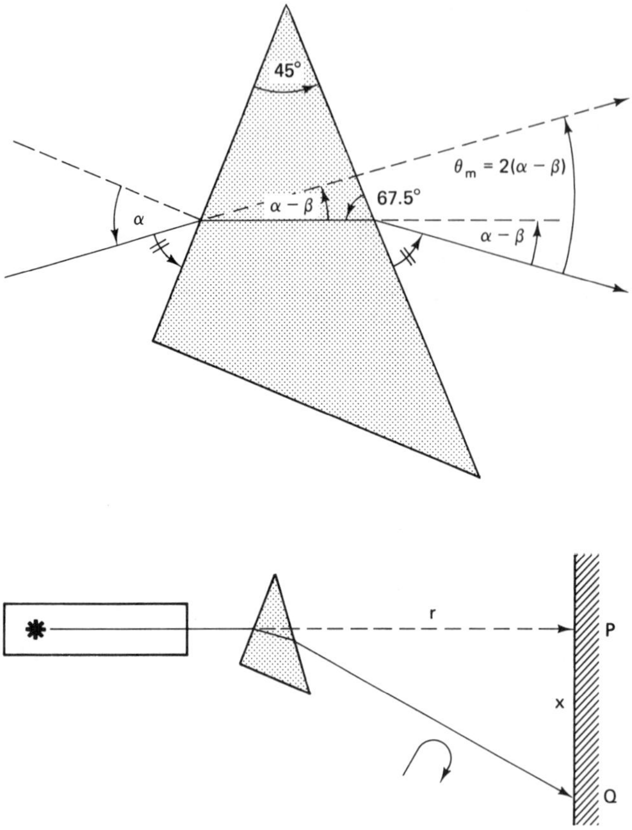

Figure 10–26 Geometry of the minimum angle of deviation.

and the refractive index is determined by the measurement of the minimum angle of deviation, θ_m.

Equipment and Materials

glass or plastic prism
monochromatic light source
polystyrene foam boards (building insulation) or some other convenient material for screens and for supporting the equipment
tape measure
protractor

Caution. If a laser is used as a light source, it should be the low-power type suggested for classroom use. Besides the usual safety procedures for shop work in the presence of electrical hazards, there are some precautions unique to optics. There is a human tendency to look into things that do not work, like looking into the barrel of a firearm that has misfired. Never look into the end of a laser or a fiber optic cable to see if it is working, just as you would never look into an open waveguide connected to a generator.

Prevent accidental intrusion of visitors into the working area; arrange the equipment well below eye level; avoid glints. Use safety glasses.

Procedure

Arrange the equipment as shown in Fig. 10–26. Mark the point P of the undeviated ray on the screen, then put the prism in place and rotate it for minimum deviation, that is, for minimum distance x.

The minimum angle of deviation can be measured with a protractor; however, if the distance r is 2 m or more, the angle can be determined more accurately as $\theta_m = \tan^{-1}(x/r)$. An average of several measurements will permit determination of the index of refraction to within a few tenths of a percent.

APPLICATION 10–3: Refractive Index from the Critical Angle

This measurement is conveniently included with those of Application 10–2 since the same equipment is required. The same safety precautions are pertinent.

We can determine the refractive index of a prism by rotating it until the emerging ray is grazing at 90°. A slight additional rotation shows that this is the angle at which total internal reflection begins.

A suggested experimental geometry is shown in Fig. 10–27. The angle of refraction of the incident ray, θ_r, and the angle of incidence of the emerging ray, β, are related by the theorem that the sum of the interior angles of a triangle is 180°: $\beta + \theta_r + 135° = 180°$, or

$$\theta_r = 45° - \beta \tag{10–19}$$

At the critical angle we have, from Snell's law, $\eta = \sin 90°/\sin\beta$, or

$$\eta = \frac{1}{\sin\beta} \tag{10–20}$$

For the incident ray we have, from Snell's law and Eq. (10–19), $\eta = \sin\theta_i/\sin\theta_r = \sin\theta_i/\sin(45° - \beta) = \sin\theta_i/(\sin 45° \cos\beta - \sin\beta\cos 45°)$, or

$$\eta = \frac{2^{1/2}\sin\theta_i}{\cos\beta - \sin\beta} \tag{10–21}$$

By equating Eqs. (10–20) and (10–21) and solving for β, we find

$$\beta = \tan^{-1}\frac{1}{1 + 2^{1/2}\sin\theta_i} \tag{10–22}$$

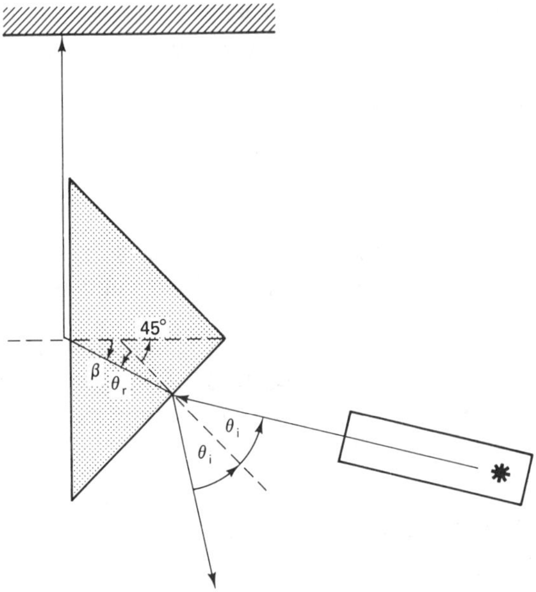

Figure 10–27 Geometry of the critical angle.

If we measure θ_i, we can find β from Eq. (10–22) and then determine the refractive index from Eq. (10–20).

Procedure

Arrange the equipment as shown in Fig. 10–27. Rotate the prism for grazing emergence or the beginning of total internal reflection. The angle of incidence can be measured directly; or, more accurately, by measuring the sum of the angle of incidence and reflection and dividing by 2.

Compare your result for the index of refraction with that determined in Application 10–2. If the same prism and light source are used, the results should agree within a few tenths of a percent.

APPLICATION 10–4: Microwave Grating

Quality optical gratings were made by H. A. Rowland near the end of the nineteenth century. Rowland devised a ruling machine capable of ruling 6000 lines per centimeter across an aperture of 15 cm. His machine is an example of the basis of our technology: A machine can be used to make a more accurate machine. Inexpensive photographic replicas of gratings are available, but because of shrinkage and warping they are not useful for precision measurements. In this application we shall build a microwave grating, a much easier task than ruling an optical grating, and use it to demonstrate spectrum analysis.

Equipment and Materials

- two 10-GHz microwave horn antennas
- X-band signal generator
- X-band receiver or VSWR amplifier/meter
- string, protractor, and tape measure
- aluminum foil-covered polystyrene or polyurethane foam insulation board

Procedure

The grating can be constructed by removing the aluminum foil from one side of the insulation board and cutting slits in the foil on the other side. Make the slits 6 cm wide and space them 11.6 cm, center to center. If the board is 0.8 m wide, there should be approximately seven slits. The other dimension of the board is not critical; it can be 0.6 to 0.8 m.

The experimental geometry is shown in Fig. 10–28. The distance from the transmitting antenna to the grating should be based on the radiation pattern of the horn; all of the grating should be illuminated but without appreciable spillover. The range from the grating to the observation point should be about 3 or 4 m, so that $R >> s$, where s is the larger dimension of the horn aperture.

Measure the angle θ for the first-, second-, and third-order maxima in the spectrum; also, measure the half-power beamwidths of the maxima by rotating the receiving horn until the received power drops to half-value. The angle θ can be measured more accurately by measuring to either side of the centerline and dividing by 2. In your report, compare the dispersion angles for the various spectrum orders and the beamwidths with what would be expected from the equations given in the text.

Rotate the grating until the receiving antenna is in the end-fire direction. Vary the angle of incidence by moving the transmitting antenna until a maximum signal is received. Now vary the angle θ by moving the receiving antenna and observe that the various orders of the grating spectrum are the sidelobes of a Yagi-type end-fire antenna.

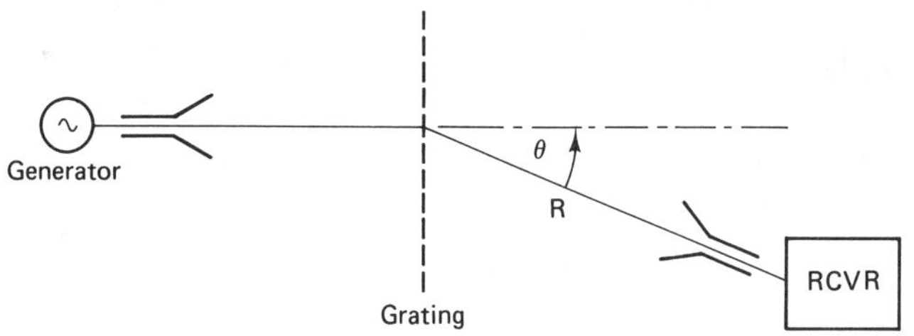

Figure 10–28 Experimental arrangement for observing the spectrum formed by a microwave grating.

APPLICATION 10–5: PIN Diode as a Variable Resistor

Since a PIN diode is a current-controlled variable resistance, it can be used to modulate a signal. In this application we observe AM and FM modulation of an oscillator.

Equipment and Materials

555 IC chip
PIN diode, for example, a Hewlett-Packard 5082–3041
dc power supply
audio signal generator
oscilloscope
assorted resistors, capacitors, and hand tools

Procedure

Use the 555 chip to construct a ½-MHz oscillator. Follow the manufacturer's recommendations, or copy a circuit from a handbook, or use the circuit shown in Fig. 10–29.

Adjust the audio-frequency signal generator output to $\pm 3\ V$ at a convenient frequency, say 1 kHz, and observe the oscillator output with an oscilloscope.

In your report describe the modulation. Is it AM, FM, or both? Why?

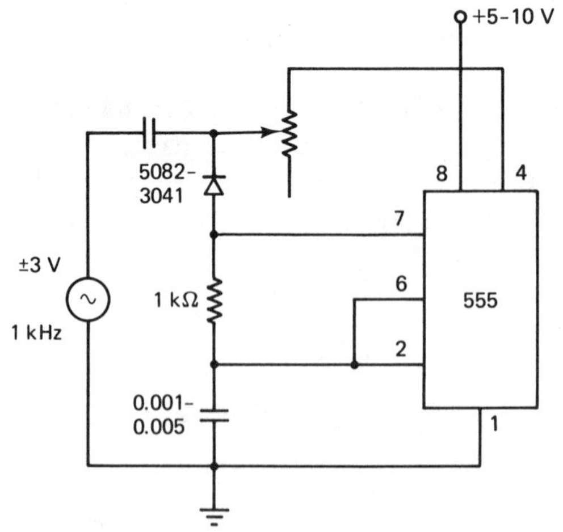

Figure 10–29 PIN-modulated oscillator circuit.

SUMMARY

The refraction of a beam of light is somewhat like a column of soldiers making a turn. The soldiers on one side march more slowly to establish a new direction. The mathematical expression relating the old and new directions of a refracted light beam is Snell's law,

$$\frac{\sin \theta_1}{\sin \theta_2} = \frac{\eta_2}{\eta_1}$$

from which we find the critical angle, the angle at which total reflection begins:

$$\theta_t = \sin^{-1} \frac{\eta_1}{\eta_2}$$

The refractivity of a dielectric can be determined accurately if the dielectric is formed into a prism. One method is the measurement of the angle of minimum deviation (Application 10–2); another is the measurement of the critical angle (Application 10–3).

Light propagates in fiber optic cables by internal total reflections. If a ray is incident on the end of a fiber optic cable core, it is totally reflected internally if its angle of incidence at the end of the core is less than θ, where angle θ satisfies the relation

$$\sin \theta = (\eta_2^2 - \eta_1^2)^{1/2}$$

The acceptance angle is 2θ, the numerical aperture is NA $= \sin \theta$.

Modal dispersion is measured by the difference in propagation times for various internally reflected modes and is estimated as

$$\frac{\Delta t}{s} \approx \frac{\frac{1}{2}(\mathrm{NA})^2}{c\eta}$$

The corresponding expectation value of the bandwidth is

$$\Delta f \approx \frac{1}{s}\left(\frac{s}{\Delta t}\right)(2)^{-1/2}$$

Dispersion-limited bandwidth is the frequency range over which the power level of a pulse drops to half value, which is approximately the reciprocal of the duration of an optimum pulse,

$$\mathrm{BW} \approx \frac{1}{\Delta t_0}$$

Chromatic dispersion can result from waveguide dispersion and material dispersion. Material dispersion is usually associated with absorption bands in the substance and can be described by a complex refractive index. In the classical theory of optics the complex refractivity is

$$\eta \approx 1 - \frac{jQ}{1 + jQ(f - 1/f)}$$
$$= \eta' - jK$$

where $K = \alpha\lambda/2$ is the extinction coefficient and α the absorption coefficient. In fiber optics technology it is customary to take 1 km as a unit of distance to define the attenuation coefficient, and expressions for attenuation are

$$\frac{p}{p_0} = e^{-\alpha x}$$

or

$$\frac{p}{p_0} = 10^{-0.434\alpha x}$$

At frequencies well below an absorption region the extinction coefficient is approximately $K \approx f^2 Q$, and the intensity decrease owing to scattering varies with the fourth power of the frequency,

$$\frac{p}{p_0} \approx \frac{f^4}{Q}$$

In optics this is called Rayleigh scattering. In antenna theory we recognize a similar effect as the fourth-power frequency dependence of the radiation resistance of a dipole.

Lasers are frequently used as light sources. Lasing occurs when the radiation in an optically pumped gas, liquid, solid, or semiconductor is partially confined. If the subsequent decay is by stimulated emission, a coherent light source is created. Coherence is measured as coherence time or coherence length. In laser diodes the partial confinement of the radiation can be achieved by gain guiding or index guiding. The gain or optical efficiency of an LD is the ratio of the power output to the drive current.

Light detectors depend on the electrical or chemical effects produced by photons interacting with matter. Ionization potential and work function are measures of such interactions. In electronics, commonly used detectors are photoelectric cells, electron multipliers, avalanche photodiodes, and PIN diodes.

A prism can be used as a spectrum analyzer, but a grating usually is preferred. The grating equation is

$$d(\sin\alpha + \sin\theta) = n\lambda \qquad n = 0, 1, 2, \ldots$$

where the slit spacing d is called the grating constant. End-fire radiations from a grating are called Wood's anomalies. A particularly simple grating with only two slits or sources is called an interferometer.

When several optical devices or functions are combined on a single chip the resultant optical/electronic system is called integrated optics.

PROBLEMS

10–1. Calculate the wavelength of radiation from a GaAs diode. Is it in the visible range?

10–2. The power delivered to a load is t^2/z, where t is the transmission coefficient and z the normalized impedance. For real impedances we have, from Eq. (5–45),

$$p_{\text{del}} = \frac{1}{z}\left(\frac{2z}{z+1}\right)^2 = \frac{4}{z + 1/z + 2}$$

Since the refractive index is the inverse of the normalized impedance, we obtain the same equation for the power delivered to a real dielectric,

$$p_{\text{del}} = \frac{4}{\eta + 1/\eta + 2} = \frac{4}{\eta_1/\eta_2 + \eta_2/\eta_1 + 2}$$

When expressed in dB, the expression above is called the *Fresnel loss*:

$$\text{Fresnel loss} = -10 \log \frac{4}{\eta_1/\eta_2 + \eta_2/\eta_1 + 2} \qquad (10\text{–}23)$$

Calculate the Fresnel loss for light propagating from glass, $\eta_1 = 1.5$, into gallium arsenide, $\eta_2 = 3.6$, at normal incidence. Since $1/\eta = z$, you can check your answer with a Smith chart, where Fresnel loss is called "reflection loss in dB."

10–3. Plastic fiber optic cable has, according to the manufacturer's specifications, a travel-time constant of 5.0 ns/m. Estimate the "effective" refractive index of the plastic fiber. Neglect dispersion.

10–4. An optical grating is ruled 15,000 lines per inch. If it is used to analyze a sodium spectrum, what is the angular separation of the sodium D-lines? Refer to Table 10–1.

10–5. If plastic fiber optic cable has a loss of 0.3 dB/m optical, what is the loss in dB electrical?

10–6. What is the critical angle in flint glass, $\eta = 1.7$?

10–7. Butt joints between glass fibers are often closed with a fluid or plastic with a refractive index approximately equal to that of the glass. If this is not done, there may be a small air gap in the joint. Estimate the power loss due to a small air gap. Assume that the gap is comparable to a wavelength and neglect multiple reflections; also, assume normal incidence. The refractivity of the glass is 1.5.

10–8. What is the numerical aperture of a fiber optic core of refractive index 1.500 if the cladding index is 1.487? What is the acceptance angle?

10–9. What is the minimum wavelength of a photon capable of ejecting an electron from the surface of a layer of cesium? Is the photon wavelength in the visible range?

10–10. When examined by casual inspection, a half-mirror appears yellow by reflected light. What color is the transmitted light? Why?

Index

A

B

C

D

M

N

O

P